From Abortion to Contraception

Map of Central and Eastern Europe.

From Abortion to Contraception

A Resource to Public Policies and Reproductive Behavior in Central and Eastern Europe from 1917 to the Present

Edited by **Henry P. David**
with the assistance of **Joanna Skilogianis**

Foreword by **Anastasia Posadskaya-Vanderbeck**

GREENWOOD PRESS
Westport, Connecticut • London

Library of Congress Cataloging-in-Publication Data

From abortion to contraception : a resource to public policies and reproductive behavior in Central and Eastern Europe from 1917 to the present / edited by Henry P. David with the assistance of Joanna Skilogianis ; foreword by Anastasia Posadskaya-Vanderbeck.

p. cm.

Includes bibliographical references and index.

ISBN 0-313-30587-0 (alk. paper)

1. Birth control--Europe, Eastern. 2. Europe, Eastern—Population policy. I. David, Henry Philip, 1923- . II. Skilogianis, Joanna.

HQ766.5.E872F76 1999

363.9'6'0947—dc21 98-51634

British Library Cataloguing in Publication Data is available.

Library of Congress Catalog Card Number: 98-51634
ISBN: 0-313-30587-0

First published in 1999

Greenwood Press, 88 Post Road West, Westport, CT 06881
An imprint of Greenwood Publishing Group, Inc.
www.greenwood.com

Printed in the United States of America

The paper used in this book complies with the Permanent Paper Standard issued by the National Information Standards Organization (Z39.48-1984).

10 9 8 7 6 5 4 3 2

In order to keep this title in print and available to the academic community, this edition was produced using digital reprint technology in a relatively short print run. This would not have been attainable using traditional methods. Although the cover has been changed from its original appearance, the text remains the same and all materials and methods used still conform to the highest book-making standards.

Equal relationships between men and women in matters of sexual relations and reproduction, including full respect for the physical integrity of the human body, require mutual respect and willingness to accept responsibility for the consequences of sexual behavior.

Paragraph 7.34,
Programme of Action of the International
Conference on Population and Development, 1994

Reproductive health is a right for both women and men . . . every individual has the right to decide the size and spacing of the family, and to have the means and information to do so . . . there must be no coercion, either to have or not to have children . . . these rights are part of the international structure of human rights, which has as its foundation the concept that all men and women are equal.

Nafis Sadik, Executive Director,
United Nations Fund for Population Activities,
28 May 1997

High abortion rates do not represent women's equality; they represent a failure on all our parts to help women prevent unwanted pregnancies in the first place.

Hillary Rodham Clinton,
International Women's Day,
12 March 1997

Contents

Tables and Figures

TABLES

FIGURES

Abbreviations

BFPA	Bulgarian Family Planning Association
BOS	Bureau of Statistics
CBS	Central Board of Statistics
CCSS	Computing and Health Statistical Centre
CDC	Centers for Disease Control and Prevention
CEC	Central Executive Committee
CEE	Central and Eastern Europe
CHSC	Computing and Health Statistics Centre
CIS	Commonwealth of Independent States
COE	Council of Europe
COMECON	Council of Mutual Economic Assistance
CPR	Contraceptive Prevalence Rate
CSB	Central Statistical Board
CSO	Central Statistical Office
DB	Data Base
ESCAP	Economic and Social Commission for Asia and the Pacific
EURO	European Regional Office
FPA	Family Planning Association
FRG	Federal Republic of Germany
FRY	Federal Republic of Yugoslavia
FSRY	Former Socialist Republic of Yugoslavia
FSO	Federal Statistical Office

GDR	German Democratic Republic
GPC	Government Population Commission
HWA	Hungarian Women's Association
IDB	International Data Base
IHS	Institute for Health Statistics
IMR	Infant Mortality Rate
IPPF	International Planned Parenthood Association
ISC	Interstate Statistical Committee
ISH	Institute for Social Hygiene
IUD	Intra Uterine Device
IUSSP	International Union for the Scientific Study of Population
MCGS	Moscow Center for Gender Studies
MMR	Maternal Mortality Rate
MOH	Ministry of Health
MOHE	Ministry of Health and Environment
MOL	Ministry of Labor
NCD	National Commission of Demography
NCS	National Center for Statistics
N.I.S.	Newly Independent States
NSC	National Statistical Committee
NSI	National Statistical Institute
PCH	Peoples' Commissariat of Health
PD	Population Division
PHSA	Public Health Statistics Annual
RFPA	Russian Family Planning Association
RLMS	Russian Longitudinal Monitoring Survey
RSFSR	Russian Soviet Federated Socialist Republic
RWRHS	Russian Women's Reproductive Health Survey
SCS	State Committee on Statistics
SECS	Society for Education in Contraception and Sexuality
SIECUS	Sexuality Information and Education Council of the United States
SFRY	Socialist Federal Republic of Yugoslavia
SO	Statistical Office
SPPFD	Society for Planned Parenthood and Family Development
SPRSV	Association for Family Planning and Sex Education
SSC	State Statistical Committee
SSO	State Statistical Office
TCS	Knowledge-Attitudes-Practices Survey

TDB	Transmonee Data Base (UNICEF)
TFR	Total Fertility Rate
TPR	Family Planning Association
TRR	Family Development Association
U.N.	United Nations
UNESCO	United Nations Educational, Scientific, and Cultural Organization
UNFPA	United Nations Fund for Population Activities
UNICEF	United Nations Children's Fund
U.S.	United States
USBOC	U.S. Bureau of the Census
USSR	Union of Soviet Socialist Republics
VCIOM	Russian Centre of Public Opinion and Market Research
WGNRR	Women's Global Network for Reproductive Rights
WHO	World Health Organization

Foreword

Anastasia Posadskaya-Vanderbeck

There are several reasons why it is my special pleasure to contribute a Foreword to *From Abortion to Contraception: A Resource to Public Policies and Reproductive Behavior in Central and Eastern Europe from 1917 to the Present.* These reasons are summarized in terms of approach, scope, and spirit.

Presently there is no other comprehensive resource on the "region in transition" that so ably integrates woman-centered political, historical, demographic, and psychosocial approaches. It overcomes the limits of "pure" demographic analysis found so frequently in the literature. For the first time the interaction between public policies and private reproductive behavior in the formerly socialist countries is considered within the context of a women's rights perspective. However, the volume moves beyond the typical "women and postsocialist transition" critiques of cultural, economic, and social trends to a more sophisticated analysis of reproductive behavior as evidenced by the meticulously compiled data on access to safe abortion and modern contraceptives. The editorial approach renders the issues of women's status, rights, and empowerment central to the understanding of public policies and private behavior of individuals and couples.

In moving from demographic aspects to improved reproductive and sexual health, including governmental support for family planning, the overviews and individual country chapters further contribute to the global shift in the framework within which the world community agreed to consider and advocate the issues of population and socioeconomic development at the major United Nations conferences of the 1990s. These included the 1992 Rio de Janeiro Earth Summit, the 1993 Vienna World Conference on Human Rights, the 1994 Cairo International Conference on Population and Development, and the 1995 Beijing World Conference on Women. The reshaping of population-related perspectives and priorities was largely the result of activism led by women's groups and gender-sensitive nongovernmental organizations worldwide. That is a further reason why the analysis of political theory in terms of daily life realities for women is such an integral part in understanding the current situation in the countries experiencing postsocialist transitions.

Another unique feature of the volume is its scope, covering 28 countries including the USSR and Yugoslav successor states. Each country chapter offers a concise analysis of historical and political development, population policy, the woman question, sexuality education, contraception, and abortion. Anyone with experience in trying to obtain information on reproductive issues and behavior in these countries knows how daunting this task is and how confusing some of the resources can be. The contributors to this volume, and especially the editor, have gone to extraordinary lengths to verify information, caution readers about data reliability, and cite original sources. The book will be an indispensable resource for current and future scholars of public policy, reproductive behavior, and gender studies. The more than 1,290 references alone are worth the price of the book.

In writing this brief Foreword, I want to conclude with a few personal remarks. The book reflects the spirit of Henry P. David. He began his cooperative interdisciplinary studies in Central and Eastern Europe in the 1960s while serving as Associate Director of the World Federation for Mental Health in Geneva, Switzerland and as a consultant to the World Health Organization. I first met him in 1993 when he visited, as part of a MacArthur Foundation delegation, the Moscow Center for Gender Studies, of which I was then the director. A brilliant scholar with practical ideas, Henry came to occupy a very special place in my life as he has for many colleagues in Russia and in other countries of Central and Eastern Europe. His generosity in giving of himself, in sharing his knowledge and expertise, and most important, in providing moral support, empathy, and warmth in at times difficult circumstances and occasionally at some personal risk, go far beyond what can reasonably be expected from a consultant or colleague. It is Henry's ability to see beyond what is possible or reasonable today that motivates his passionate and optimistic commitment to the future.

From Abortion to Contraception is an embodiment of Henry David's vision, dedication, and will. Overcoming a serious illness, he continued to motivate the contributors, extensively edited every chapter, and moved the manuscript forward. I am delighted to recommend this unique and insightful book, which will stand for years to come as a comprehensive, archival, and historical record of public policy and reproductive behavior over more than 70 years of socialist rule and nearly a decade of postsocialist transition in Central and Eastern Europe.

Preface

Henry P. David

This book evolved from my cooperative endeavors with colleagues in the socialist countries of Central and Eastern Europe (CEE) over more than 30 years, beginning in the mid-1960s when I was Associate Director of the World Federation for Mental Health, based in Geneva, Switzerland. It builds on two prior monographs, both long out-of-print (David, 1970; David and McIntyre, 1981). The Transnational Family Research Institute (TFRI), which I serve as director, is a nonprofit and nongovernmental organization with an extensive history of cooperative interdisciplinary and international research in reproductive behavior. Perhaps best known is the still continuing study of persons born in 1961–63 to Prague women twice denied abortion for the same pregnancy and pair-matched controls (David et al., 1988; Kubicka et al., 1995), summarized in Chapter 6.

The title *From Abortion to Contraception* owes much to the initiative of Daniel Pierotti, who conceived of an international conference convened under that theme in Tbilisi, Georgia in October 1990. It was cosponsored by the United Nations Fund for Population Activities, the World Health Organization, and the European Office of the International Planned Parenthood Federation. It was my privilege to serve as one of the consultants in planning that conference and to present a keynote address.

Different scholars have named the region "Eastern Europe" (Blayo, 1991; Stloukal, 1995) or "East-Central Europe" (Kosinski, 1977b; Batt, 1991), or divided it into "Eastern Europe" and "Southern Europe" (U.N., 1991). Others have preferred "Central and Eastern Europe" (David, 1970; David and McIntyre, 1981), differentiating between countries that preserved major elements of Western traditions or belonged to the Habsburg empire and those that readily identified with Eastern cultures.

With contemporary national borders reflecting diverse ethnic realities, the old political terminology has lost some of its meaning. For the purposes of this volume, the terms employed to identify the region are less important than the experience of the populations in the countries as they traversed the road from an autocratric communist regime through a postsocialist transition that included, in some cases, separation into independent successor states.

With the generous support of the Moriah Fund and the encouragement of colleagues in the CEE countries, it was decided to explore the interaction between population policies and private reproductive behavior and couple decision making from 1917, the time of the Russian Revolution, through 1997. I am most grateful to the colleagues who wrote the original country reports and who patiently accepted the many editorial revisions based on additional material from the TFRI archives and other sources. The result is a problem-focused rather than discipline-oriented resource book. Several readers reviewed the revised chapters, adding additional incisive comments. Their names are cited in my editorial introductions with gratitude for their generosity in time and spirit. Full responsibility for any remaining errors and misunderstandings falls on my shoulders.

Throughout the development of the monograph much valuable assistance was generously given by Carl Haub of the Population Reference Bureau. Additional support was graciously provided by John Dunlop, Ward Kingkade, and Lorraine Wright of the International Program Center of the U.S. Bureau of the Census, by Timothy Heleniak of the World Bank, and by Anne Marie Amantia, the ever-resourceful senior Librarian of Population Action International.

Particularly welcome were the many excellent suggestions of Tomas Frejka, Stanley Henshaw, and Libor Stloukal for improving my understanding of demographic trends. Gail Kligman and Elaine Murphy were especially helpful in reviewing segments of the draft. While completing her Ph.D. dissertation and having a baby, Joanna Skilogianis skillfully critiqued all the chapters and prepared every table and figure (with innumerable revisions). The manuscript, in its myriad formats, was expertly prepared by Lena Taggart. The map of Europe was kindly contributed by Heather Lilley of the Population Reference Bureau. Nita Romer, my Greenwood Press editor, encouraged the project and provided strong support all along the way from an idea to its realization. The positive attitude of my very conscientious copy editor, Terri Jennings, made a sometimes tedious task easier to bear.

I am most grateful to my life companion, Tema, for her loving support at a time of major illness, for encouraging me to complete the task, and for sharing with me a richly rewarding family and international life. It is with the hope of further enhancing transnational communication of experience that this monograph is presented.

Bethesda, Maryland
January 1999

part I

INTRODUCTION

chapter 1

Overview

Henry P. David

INTRODUCTION

The collapse of the communist era at the end of the 1980s and the turmoil of subsequent events offer an opportunity to reflect on the influence of public policies in the formerly socialist countries of Central and Eastern Europe (CEE) on private reproductive behavior. The aim of this volume is to review each country's unique experience within a broader historical, psychosocial, and public health context. Consideration is given to a range of factors impinging on sexual and reproductive behavior. These include (1) societal values and traditions, (2) Marxist theory, (3) socialist and patriarchal perceptions of gender roles, (4) the status of women as producers of labor and reproducers of families, (5) changes in legislation facilitating or constraining access to modern contraceptives, (6) pronatalist incentive influences on demographic trends, (7) attitudes of health service providers, (8) views on sexuality education, (9) adolescent sexual behavior, and (10) emerging roles of public services and nongovernmental organizations. Motivations for changes in public polices are discussed as are fluctuations in the interpretation and implementation of laws and decrees. Each country chapter attempts to explore the complex tapestry of historical and political development within evolving trends in fertility regulation and the interrelationship between abortion and contraception from the period before the establishment of communist regimes to the present.

This overview, which continues in chapters 2 and 3, focuses on major themes variously noted in the separate country chapters. In Chapter 2 Stloukal discusses psychosocial and cultural aspects of the abortion experience, the effects of pronatalist policies, and the contraceptive fatalism accepted by many Central and Eastern Europeans. Consumerism, inertia, and alienation dominated lives. Abortion was not only the product of erratic contraceptive supplies, but also of evolving value systems and social relations. There was a close association between the lack of pregnancy prevention, induced abortion, and the perception of social welfare entitlements as compensating for the shortcomings of daily life. Individuals were relieved of initiative and personal responsibility by accepting dependence on a paternalistic state. Institutional adjustments and nonadjustments often determined the roles and functions of health services, sexuality education, and the mass media.

Chapter 3, The Woman Question, traces the historical development of socialist views, noting trends before and after the October 1917 Revolution. It considers divergences between rhetoric and realities of daily life, depending on local culture and political mores. It then examines effects of postcommunist transitions from central planning to market economies on such issues as production and reproduction, reproductive health, and feminism. In several countries, steps taken to reduce women's multiple burdens of worker, wife, and mother delivered low levels of benefits and services.

The delays in producing consumer goods, the long hours required for shopping, the lack of housing, and the need for two full-time incomes persuaded many couples to reduce family size and terminate unwanted pregnancies. However, in some countries (Albania and Romania) the right to control fertility became a casualty of the drive to meet economic targets. Restrictions on abortion were justified on the ideological basis that family planning was not a human right but rather an inextricable part of socioeconomic planning for which all individuals were responsible (Stloukal, 1995).

Women are confronted by special problems in the transition process. Access to modern contraception varies, domestic production is often poor or nonexistent, and costly imported products are beyond the economic reach of many. Women's preferences are rarely considered. Paying under the table is part of daily life, alternative services are seldom found, and women's organizations without an obscure political past are only slowly evolving. In several countries, the waning influence of communism is associated with the rise of a new conservatism, often supported by revitalized religious organizations. Ideological opponents of family planning argue that liberal abortion legislation is a relic of the communist past, that "planning" has been discredited, and that "traditional" values should be restored. Poland has experienced several changes in abortion legislation, most recently again placing severe restrictions on access to legal abortion while modern contraceptives remain in short supply, especially in rural areas.

COUNTRY CHAPTERS

Part II presents nine country chapters in alphabetical order, based on their pre-1989 status plus their successor states, totaling 28 separate national entities as of 1999. Each chapter begins with an editor's introduction, commenting on the historical context, and citing topics reviewed. The nine original socialist countries generally shared the goals of a centrally managed economy (with the exception of Yugoslavia), a one-party political system, and the sometimes submerged but seldom forgotten social ideal of gender equality. While heavily influenced by the Soviet Union, the CEE countries did not constitute a monolith (David, 1982; Turnock, 1989). On the contrary, each had its own historical, cultural, and religious values, its own ethnic traditions and conflicts, and its own political and socioeconomic development experience. Considerable differences continue to exist in population size, urbanization, and living standards.

The determinants of fertility behavior varied greatly between and within countries. The Soviet Union was extremely heterogeneous with each of its 15 republics at different stages of economic development and with considerable ethnic and cultural dissimilarities. Albania is the only country in the European region with a predominantly Muslim population, although parts of Bulgaria and Bosnia-Herzegovina also have a large proportion of Muslims as do a number of USSR successor states.

Chapter 6 on the Czech and Slovak Republics covers Czechoslovakia and developments after the country was peacefully divided in January 1993. Chapter 11 on the Russian Federation notes trends from before the October 1917 Revolution through the 1991 breakup of the Soviet Union. Mention is made of subsequent developments in the Baltic states of Estonia, Latvia, and Lithuania; in Belarus, Ukraine, and Moldova; in the Caucasian republics of Azerbaijan, Armenia, and Georgia; and in Kazakstan and the Asian lands of Kyrgyz Republic, Tajikistan, Turkmenistan, and Uzbekistan. Similarly, Chapter 12 reports on the Socialist Federal Republic of Yugoslavia until its dissolution in June 1991. Subsequent events in the independent republics of Yugoslavia (Serbia and Montenegro), Bosnia and Herzegovina, Croatia, Macedonia, and Slovenia are cited in relevant sections.

In all, Part II presents historical and current information on the influence of public policies on private reproductive behavior within the societal context of 28 formerly socialist countries. To the extent available, statistical data tabulated include births, legally induced abortions, crude birth rates, legal abortion rates, legal abortion ratios, and total fertility rates. Cautions about the quality of collecting information and "optimistic reporting" of official data are indicated. Also mentioned, usually in the text, is the incidence of maternal mortality due to abortion. Deaths from the consequences of abortion performed under unsafe clinical conditions still account for one in four maternal deaths (Brandrup-Lukenow and Mangiaterra, 1998). In addition, every chapter traces the influence of cultural traditions on the evolution of current concepts of sex and society, including abortion and contraception. Trends in adolescent sexual behavior are noted. Concerned teenagers often have only few or no services to which they can turn for advice about reproductive health or contraception.

RESOURCES

To reduce repetition and facilitate bibliographical searches, the nearly 1,300 references have been combined into a single chapter with listings by author in alphabetical order. Titles are presented in English with the language of original publication stated in parentheses. Accents and diacritics have been omitted. A general subject index and an About the Contributors chapter are provided.

FROM DEMOGRAPHY TO REPRODUCTIVE HEALTH

In the late 1950s influential demographers began to question the macro-theories of demographic transition (Hauser and Duncan, 1959). During the 1960s, researchers gradually shifted attention to micro-factors that operate on community and individual levels (Bogue, 1969; Coale, 1969; Coale and Watkins, 1986). Within that context, demography opened its doors to psychosocial and anthropological approaches to explain reproductive behavior (Luker, 1975; Miller, 1983; Newman, 1985; and Presser, 1997).

The 1994 International Conference on Population and Development (UNFPA, 1995b) confirmed a significant change in policy perspectives. Instead of concentrating on numbers of people, demographic targets, and macro-level population control, the focus shifted to the micro-level needs of individuals. There was a move away from controlling women's fertility to assisting individuals in making fully informed choices. Emphasis was placed on raising awareness of reproductive rights and challenging the social norms that govern gender relations (Murphy, 1998). The conference reaffirmed the right of couples and individuals to decide freely and responsibly whether to have children and when to have them. It adopted the concept of reproductive health (WHO, 1992), deeming it to be determined by socioeconomic development, lifestyles, quality and accessibility of health services, and by the status of women but "more than anything else, by the freedom to make choices" (UNFPA, 1995b, 33). Sexual health was considered an important component of reproductive heath along with family planning. By joining sexual health with reproductive health, the ambivalence and taboos surrounding the concept of sexuality were reduced, particularly as related to sexuality education and adolescent sexual behavior.

The country chapters try to present both a demographic/public policy and a reproductive health perspective. Private reproductive behavior involves a series of individual decisions reflecting historical, political, economic, social, cultural, and other values and traditions. Individual reproductive decisions are viewed in a process-context approach over the life course and over time (Greenhalgh, 1994; McNicoll, 1944; and Hutter, 1996). While abortion decisions are usually made jointly by partners in an emotionally secure relationship, discussion of the "right" of the male partner to have a voice in the decision emerges primarily in the context of forcing a woman to continue a pregnancy she does not want (Nolan, 1998).

DECISION MAKING

Throughout their fertile years individuals are faced with many choice points and alternative courses of action likely to determine the success of their efforts to control their fertility. The awareness of these choice points, the extent to which alternative courses of action are recognized, and the degree to which choices are based on realistic appraisals of benefits, costs, and consequences are the cornerstones of rational reproductive behavior (David and Johnson, 1977).

Successful fertility regulation generally requires the coordination of three distinct human forces: (1) the desire to have sexual relations; (2) the wish to have or not to have a child; and (3) the will to practice consistent contraception (David and Friedman, 1973). While these forces may be logically linked, they are not necessarily psychologically related. Coordination requires considerable and ever-vigilant effort, typically with a longer term reward that appears only in the guise of the absence of an event (pregnancy).

Although improvements have been made in delivery systems, regular availability and wide method acceptability remain elusive goals in some of the CEE countries. Contraceptives reduce but do not eliminate the need for abortion. Many couples cannot achieve a small family with existing contraceptive methods, owing to poor services, side effects, or misperceptions about risks (Kulczycki et al., 1996).

When pregnancy prevention fails, the management of an unintended and unplanned pregnancy becomes an urgent personal concern. An unwanted pregnancy threatens the well-being of the woman, the partner, the family, or other children. In the CEE countries numerous women have at least one unintended pregnancy by the time they reach menopause. In countries lacking ready access to contraception, the number of unintended pregnancies is higher. Most women are fertile for nearly half of their lives. They may well be sexually active for several years before desiring to become pregnant and give birth for the first time. Unless protected by surgical sterilization or infecundity, a woman is able to conceive for many years after she has had all the children she wants.

The decision to use or not to use a fertility-regulating method is often associated with the perceived risk of pregnancy, the openness of partner communications, the support of parents and peers, societal desensitization of sex as a taboo topic, the community's attitude toward sexuality education, and the influence of specific guidance by health clinic staff. One difficulty is that many societies are not open about sex, far too many parents are afraid to talk about fertility regulation with their own children, and too few schools are prepared to broaden their biology-oriented courses to include practical information about contraceptives. Policymakers tend to shy away from what they perceive to be politically sensitive topics, and governments are reluctant to initiate adolescent pregnancy prevention programs for fear of unfavorable public reaction. It is seldom acknowledged that teenagers will have sex, with or without contraceptive information and services.

ABORTION POLICIES

Throughout recorded history, women have resorted to abortion to terminate unwanted pregnancies, regardless of religious or legal sanctions (David, 1972, 1981). On 18 November 1920 the Soviet Union became the first country in the world to legalize abortion on request of the woman in the first trimester of pregnancy. The subsequent history of abortion legislation, decrees, and court decisions is detailed in each of the country chapters. Even within the same country (Romania) experience has ranged from very permissive to rigidly repressive.

Little is known about the genesis of abortion policies before they were promulgated (Macura, 1974b). As will be seen in the country chapters, liberal policies were officially justified on the basis of public health concerns (Klinger, 1971; Potts et al., 1977) and a desire to grant greater social freedom to women (Field, 1956; Berent, 1970; David, 1974; Macura, 1974b). They served other purposes as well The post-Stalin liberalization of abortion policies may have reflected a policy preference for a low cost means of fertility regulation without expending hard currency on importing contraceptives (Callahan, 1970; Besemeres, 1980) as well as an effort to put Marxist ideology and theory into practice (Macura, 1974b). However, governmental attitudes and practices varied, as did the timing and implementation of measures enacted (Macura, 1981). The vacillating views on oral contraception in the former USSR are noted in Chapter 11.

As mentioned in Chapter 2, Understanding the "Abortion Culture" in Central and Eastern Europe, and particularly in Chapters 9 and 11, on Poland and the Russian Federation, abortion was accepted as the method of choice because there seldom was a regularly available and reliable alternative. As shown in several country reports, *coitus interruptus*, the oldest but not very reliable method of fertility regulation, was and remains a major method of pregnancy prevention, backed by abortion. Except for the German Democratic Republic, abortion became freely available in the CEE countries before the introduction of modern contraceptives. For many couples, the one-event abortion appeared cheaper and more reliable than dependence on irregular contraceptive supplies. As noted in Chapter 11, on the Russian Federation, once abortion was legalized and widely utilized much of the ethical resistance to this form of fertility regulation was removed. The use of contraceptives was seldom actively encouraged, supplies were limited, and inaccurate reports of the health risks were widely circulated. The abortion culture evolved because women had no realistic alternative.

The evolution of abortion policies is entangled with a host of political, religious, and moral concerns. Current political discussions often turn on religious beliefs on when life begins, but "embryology can no more establish when life begins than astronomy can determine whether heaven exists" (Kulczycki et al., 1996: 1667). A first-trimester abortion is generally available on some medical, social, and juridical grounds in all the CEE countries, except in Poland where access has been increasingly difficult for economically disadvantaged women (Rahman et al., 1998). Resistance by the medical community may make abortion services scarce even in circumstances where it would be legally permitted, as has occurred in Romania and in Poland.

CONTRACEPTION

Wide acceptance of modern contraceptives requires (1) availability, (2) accessibility, and (3) a general perception that efficient contraceptive practice is preferable to reliance on abortion. This combination of factors is gradually achieving headway. While availability has greatly improved, accessibility and acceptance are increasing less rapidly. In some countries, such as Bulgaria and Romania, dispensing pills without prescription is tolerated when supplies are available (Blayo and Leitao, 1993). However, information is not always easily shared and the interests of women do not rank high among the priorities of health professionals (Pierotti, 1993).

There are vast differences between and within countries in integrating awareness and communicating knowledge on how to improve contraceptive practice. In several lands, a substantial proportion of (mostly rural) women and health professionals believe erroneously that the pill constitutes a significant risk to health. Considerable reliance continues to be placed on less effective traditional methods. Provision of abortion services is not always associated with contraceptive counseling designed to prevent repeat abortions.

REPORTED ABORTION RATES AND RATIOS

The abortion rate, representing the reported number of abortions per 1,000 women of reproductive age (usually defined as 15–44 years) in a given year permits comparisons over time between and within countries. The abortion ratio, representing the reported number of abortions per 1,000 live births in a given year, permits similar comparisons. The rate is more useful for describing the level of abortion incidence in a given female population. Since the ratio illustrates how the number of abortions compares to the corresponding number of live births in a given year, it can be more readily affected by a rise or decline in fertility. The abortion ratio may rise even as the abortion rate declines and thus needs to be interpreted carefully (Singh and Henshaw, 1996; Kulczycki, 1996). The absolute level of abortions is best measured by the absolute number of abortions.

The quality of reported abortion data in the CEE countries varies according to the type of legislation, adequacy of services, completeness of the reporting system, and how much (if any) social stigma is attached or believed to be attached to the procedure (Blayo, 1989, 1991; Singh and Henshaw, 1996). Czechoslovak and Hungarian data are among the most complete national sets. Information from other CEE countries varies in coverage and reliability, as noted in the separate country chapters. Reported data from the former Soviet Union are particularly difficult to interpret (Avdeev, 1998). With the post-1989 advent of private medical practice and the establishment of commercial abortion clinics, reporting has become more lax. Except to express an opinion, no attempt has been made to compensate for over- or underreporting. Presently least reliable are abortion data from Poland where access is severely restricted and many women obtain abortions clandestinely in or outside the country. Since complete accuracy will probably remain elusive, qualitative studies are needed to provide a more accurate reflection of the meaning of abortion for coping with an unwanted pregnancy under conditions of severe restrictions on access to legal abortion (as noted in the chapters on Romania and Poland). When data sources are shown in the tables, they are not always repeated in the text. Few estimates are made about reported illegal abortion because they are generally based on hospital admissions that fail to reflect the large number of women who do not come to medical attention after a clandestine procedure.

Additional problems arise from difficulties in terminology and from changes in reporting systems, as for example, in the Russian Federation (Chapter 11). Broad agreement on nomenclature has yet to be reached on what constitutes a specific type of abortion, whether induced at what particular stage of gestation, and under which circumstances. In this volume, unless otherwise noted, the term "abortion" refers to a legally induced procedure. It remains to be seen whether any definition of abortion will become widely acceptable for universal statistical utility (Tietze, 1983; Stloukal, 1995).

The number of abortions (whether legal or clandestine) in a given country, the extent to which pregnancy termination is used to regulate fertility, and the demographics of the population subgroups relying most heavily on abortion are shaped by social and economic pressures to limit or delay childbearing, by effective contraceptive practice, and by the prevailing legal, cultural, and political climate (Frejka, 1985). In some countries, such as Romania in 1990, the initial rise in reported abortions after liberalization reflected an actual increase in numbers plus official recording of pregnancy terminations previously unreported. In other countries, such as Czechoslovakia, the effect of the 1957 abortion legalization was both a rise in registered abortions and a significant acceleration of fertility decline. The decreasing level of abortion-related mortality in most CEE countries is another indicator of the gradual decline of procedures performed by untrained providers in unsafe settings.

The incidence of abortion and the ratio of abortions to births vary among population subgroups for a variety of reasons, usually related to local service conditions and differences in cultural traditions. The rate of unintended pregnancy depends on the extent of exposure to intercourse at a time when children are not desired, the use of modern contraceptives, and rates of infecundity (Henshaw et al., 1990, 1999). Unmarried women frequently resort to abortion to delay birth of the first child, whereas married women with children seek abortion to end childbearing. Abortion used for spacing children is common in those countries where contraceptive use is comparatively low because of irregular availability of contraceptive supplies. Abortion to delay childbearing is common where women are sexually active for several years before marriage. Although the use of abortion for delaying, spacing, or ending childbearing exists in all countries, the relative importance of each in a given country may be inferred from the age distribution, marital status, and the parity of women having abortions. However, the interpretation of such findings needs to be considered within the sociocultural context of each country. Earlier reviews of the literature (David, 1992a; Stloukal, 1995) suggest that in the CEE countries abortions were used mainly to space births and end childbearing rather than delay the start of childbearing. Indirect characteristics, such as a women's socioeconomic situation, value orientation, psychological disposition, or approach to fertility regulation are critical factors in assessing reliance on induced abortion. Rates tend to peak among older married women who are likely to use less efficient contraception and seldom rely on sterilization (Frejka, 1985).

While specific data are discussed in the individual country chapters, the continuing decline in birth rates in Central and Eastern Europe is shown in Table 1.1. A comparison of the latest available birth rates is depicted in Figure 1.1. Total fertility rates are similarly presented in Table 1.2 and Figure 1.2. Quite a few CEE countries have fallen well below the replacement level of about two births per woman. Abortion rates and ratios are shown in Tables 1.3 and 1.4 as well as in Figures 1.3 and 1.4, respectively. While a declining trend in abortion rates and ratios is apparent, abortions continue to exceed births in Bulgaria, Romania, the Russian Federation, Estonia, Latvia, Belarus, and Ukraine.

Officially reported country data are cited in the tables. When official data were not available, this is indicated by the letters "na." Figures show reported country data by order of magnitude. For comparison purposes, abortion rates in the Netherlands and the United States are included in Figure 1.3. More extensive discussion of the data is provided in the country chapters. When data sources are cited in the tables, they are not always repeated in the text.

Worldwide, among countries reporting reliable statistics, the Netherlands had the lowest abortion rate, 6.5 per 1,000 women 15–44 years of age in 1996. For native Dutch women it was even lower. The 1996 rate reported from the United States was 22.9 per 1,000 women of reproductive age (Henshaw, 1998). Among the reasons for the low rate in the Netherlands is the excellent sexuality education program and the congruence of public policy with private behavior in reproductive and sexual health (David and Rademakers, 1996). The highest rate in the CEE countries was probably recorded in Romania, 252 in 1965 (not shown), the year before severe restrictions were placed on access to legal abortion. Abortion rates declined most rapidly in those CEE countries where abortion is part of a comprehensive reproductive health program, where a full range of contraceptive methods is available free of charge or at low cost, and where services are responsive to client needs within a sociocultural context fostering responsible sexual behavior.

The low fertility and high abortion rates in most CEE countries reflect a preference among the vast majority of women for a small family. By the time these women have two children they are ready to terminate childbearing. Low levels of desired childbearing, especially within a context of limited availability of effective long term contraception and an early start (and finish) of childbearing, enhances the probability of unintended and unwanted pregnancies and subsequent abortion.

There is little doubt that abortion liberalization had a substantial fertility-suppressing effect (Blayo, 1991). However, a classic analysis of the proximate determinants of fertility, using Czech, GDR, and Hungarian data on abortion, fertility, contraception, and nuptiality during the 1960s and 1970s, showed that "the fertility inhibiting effect of induced abortion has, as a rule, been less than that of contraception" (Frejka, 1983: 517). One difficulty of assessing the demographic impact is that in most countries the fertility decline began before the introduction of liberal

Table 1.1
Birth Rates in Central and Eastern Europe: 1950–96

	1950	1960	1970	1980	1990	1991	1992	1993	1994	1995	1996
Albania	38.9	43.4	32.5	26.5	25.2	23.8	23.7	22.4	21.9	22.7	17.8
Bulgaria	25.2	17.8	16.3	14.5	11.7	11.1	10.5	10.5	10.0	8.6	8.6
Czech Republic	21.1	13.3	15.1	14.9	12.6	12.5	11.8	11.7	10.3	9.3	8.8
Slovak Republic	28.8	22.1	17.8	19.1	15.1	14.9	14.1	13.8	12.4	11.5	11.2
GDR[a]	16.5	17.0	13.9	14.6	11.1	6.8	5.6	5.1	5.1	5.4	6.0
Hungary	20.9	14.7	14.7	13.9	12.1	12.3	11.8	11.4	11.3	11.0	10.3
Poland	30.7	22.6	16.8	19.5	14.3	14.3	13.5	12.8	12.5	11.2	11.1
Romania	26.2	19.1	21.1	18.0	13.6	11.9	11.4	11.0	10.9	10.4	10.2
FRY[b]	29.5	21.6	17.7	17.7	14.7	14.6	13.5	13.4	13.1	13.3	13.0
Croatia	24.9	18.4	13.9	14.9	11.6	10.8	9.8	10.2	10.3	10.8	11.7
Slovenia	24.5	17.6	15.9	15.7	11.2	10.8	10.0	9.9	9.8	9.5	9.4
Russian Federation	26.9	23.2	14.6	15.9	13.4	12.1	10.7	9.4	9.6	9.3	8.9
Estonia	18.4	16.6	15.8	15.0	14.2	12.3	11.7	10.0	9.5	9.1	9.0
Latvia	17.0	16.7	14.6	14.1	14.2	13.0	12.0	10.3	9.5	8.6	7.9
Lithuania	23.6	22.5	17.6	15.1	15.2	15.0	14.3	12.5	11.5	11.1	10.5
Belarus	25.5	24.4	16.2	16.0	13.9	12.9	12.4	11.3	10.7	9.8	9.3
Ukraine	22.8	20.5	15.2	14.8	12.7	12.1	11.4	10.7	10.0	9.6	9.1
Moldova	38.9	29.3	19.4	19.8	17.7	16.5	16.0	15.2	14.3	13.0	12.0

[a]German Democratic Republic. [b]Federal Republic of Yugoslavia.

Sources: See individual chapters and tables.

Table 1.2
Total Fertility Rates in Central and Eastern Europe: 1960–96

	1960	1970	1980	1990	1991	1992	1993	1994	1995	1996
Albania	6.85	5.16	3.61	3.03	2.80	2.38	2.08	1.98	2.32	na
Bulgaria	2.31	2.18	2.05	1.81	1.65	1.54	1.46	1.37	1.23	1.24
Czech Republic	2.09	1.93	2.07	1.89	1.86	1.72	1.67	1.44	1.28	1.18
Slovak Republic	3.07	2.40	2.32	2.09	2.05	1.98	1.92	1.66	1.52	1.47
GDR[a]	2.33	2.19	1.94	1.52	0.98	0.83	0.78	0.77	0.84	0.95
Hungary	2.02	1.97	1.92	1.84	1.86	1.77	1.69	1.64	1.57	1.46
Poland	2.98	2.20	2.28	2.04	2.05	1.93	1.85	1.80	1.61	1.60
Romania	2.33	2.89	2.45	1.83	1.56	1.51	1.44	1.41	1.34	1.30
FRY[b]	2.59	2.28	2.26	2.08	2.08	1.92	1.91	1.85	1.87	1.87
Croatia	2.20	1.80	1.92	1.63	1.53	1.48	1.52	1.47	1.58	1.67
Slovenia	2.18	2.10	2.11	1.46	1.42	1.34	1.34	1.32	1.29	1.28
Russian Federation	2.54	2.01	1.89	1.89	1.73	1.55	1.39	1.40	1.34	1.28
Estonia	1.97	2.16	2.02	2.05	1.79	1.69	1.45	1.37	1.32	1.30
Latvia	1.94	2.01	1.86	2.02	1.86	1.73	1.51	1.39	1.25	1.16
Lithuania	2.57	2.40	2.03	2.03	2.01	1.90	1.67	1.54	1.49	1.43
Belarus	2.76	2.33	2.03	1.91	1.80	1.75	1.61	1.57	1.39	1.31
Ukraine	2.23	2.09	1.95	1.85	1.81	1.72	1.55	1.46	1.40	1.38
Moldova	3.36	2.58	2.36	2.39	2.26	2.22	2.10	1.95	1.76	1.60

[a]German Democratic Republic. [b] Federal Republic of Yugoslavia.

Sources: See individual chapters and tables.

Table 1.3
Abortion Rates[a] in Central and Eastern Europe: 1960–96

	1960	1970	1980	1990	1991	1992	1993	1994	1995	1996
Albania	na	na	na	28.4	39.9	37.0	45.0	49.7	51.1	35.8
Bulgaria	29.0	63.2	76.7	67.8	65.6	67.5	52.1	50.0	50.2	51.3
Czech Republic	35.7	34.2	32.2	49.3	47.1	41.8	31.3	24.3	22.0	20.6
Slovak Republic	25.1	28.2	28.7	40.9	38.4	35.8	31.8	28.3	21.0	17.2
GDR[b]	0.2	6.2	25.3	20.4	15.5	15.0	11.0	9.0	8.3	10.1
Hungary	76.7	83.5	36.3	41.2	40.7	38.9	33.7	33.3	34.5	34.5
Poland	24.2	19.9	16.7	7.0	3.6	1.3	0.1	0.09	0.06	0.06
Romania	186.1	64.0	90.2	199.3	172.3	139.1	166.2	105.1	99.5	90.2
FRY[c]	na	na	na	na	70.7	62.4	51.4	52.8	44.9	na
Croatia	na	na	na	na	na	25.8	25.0	19.1	13.9	12.4
Slovenia	na	na	na	na	na	na	27.4	25.1	23.9	22.7
Russian Federation	161.2	156.3	140.2	125.6	109.5	103.7	97.7	92.1	83.3	79.8
Estonia	na	131.5	110.7	87.9	87.6	84.9	78.6	70.1	64.5	53.6
Latvia	na	123.0	107.8	87.2	69.1	61.2	57.2	49.8	48.9	46.1
Lithuania	na	63.4	59.3	61.7	50.2	50.1	43.0	37.1	38.2	34.0
Belarus	na	91.3	94.0	116.9	104.9	103.6	93.2	90.5	81.9	73.6
Ukraine	na	103.1	106.8	96.5	86.9	84.0	76.7	71.1	66.5	62.3
Moldova	na	101.9	101.2	83.1	73.5	70.9	63.5	58.0	56.7	45.8

[a]Per 1,000 women age 15–44. [b] German Democratic Republic. [c] Federal Republic of Yugoslavia.

Sources: See individual chapters and tables.

Table 1.4

Abortion Ratios[a] in Central and Eastern Europe: 1960–96

	1960	1970	1980	1990	1991	1992	1993	1994	1995	1996
Albania	na	na	225	318	393	390	454	438	459	405
Bulgaria	364	863	1,063	1,213	1,287	1,334	1,084	1,102	1,211	1,233
Czech Republic	524	486	448	820	797	768	574	504	503	513
Slovak Republic	234	346	328	605	584	571	529	526	479	419
GDR[b]	3	86	376	373	462	496	400	333	288	316
Hungary	1,107	1,267	544	719	707	716	644	644	687	728
Poland	225	271	192	108	56	22	2	2	1	1
Romania	2,186	685	1,036	3,153	3,149	2,657	2,343	2,149	2,125	1,972
FRY[c]	na	na	1,212	1,263	1,034	1,011	846	720	690	na
Croatia	92	606	701	697	778	557	520	405	285	229
Slovenia	295	310	646	656	648	650	616	580	568	543
Russian Federation	1,653	2,541	2,046	2,063	2,011	2,165	2,320	2,173	2,028	2,033
Estonia	na	na	na	1,318	1,523	1,578	1,684	1,584	1,507	1,271
Latvia	na	na	na	1,293	1,121	1,085	1,168	1,103	1,199	1,222
Lithuania	na	na	na	879	726	763	754	710	760	709
Belarus	na	na	na	1,825	1,783	1,830	1,812	1,878	1,866	1,753
Ukraine	na	na	na	1,590	1,517	1,562	1,544	1,531	1,501	1,470
Moldova	na	na	na	1,061	1,021	1,024	970	947	1,014	886

[a]Per 1,000 live births. [b] German Democratic Republic. [c] Federal Republic of Yugoslavia.

Sources: See individual chapters and tables.

Figure 1.1
Birth Rates in Central and Eastern Europe: 1996

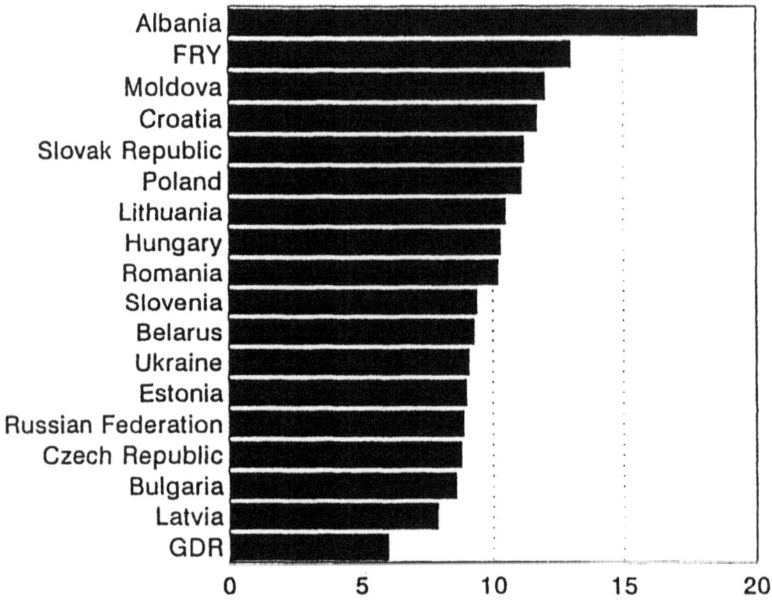

Figure 1.2
Total Fertility Rates in Central and Eastern Europe: 1996 (or latest year available)

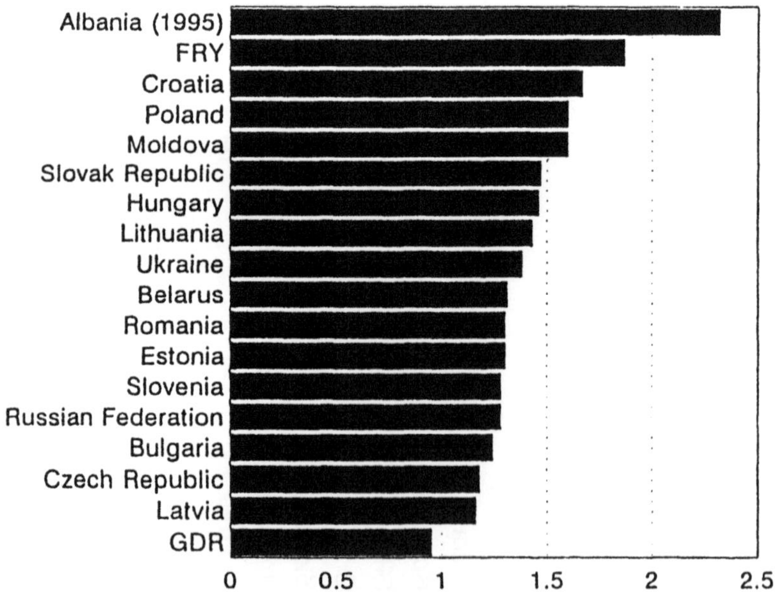

Figure 1.3
Abortion Rates* in Central and Eastern Europe, the Netherlands, and the United States: 1996 (or latest available year)

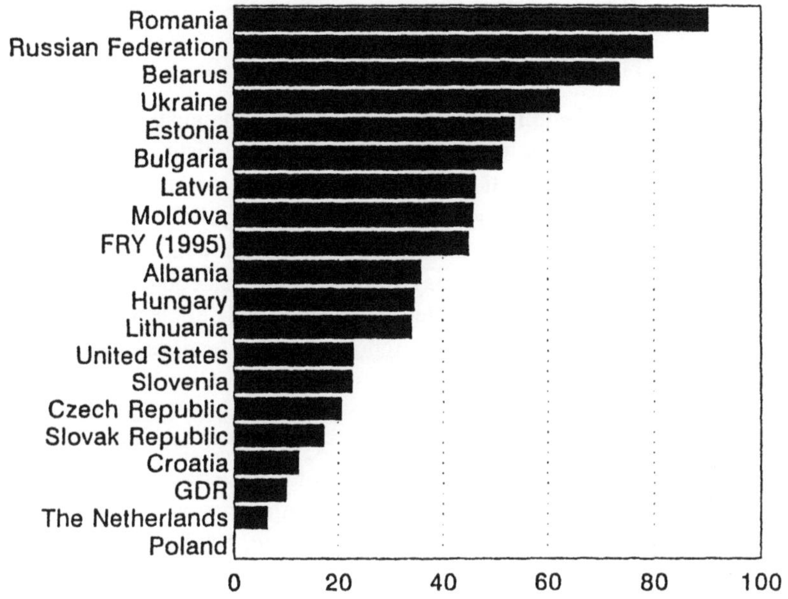

*Per 1,000 women age 15–44.

Figure 1.4
Abortion Ratios* in Central and Eastern Europe: 1996 (or latest available year)

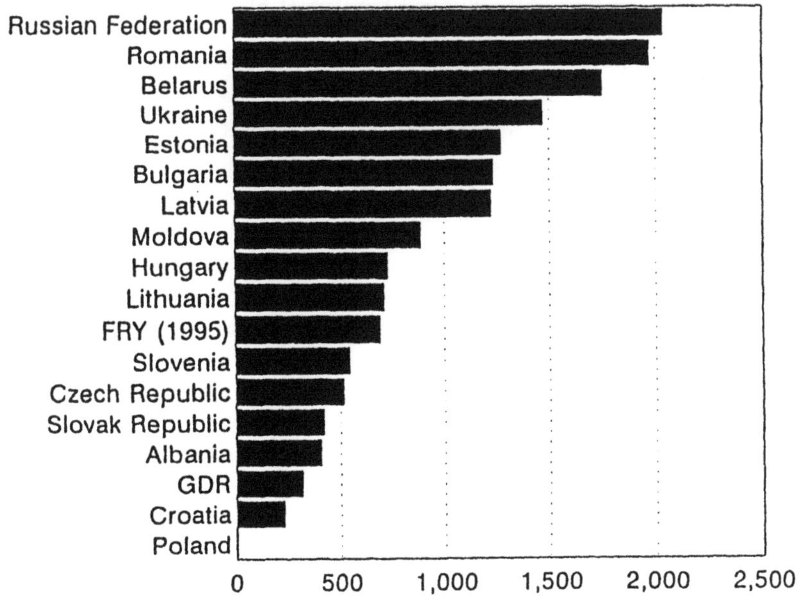

*Per 1,000 live births.

abortion laws. Changes in abortion legislation tended to accelerate rather than initiate the fertility decline (Stloukal, 1995).

POSTABORTION PSYCHOLOGICAL RESPONSES

Of all the potential complications of abortion, psychological responses are the most difficult to assess—far more so than mortality statistics. It is a question increasingly raised by antiabortion activists in the CEE countries, usually citing anecdotal observations. Assuming that psychiatric and psychological morbidity are real and measurable phenomena, the explanation for the wide range of opinions expressed in the Western literature may well lie in the inadequacy of much of the published work (David, 1985). Included in the scientific deficiencies are an overemphasis on clinical case histories that ignore the large majority of women who terminate unwanted pregnancies and never seek postoperative mental health consultation; the absence of standardized follow-up procedures; failure to reach consensus on diagnostic psychological criteria; or disagreement on psychological variables related to the sociocultural context within which the abortion decision occurs. Differing political, moral, ethical, and religious perspectives impinge on how abortion is perceived by diverse observers (David, 1997).

In 1989 the American Psychological Association convened an expert panel to examine psychological factors (Adler, et al., 1990, 1992). It focused on U.S.studies with the most rigorous research designs, reporting findings on the psychological status of women who had abortions under nonrestrictive legal circumstances, that is, on request in the first trimester and not solely on grounds of physical or mental health.

The panel found that psychological distress is generally greatest *before* the abortion when the woman has to decide how to resolve an unwanted pregnancy. Responses after abortion reflect the range of psychological experience and the resources a woman has for coping with negative life events. While there may be temporary feelings of regret, sadness, or guilt, the weight of the evidence indicates that legal abortion of an unwanted pregnancy in the first trimester does *not* pose a severe psychological hazard for the vast majority of women. Indeed, most women report experiencing a feeling of relief—of anxiety lifted (Russo, 1992).

In a longer term study Russo and Zierk (1992) noted women who had an abortion eight years earlier had a statistically significant higher global self-esteem rating than women delivering unwanted pregnancies (who had the lowest self-esteem). Women with repeat abortions did not differ in self-esteem from women who had never had an abortion. In all, the evidence confirmed earlier findings that factors other than the abortion experience itself determine postabortion emotional status, particularly how a woman perceives her pregnancy and how she believes it to be perceived in her immediate social environment (Adler, 1975). Some women continually reconstruct and reinterpret past events in the light of subsequent experience and can be pressured into feeling guilt and shame long afterward (Russo and Zierk, 1992).

Severe psychological reactions after abortion are infrequent. Psychoses are very uncommon. While individual case studies and anecdotal reports of severe stress or psychopathology following abortion abound in the literature there is no clear evidence of causal linkage to abortion. Although such responses can be emotionally overwhelming for the woman concerned and for her family, the number of such cases is very small, and has been characterized by former U.S. Surgeon General C. Everett Koop (1989:14) as "minuscule from a public health perspective." There is, as yet, no credible evidence for the existence of a postabortion syndrome (Stotland, 1992).

Women identified in Western literature as being at some risk for negative psychological reactions, and in potential need of special counseling, are those who (1) terminate a very much wanted pregnancy for medical reasons; (2) lack support from partners or parents for their decision; (3) are coerced into making a decision they subsequently regret; (4) are conflicted about deeply held religious values; (5) are uncertain of their coping abilities beforehand; (6) blame themselves for the pregnancy; (7) delay the decision into the second trimester; or (8) had a previous psychiatric episode (David, 1997).

There are few scientific reports comparing women who have had an abortion with women who have delivered a pregnancy. An extensive Danish study showed that, when controlled for previous psychiatric history, the risk of admission to a psychiatric hospital three months postabortion or postpartum was about the same for women who were married or living in stable partner relationships (David, 1985). A unique longitudinal study reported from the Czech Republic (Chapter 6) demonstrates that denial of an abortion request in cases of unintended conception and unwantedness in early pregnancy can result in detrimental psychosocial development of children born under such circumstances (David et al., 1988; Kubicka et al., 1995).

PRONATALIST INCENTIVES

As reported in the country chapters, in the late 1960s and 1970s the official views of CEE governments gradually reflected alarm about declining birth rates, prompting some (at times temporary) tightening of abortion laws while implementing pronatalist policies. These measures were diverse in nature and differed in terms of the strength and determination with which they were employed. Financial incentives gradually lost their value. Overall, they had the effect of encouraging earlier marriage and earlier childbearing, generally resulting in a temporary improvement of the birth rate as, for example, in the German Democratic Republic (Chapter 7). However, they seldom resulted in the birth of a third child, which was deemed to be needed for achieving the desired pattern of population reproduction. Nevertheless, it can be argued that without such policies, the birth rates might have been even lower (Frejka, 1998).

SEXUALITY EDUCATION AND GENDER ROLES

The individual country chapters provide ample evidence that in the first 15–20 postwar years, sexuality education did not exist in the CEE countries either in schools or in the media. After an initial dose of liberalism, Soviet-style communism turned puritanical and opposed sexology (see Chapter 11). While the German Democratic Republic had perhaps the most advanced program, a low level of sexuality education prevailed and persists in most CEE countries, reflecting diverse religious and cultural influences.

In addressing gender differences and sex roles, teachers directed most remarks toward girls. Fathers were generally depicted as important but mothers were regarded as irreplaceable. Men were perceived as providers and women as homemakers (Attwood, 1987). Where permitted, discussions of fertility regulation were left mostly to school physicians. All too often patriarchal concepts prevailed.

HEALTH PROFESSIONALS

The attitude of health professionals is central for the administration of laws. As "gate keepers," they often control access to modern contraceptives, abortion, and effective implementation of a women's decision (David, 1992a). Conditions of practice, perceptions of professional roles, and the economic status of the woman are usually interrelated. Historically, women with greater economic resources have had fewer difficulties obtaining desired services from qualified physicians than have economically disadvantaged women, whether such services were legal or not.

As will be noted in the country chapters, medical professionals played an important role in perpetuating the "abortion culture." This may be explained by isolation from contraceptive developments in Western countries, by severe shortages in the latest pharmaceutical products, and by lack of official interest in encouraging more effective contraception. Indeed, at times barriers were erected, as in the USSR for example. Abortion was generally a better and more regular source of income than the provision of contraceptive services, a situation that still prevails in some countries of the region (Pierotti and Blayo, 1994; Serbanescu et al., 1995).

The World Health Organization (1995) recommends that abortions be provided at the lowest appropriate level of the health care system with complex cases referred to higher level facilities. Appropriately trained midlevel clinicians possess the skills to perform first-trimester abortions and have been providing them safely where laws do not prohibit the practice (National Abortion Federation, 1997). However, the organized medical community does not favor training physicians' assistants, nurse practitioners, or nurse-midwives to provide first-trimester abortions (Kulczycki et al., 1996). Still, there is a perceptible trend toward better training in reproductive health that is sensitive to women's needs regardless of economic circumstances. International donors, nongovernmental organizations, and women's groups are raising social awareness and are working with sympathetic health providers and public facilities to extend the range and quality of specialized services.

FUTURE PERSPECTIVES

One of the major lessons to be learned from the experience of the CEE countries is that there will always be a need for safe abortion. However, this need will decrease and may approach that of the Netherlands depending on local circumstances. Even if contraceptives were widely available and guaranteed to be 100 percent fail-safe, unplanned pregnancies would continue to occur. Some women cannot always negotiate the use of contraception with their partners (Furedi, 1993; Rogow, 1993). The role of men in improving reproductive health is only beginning to be explored (Mundigo, 1998). Meanwhile, women will do whatever is necessary to terminate an unwanted pregnancy.

Although the availability and costs of modern contraceptives vary in each of the CEE countries, the choice of contraceptive methods has increased. As a result of international donor assistance and governmental programs, there is a greater supply of hormonal contraceptives, including injectables and implants, IUDs, condoms, and spermicides Emergency contraception is beginning to be available. Acceptability of contraception is improving, but, the high cost of imported products in the marketplace prevents wider usage by all women at risk for an unwanted pregnancy.

As noted in the country chapters, the officially reported number of legal abortions has declined and is continuing to decline nearly everywhere. While in part due to unreported procedures in commercial clinics or to increased costs for the woman, it is to a greater extent the result of an authentic shift from abortion to better contraceptive practice. There may well be a fundamental behavioral change from an "abortion culture" to a "contraceptive culture."

Few reasonable people would disagree with the public health policy goal of preventing unintended pregnancies and reducing the number of abortions, while ensuring that the vast majority of abortions are performed in the first trimester, when they are safest. There is also a growing awareness that abortion is a social reality that cannot be obliterated by legal codes. It is closely related to the empowerment of women and can no more be eliminated than the controversy surrounding it can be silenced.

As contraceptive practice improves with better access to a steady supply of modern methods, the incidence of abortion and repeat abortion can be expected to decrease further. While each country will have its own abortion/contraception experience, policymakers, service providers, and the public need to understand that some women are more fecund than others and that some contraceptive failures and human error are inevitable. Reproductive behavior and fertility regulation are matters of cultural tradition and evolving perceptions.

The 1993 World Conference on Human Rights reaffirmed that women's rights are an integral and inalienable part of all human rights. This principle was reiterated at the 1994 International Conference on Population and Development in Cairo and at the 1995 Fourth World Conference on Women in Beijing. In 1998, the fiftieth anniversary of the Universal Declaration of Human Rights, women's health, including sexual and reproductive health, was acknowledged as a human right (Sadik, 1998). While much work remains to translate reproductive rights into day-to-day realities, it is important to recognize that within less than a decade most of the CEE countries and their successor states moved from tightly controlled authoritarian political systems to more pluralistic forms of government with institutions, political parties, a semblance of free media, and evolving market economies (Quinn, 1998). At the same time, women are confronting harder conditions with less support from state allowances, daycare, and housing, motivating a desire for smaller families.

When viewed in historical and sociocultural perspective, fertility regulation, whether by pre- or postconceptive means, is not a new idea. Only the technical possibilities and social attitudes have changed over time. The evidence is persuasive that people can be motivated to change reproductive behavior and prevent unwanted pregnancies when they perceive themselves as playing an active role in determining their own future and in improving their family's quality of life. The experience of the CEE countries demonstrates that the extent to which ambivalences are resolved and equal access to modern and reasonably priced contraceptives and safe abortion is provided to women of all income levels will, to a large degree, be determined by whether public policy concurs or conflicts with private values (David, 1994b).

chapter 2

Understanding the "Abortion Culture" in Central and Eastern Europe

Libor Stloukal

EDITOR'S INTRODUCTION

Researchers of reproductive behavior in Central and Eastern Europe (CEE) are often asked for some explanation of the high incidence of induced abortion in the formerly socialist countries. Libor Stloukal suggests that the heavy reliance on abortion in the post–World War II period reflects political developments and prevailing social values. The circumstances of personal life amidst the realities and limitations imposed by the socialist regimes in CEE countries generated a cultural milieu in which people shifted from a sense of personal initiative and responsibility for their well-being to a dependence on the paternalistic state. The national health system provided free or low cost abortions while access to (often expensive and imported) modern contraceptives was erratic.

Libor Stloukal's chapter is based on research conducted in 1992–95 as a Ph.D. candidate in demography at the Australian National University in Canberra and builds on material originally presented at the Seminar on Sociocultural and Political Aspects of Abortion from an Anthropological Perspective, organized by the International Union for the Scientific Study of Population in Trivandrum, India in March 1996. His contribution is unique in developing an overview encompassing sociopolitical, cultural, and psychological trends from the perspective of a Czech observer.

The Introduction is followed by a discussion of the legalization of abortion and pronatalism as policies designed to facilitate social control based on male-centered concepts. Institutional adjustments and nonadjustments to social policies are considered in terms of health services, sexuality education, and the roles of the mass media. Behavioral adjustments and social values are noted within a context of the impact of contraceptive fatalism, inertia, and alienation on reproduction. The chapter concludes with some speculations on the influence of recent policy changes on future perspectives.—H.P.D.

INTRODUCTION

During the period 1960–90, the officially reported levels of legally induced abortion in the CEE countries were among the highest in the developed world. There are good reasons to believe that the true incidence of abortion in some of these states, notably Albania, Poland, Romania, former Yugoslavia, and especially the former Soviet Union, was even higher than the official figures. Only a few other developed countries, such as Japan in the early 1950s and Cuba since the mid-1960s, have experienced similarly high levels of abortion.

There are several competing explanations for the relatively high abortion levels in Eastern Europe (e.g., Callahan, 1970; David and McIntyre, 1981; Mazur, 1981; Blayo, 1991, 1993; and David, 1992a). According to the "low-cost-of-abortion" explanation, high abortion levels were a result of the liberal abortion laws implemented almost universally throughout the region during the period 1955–60; that is, before the mass introduction of modern contraceptives. A "high-cost-of-modern-contraceptives" explanation would stress the lack of political commitment to promote other methods of fertility regulation; the limited availability and "bad press" of modern contraceptives; and the resulting reliance of Central and Eastern Europeans on ineffective traditional means of avoiding unwanted pregnancies. Yet another possibility is to interpret the high abortion figures as a reflection of a pattern of fertility behavior that emerged in the late 1950s and proved to be largely resistant to social change under Soviet-style socialism, a pattern occurring under conditions of suppressed social dynamism and institutional conservatism. There are elements of truth in all these propositions. Disentangling the complex determinants of reproductive behavior is difficult.

SOCIAL POLICIES

CEE socialist regimes always viewed social policy as an important political instrument for social planning and control. Everyone was entitled to certain rights: to work, to leisure, to health care, to security in old age, and so on. The task of social policy was to regulate the ways in which these rights were implemented while retaining a productive and loyal workforce. Social policy was regarded as part of social planning reflecting political ideology. It was used in a frankly discriminatory way to favor the working class, to encourage the collectivization of agriculture, and to influence the use and distribution of the labor force. Social policy was also a device to implement socialism's ideological commitment to gender equality. With such a full agenda, there was considerable danger that it would sometimes work at cross purposes.

The period immediately following the communist assumption of power in the CEE states after 1945 was characterized by an upheaval in general socioeconomic conditions. Changes were greatest in those areas of life most amenable to formal institutional control: the economy and employment. Within a relatively short time span, the rather backward CEE countries were transformed into industrialized states whose economies were free from the influence of price fluctuations and trade boycotts. Masses of people were guaranteed work and an unprecedented level of social security. The sharp increase in women's economic independence and the granting of legal equality were important components of these changes. By the mid-1950s, economic emancipation of women reached a level that the West had still to match. There was no room for the idea that socialist planning might have a built-in gender bias because it incorporated primarily male values and priorities. Gender "equality" was often interpreted in terms of gender "identity." For example, it was recognized that women who worked eight hours daily still had to cope with more drudgery at home. Certain political steps were taken to lift some of the burden from family units, designed chiefly to reduce the need for women's extra work. These policies did not have much prominence until well into the 1960s, and even then they delivered generally low levels of benefits and services (David and McIntyre, 1981).

Important areas of criticism were opened by Stalin's death in 1953. The Khrushchev era (1953–64) temporarily stirred a debate over the legitimacy of centralized communist rule. In most CEE countries, the permissive abortion laws, allowing abortion on request or broadly defined social reasons, emerged concomitantly with the start of the post-Stalinist period. There are various ways of assessing this historical coincidence. Besemeres (1980) offers a possible interpretation, one that regards the legalization of abortion not as merely simultaneous with the political thaw, but essentially as its direct, politically motivated product. By this move, Soviet authorities could show their sensitivity to current social problems as well as their adherence to more permissive social norms. Most remaining CEE governments (the only exceptions being the East German and Albanian) quickly mimicked the Soviet model, as they were used to imitating almost everything that happened in that country in those years. However, socialist policies on fertility regulation remained contradictory. The failure to develop adequate provisions for modern contraceptives made women largely dependent on abortion.

In the 1960s the price to be paid for the early success in socialist reconstruction became fully apparent (Kornai, 1992). The priorities assigned to heavy industry caused indefinite postponement of development of the service sector that, according to Marxist theory, was to take over women's household tasks. The lag in consumer goods industries and failures in distribution made shopping for essentials a chore consuming several hours daily. Wages were such that families needed two incomes, and levels of female labor force participation were approaching a demographic maximum. The housing shortage meant that few young couples had immediate prospects for a place of their own. Women were forced to adopt the male norm of full-time employment throughout their adult lives, except for short interruptions due to maternity leave. The clearest evidence that women found it hard to combine employment with motherhood was the precipitous decline in fertility levels throughout the CEE countries between 1950 and 1960 (Berent, 1970).

To ensure an expanding labor reserve, an increase in the birth rate was considered essential. There could be no question of women returning permanently to the home. Macroeconomic realities did not permit a massive reallocation of resources to child care facilities and household services. Beginning in the early 1960s, all CEE governments began to encourage births by introducing pronatalist policies: prolonging paid maternity leave or raising family allowances, or both (David and McIntyre, 1981). Although costly, these measures seemed politically promising. They were likely to be perceived by populations not as manipulations to encourage them to have children they did not want, but rather as a relief from the obstacles to childbearing they experienced. In Bulgaria, Czechoslovakia, Hungary, and Romania, the governments came to the conclusion that people's right to control their own fertility had to be sacrificed in the drive to meet economic targets. Measures were also taken to make abortions more difficult to obtain. This was justified by an ideological stance implying that family planning was not a human or legal right, but rather a part of socioeconomic planning for which all individuals shared responsibility.

To describe the CEE pronatalist policies of the 1960s and the early 1970s merely as governmental efforts to change demographic patterns in the name of the communist future would obscure the organic connection between political means and political ends. Communist policies were always purposeful before they were social. Regardless of what aims they were proclaimed to serve, their origins were essentially political. Specifically, population policy played its role in the regime's strategy to deter individual freedom and to eliminate the possibility of public unrest. Communist leaders sensed that authoritarianism in political relations is incompatible with tolerance of autonomy in matters related to citizens' personal lives in general, and their reproduction in particular. Measures adopted differed from country to country, depending on the local culture and political mores. As will be noted in Chapter 6, Czechoslovak pronatalist polices of the 1968–73 period are an example of efforts to restrict undesirable forms of behavior by strengthening the family. Romania's experience under Ceausescu (1965–89) illustrates the implementation of policies designed to enable the regime to better control and discipline its citizens (Chapter 10).

From the mid-1970s to the end of the 1980s, population policies did not undergo dramatic changes in the CEE countries. Measures became less oriented toward the transformation of society and more toward the system's conservation and preservation of its human capital base. The regimes largely succeeded by suppressing major social problems rather than by solving them. They supplied certain consumer comforts to the citizenry; supported a high degree of female participation in the labor force; offered women job protection as well as financial security in motherhood; and, except in Romania and Albania, returned to the more permissive policies allowing fairly broad access to abortion. Thus, most CEE regimes apparently came to the conclusion that it made more sense to adapt to the consequences of long-term demographic trends than to invest in trying to reverse them. Such a strategy was more likely to be successful in maintaining political stability. Society paid for its comfortable social certainties by living at the expense of the future. Moreover, the comprehensive social welfare policies relieved individuals, to a significant degree, of responsibility for providing for their own living conditions.

Ideologically, social policies continued to be based on male-centered concepts of society and the family. However generous the intentions of the men who formulated them, policies were essentially aimed at making a male-dominated society function better. As noted in Chapter 3, freeing women from housework remained among the proclaimed goals, but very little progress was achieved. The opportunity to be a full- time mother turned into the "opportunity" to be a full-time housewife. Gender equality was not facilitated by the actual policies implemented. Measures were aimed at softening the double burden on women, not on redistributing the burden. The fundamental contradiction lay in a definition of women as workers *and* mothers without any parallel reconceptualization of men's role (Heitlinger, 1979).

INSTITUTIONAL ADJUSTMENTS (AND NONADJUSTMENTS)

Health Services

During the 1950s, the area of health services underwent substantial changes. In most CEE countries, with the exceptions of Poland, German Democratic Republic, and Yugoslavia, the vast majority of health-care facilities became state-owned. All legal and procedural parameters of medical practice were dictated directly by the government. Medical personnel had few options other than to act in conformity with directives from above. The health service was thus capable of providing social control and discipline on a scale much wider than in any pluralistic society.

Soon after the new abortion laws were implemented, the whole medical system adapted itself to new conditions. In the Soviet Union, for instance, large numbers of physicians started to work as full-time abortion providers (Popov, 1993). In Czechoslovakia, expertise in basic techniques of pregnancy termination were routinely required from doctors wishing to specialize in gynecology and obstetrics (Havranek, 1982), which meant that practically no individuals with objections to abortion could enter the profession. Once abortions were made accessible on request or on easy-to-meet social criteria, similar institutional responses occurred in the health-care systems of all CEE countries. The control of the process was placed in the hands of the medical profession, whose discretion in implementing government policy was thus confirmed. This was a component of the political strategy of suppressing autonomy and promoting management from above.

Although it would be unfair to say that CEE physicians "liked" to perform abortions, their attitudes to pregnancy termination were probably much less scrupulous than those of their counterparts in the West. The surgical nature of abortion operations apparently appealed to doctors' idea of "real" medicine and complemented their view of the need for medical experts to manage women's reproduction. The evidence for this is little more than anecdotal but nevertheless illustrative of the contrasts in approaches to abortion and to contraception. Moore (1972: 4) quotes a Yugoslav doctor as saying that "the psychological rewards to the physician—the gratitude of an abortion patient rescued from a crisis situation—far exceed the routine 'thank you' the patient may offer on simply being supplied with a contraceptive prescription." Medical research tended to focus on the clinical advantages of particular abortion techniques rather than on the problem of how abortion can be replaced by contraception (Sracek, 1988). Under certain circumstances, the physician's willingness to perform an abortion might even become a form of political statement. Kligman (1992) suggests that some Romanian doctors were inclined to meet the wishes of women seeking abortion during the 1970s and 1980s, as this provided an opportunity to demonstrate private opposition to Ceausescu's demographic tyranny.

As noted in Chapter 11 on the Russian Federation, CEE medical professionals played an important role as perpetuators of what had become known as "the abortion culture." Part of this reality can be explained by the lack of access to modern contraceptives, the doctors' isolation from contraceptive developments in the West, and by the lack of medical technology. This interpretation is supported by the well documented low level of state expenditures for health care in CEE states (Field, 1994), the observations of Remennick (1991, 1993, 1996), and the personal experiences of some Western physicians (Prager, 1987).

Sexuality Education

For the first 15 to 20 postwar years, sexuality education did not exist in CEE. Puritanism was a striking feature of orthodox Marxism. Marxism was similar to the traditional Christian position on sexual matters, while diametrically opposed to all other teachings. Official hostility to sexuality education was motivated by a feeling that it might promote sexual promiscuity. It was also believed that the family could deal with the matter more effectively.

During the 1960s and 1970s, all the CEE countries, except Albania and Romania, developed elaborate schemes for sexuality education, largely to complement the pronatalist policies introduced in those years (Wynnyczuk, 1974; Kozakiewicz, 1977). Characteristically, sexuality education was subsumed under "parental" or "familial" education, education for marriage and parenthood, or education for family life. These educational strategies were designed to cover a vast number of topics, ranging from the biology of human reproduction to propaganda about the place of the family in society and the social importance of childbearing. In theory, education for parenthood was outlined as the task not only of the educational system, but also of the family, the mass media, medical institutions, youth organizations, and even the army. Inevitably, such a complicated organization proved hard to command. In reality, it was the schools that played the major role in providing education on sexuality and parenthood.

Although the accounts of sexuality education in schools given by some CEE theorists depicted an almost flawless framework for providing knowledge to the young, the *de facto* situation was quite different. As reflected in the country chapters, biological aspects of sex were typically included in natural science lessons during a course on human anatomy, physiology, and hygiene for 8 to 15 year olds. There was little about family life or about the emotional and personal value aspects of human sexuality. A narrow version of sexuality education prevailed. It approached the subject as a second rate pedagogical exercise in delivering basic facts of physiology. Classes were frequently taught by school physicians, which meant that the lessons had a health emphasis rather than being placed into a broader context of personal fulfillment.

Puritanical views and prejudices usually dominated the educational process. Muckle (1987) reports that in the standard syllabuses in Soviet schools in the late 1980s the word "sexual" was rarely used, mostly in passing, and often with the sole purpose of condemning the importance of sex. Topics such as sexual debut and methods for preventing pregnancy were largely dependent on the willingness of teachers to absorb these subjects into the curriculum. In Poland the frequency of this type of sex education ranged from never in 49 percent of urban and 70 percent of rural schools to regularly in only 19 percent of urban and 9 percent of rural schools (Kozakiewicz, 1978). Gender roles were addressed in a 1983 directive for Soviet teachers. Fathers were depicted as important but mothers as irreplaceable. Gender differences in social roles were deemed natural and inevitable with men as providers and women as homemakers (Attwood, 1987).

With respect to abortion, educators could choose between two extremes. They could either praise the provision of abortion laws as a gain of the socialist system, or they could portray the phenomenon as legal but nevertheless alien to socialist morality. Both these positions were in certain senses justifiable, but they also contradicted everyday reality. The well-known high incidence of abortions largely prevented black and white interpretations. The question was further complicated by the unpredictability of official policy. Politics was thus among the major factors dominating the classroom context.

Mass Media

The values propagated by the socialist media did not conflict with those officially promoted by governments and taught during parental education courses in schools. Media, too, had adjusted to the widespread resort to abortion by virtually ignoring it. This was because the issue of abortion was likely to damage national self-esteem and to have corrosive implications for the doctrinal legitimization of state power—overtones that would hardly fit the journalists' main task of convincing the people that they were being well-governed and need not worry. Moreover, publicizing matters related to sex might have been understood by the censors as an attempt to disseminate Western style ideas of sexual freedom. It was only when the government's line required criticism of abortion as a social evil, complicating the economic future, that abortion was given more negative publicity. At such times, the media gave center stage to moral values rather than to sexuality and the conditions of everyday life, and tended to ascribe all problems to poor interpersonal relations rather than poor governmental policies. One infamous example of such practices was the very elaborate and pervasive media campaign on behalf of increased childbearing in Czechoslovakia during 1968–73 (Besemeres, 1980). The most aggressive campaign was implemented by Ceausescu's regime in Romania when horrifying details of botched abortions were routinely sensationalized in the media to act as a deterrent (Ionescu, 1987).

Ignorance, political tendentiousness, and lack of responsibility toward the public were thus the main characteristics of communist journalists' portrayal of issues related to reproduction and sex. Over the years, the unwillingness to examine controversial questions reached dimensions of a culpable negligence (Jakubowicz, 1992). For example, AIDS was long regarded as a malady afflicting only decadent capitalist societies (Williams, 1994). Consequently, many people sought relevant information in unofficial sources, such as erotic and pornographic literature. A 1988 survey of attitudes of 95 female college students in the former USSR showed a correlation between the development of sexual awareness and early familiarity with such unofficial texts, while exposure to the official press did not have any effect on the respondents' attitudes toward sex (Borisov, 1989). Ignorance regarding sex and reproduction was widespread among lay people and, paradoxically, even among media workers. In 1994 representatives of the Bulgarian Family Planning Association still complained that many Bulgarian journalists did not understand such basic terms as contraceptives or reproductive health (Chernev and Stamencova, 1994).

BEHAVIORAL ADJUSTMENTS AND SOCIAL VALUES

The occurrence and degree of "positive" behavioral responses to social policies was an important aspect of political and social culture. Most Central and Eastern Europeans displayed a remarkable sociopsychological tolerance for induced abortion, poorly covered by hypocritical silence at both individual and public levels. In Remennick's (1991: 844) words, "abortion was perceived for decades as a routine, although certainly unpleasant, medical procedure, comparable, say, to a removal of a decayed tooth." For some observers, this may merely suggest the destruction of religious consciousness in CEE societies. Was this really the case, or was the true reason to be found somewhere else?

Perhaps the best way to fully understand the abortion experience of CEE populations is to examine the habits and attitudes that developed during the years of communist rule. When the communists came to power after 1945, they launched a frontal attack against the traditional value system. They dismantled the traditional value-generating institutions such as churches, communities, associations, and social movements. They denigrated the values of the previous sociopolitical system, including values relating to private property, personal autonomy, freedom, and responsibility. However, the idea of replacing these values with those of collective discipline, altruism, and revolutionary consciousness failed almost completely. It failed because the regimes stuck to their anachronistic revolutionary ideology in an age of socialist conservatism and cautious reforms, because the almost total inefficiency of their economies ruined working morale and generated negligence, and because their authoritarian political systems excluded people from public life and created alienated, individualistic societies.

Among the most powerful factors shaping individual behavior was the eclectic and somewhat incoherent set of social values which a group of Czech sociologists (Mares et al., 1991) characterized as follows: (a) society should not be overly differentiated, people's living standards ought to be as equal as possible; (b) everyone must work, regardless of whether one's work is meaningful or productive; (c) state institutions know best how to satisfy people's needs, people need not care excessively for themselves; and (d) the living standard need not be high, provided it is secure. Other important features of the value orientation were small intergroup and generational differences, suggesting that the pattern generally prevailing within the population was only partly modified among subpopulations; a weak correlation between values and the demographic characteristics of the respondents; and substantial stability in the distribution of responses about values over time. Reflecting on this highly homogeneous and stable structure of values, one Polish sociologist remarked that, in the CEE countries, it would probably be more appropriate to speak about one system of values than about multiple systems (Nowak, 1981).

It is not abnormal for most people in most countries to live nonpublic apolitical lives and to focus most of their attention on the material well-being of themselves and their families. In CEE, however, consumer values became extremely widespread during the years of communist rule (Matejko, 1991; Grad, 1993). From the point of view of individuals, consumerism was a reaction to the difficulties of everyday life, marked by capricious shortages and sudden supplies of goods. The gulf between consumer values and those of official Marxism-Leninism did not mean that the existence of, for instance, black markets and a popular desire for more consumer goods was necessarily a source of political instability. Many, though by no means all, unofficial economic activities actually helped the economies to work more smoothly than they otherwise would have. Moreover, an individual's preoccupation with standards of living could have been an alternative to authentic political activity- an enterprise potentially dangerous for the individual, and consequently for the system as well. Over the years, attempts to satisfy the material demands of the consumer became one of the principal ways in which CEE communist leaders could supplement a purely Marxist-Leninist appeal for support within their societies (Brown, 1979).

The spread of consumer attitudes in CEE societies had many important consequences. First, consumerism provided a source of distraction from those aspects of life that ordinary people found frustrating. It helped to push religious and other spiritual values down in the social value system. Consumerism was intimately interconnected with familism, as many people evaluated their standard of living in terms of family welfare and displayed a strong pro-family orientation. Another consequence was social atomization. In people's minds, society became a confederation of primary interest groups and closed family-like communities, identified through blood ties or shared economic goals and strong intragroup solidarity (Horowitz, 1979). Individual action was guided mainly by short term materialist interests and by respect for the existing value system and power structure. For most individuals, the struggle to secure higher consumer standards was more important than a struggle for the solution of other problems, such as gender inequality. The orientation of both genders towards the private sphere of family life made them partners, which may be a partial explanation of the virtual absence of a genuine women's movement in the CEE countries.

Hand in hand with consumerism, perhaps the most important consequence of Soviet-style socialism in the realm of social psychology was a tendency toward anonymity—the absence of any distinctive, forceful personal identity, and the avoidance of initiative and responsibility. While in a community of free persons initiative and responsibility are valued, mediocrity was ideal to survive under oppression. For most CEE citizens, initiative and responsibility were luxuries they could not afford, a potential source of conflict with the repressive political environment, and thus maladaptive. The best adaptive strategy was maximum flexibility in fitting into a context (Kohak, 1992). Such behavior was often accompanied by a learned helplessness syndrome (Ziolkowski, 1993)—the passive demand for and expectation of protection and of minimum social and economic security provided by the state regardless of an individual's efforts.

Contraceptive Fatalism

Fatalism and a sense of powerlessness are antithetical to planning for a successful future. The fatalistic person believes that future happenings are inevitable, that one is not in control of one's life and powerless to alter the course of events. Fatalism, inertia, and alienation were the way of life for many Central and Eastern Europeans.

One of the causes of fatalism was that for a large part of society individual life goals and aspirations became unattainable through legally available means. The blocking of the attainment of life goals was accompanied by a strong feeling that little could be achieved in a society whose functioning was chaotic. Powerlessness was taken for granted, often resulting in passive fatalistic alienation (Hollander, 1973).

Research focusing on the links between fatalism and reproductive behavior in CEE countries was rare and seldom met acceptable scientific standards. The few existing studies were largely empirical, showing that fatalism ranked high among reasons reported by women for not using modern contraceptives (e.g., Fukalova and Uzel, 1975).

Beginning in 1987 official Czech and Slovak abortion statistics recorded what kind of contraception a woman used immediately before she last became pregnant. They too reflect a high degree of fatalism (Czech Republic/IHS, 1995).

During 1987–90, 93 percent of abortion obtainers in the Czech Republic claimed they did not use any contraception, only 5 percent used an IUD, and just 2 percent relied on the pill. For the Slovak Republic, the corresponding figures were 96 percent, 3 percent, and 1 percent. Taking the sum of statistically recorded pregnancies as a base, the annual proportion of those conceived in circumstances of a total absence of contraception and terminated by an induced abortion was 42 percent in the Czech Republic and 36 percent in the Slovak Republic (Czechoslovakia/FSO, 1993). These results indicate that abortions were predominantly the result of contraceptive nonuse rather than of contraceptive failure. Similar findings have been reported from other CEE countries. Popov (1990), reviewing the official Soviet statistics for the 1980s, estimated that as many as 25 percent of the female population of Moscow adopted abortion-oriented attitudes and behavior. The prevailing fatalism in reproductive behavior of Soviet couples is also emphasized by Remennick (1991) and Kon (1993). Interestingly, resort to abortion was likely to persist even when women born in Eastern Europe were subsequently exposed to an environment with more developed contraceptive resources. For example, among Russian-born Jewish female immigrants to Israel, age-specific rates of applications for abortion were significantly higher than among long-resident Jewish women, and continued to remain so even after 10 to 15 years of residence in Israel (Sabatello, 1992).

While these findings are indicative in themselves, they need interpretation. One way is to relate them to existing theories about why women have unwanted pregnancies. According to Luker (1975), there are two grand theories. The first, commonly associated with the classic concept of the demographic transition suggests that women have unwanted pregnancies because they lack the contraceptive experience or technology to prevent them. The second proposes that couples often have the contraceptive skills to prevent pregnancies but encounter psychological resistance or lack of motivation when it comes to using them. Psychic conflict regarding contraception is in a certain sense natural, as conscious and unconscious reasons for and against its use often exist simultaneously. For instance, some women may experience conflict between sexual desire and not wanting a child or being uncertain about the partner's commitment. Similarly, there may be a conflict between the perceived inconvenience of a contraceptive technique and the relative convenience of spontaneous (unprotected) sexual intercourse. Indeed, for some individuals certain disadvantages of contraception may outweigh certain advantages of abortion (Moore, 1971). Contraceptive decisions (and nondecisions) may thus be seen as being made on the basis of balancing the costs and benefits of risky but spontaneous sexual intercourse versus the costs and benefits of pregnancy and its possible termination by an abortion.

Seen from this theoretical perspective, the widespread resort to abortion in CEE countries becomes somewhat easier to understand. It was not only an outcome of irrational behavior, or of erratic contraceptive supplies (though both played a role), but also an outcome of prevailing value systems and social relations. Because abortion was relatively easy to obtain and largely socially acceptable, engaging in sexual intercourse without much regard for the consequences to oneself or one's partner was psychologically not so distinct from other kinds of behavior in which people relied, consciously or subconsciously, on the protective capacity of the paternalistic state. Inexpensive and easily available abortion services (in most countries), heavily subsidized food and rents, full employment, and provision of free or cheap health care and education, all represented a type of social contract between the party-state regime and the people (Silver, 1987). And, the more individuals relied on "society" as the guarantor of social policy achievements, the less space remained for articulating and working through conflicts involving questions such as fertility regulation. Many abortion clients behaved more as objects of provisions rather than as active subjects capable of defining their needs. Social benefits and services were so taken for granted that they had no stimulating or motivating effects, and often resulted in the passivity and resignation of the majority of the population. The opportunity for an individual to exercise influence, either through the marketplace or the ballot box, was entirely absent.

Thus, there was an intimate connection between the lack of pregnancy prevention, abortion, and the perception of social welfare entitlement as compensating for the shortcomings of daily life. However, the state-father was not perceived equally by all people. There were opportunities to exercise individual initiative and to make choices. This is reflected by the substantial intercountry and intergroup differences in abortion rates and ratios, discussed in the relevant literature (e.g., David, 1970; David and McIntyre, 1981; Frejka, 1983; Blayo, 1991; Anderson, 1987, 1991; Anderson et al., 1992; and Stloukal, 1995). It might be hypothesized that underlying the differences in resort to various strategies of fertility regulation were differing attitudes and interpersonal relations. More specifically, it seems that the impact of communist rule on individual reproductive behavior was partly a function of factors such as religion, education, social status, and so on. Selective attention to the risk of unwanted pregnancy and preferences for different types of risk-taking (or risk-avoiding) behavior corresponded to cultural biases—to world views, deeply held values, beliefs, and normative presuppositions defending different patterns of action. A broader quantitative analysis of the various social outcomes from Soviet-style socialism, and their differential effects by class, country, and so on, would be necessary to elaborate this hypothesis further.

FUTURE PERSPECTIVES

The collapse of the communist governments across the CEE countries 1989–91 brought to an end a particular type of political, economic, and welfare system. Most of the previously Soviet-dominated countries underwent unprecedented changes in the years immediately following the breakdown of the old regime. The transition was of a magnitude and importance rarely experienced in human history. Within a few months, or even days, in most countries of the region one-party regimes were replaced by multiparty democracies. By and large, the changes occurred peacefully;

it was only in a few instances that the reform forces clashed violently with proponents of the status quo. However, while the early 1990s brought about significant political and socioeconomic changes that many Central and Eastern Europeans had dreamed of for decades, the period also gave rise to a plethora of fresh, previously nonexistent or hidden problems, such as nationalism and ethnic hatred. Of great significance was the dissolution of the multiethnic federations—Czechoslovakia, the Soviet Union, and Yugoslavia—and the subsequent creation of more than 20 descendant states. Several of these newly established states have experienced civil or intercountry wars on their territories, ranging from temporary conflicts of minor importance to devastation on a scale not experienced in Europe since World War II.

One of the principal reasons for the collapse of Soviet style socialism was its inability to reform the national economies by freeing them from the pitfalls of central planning. It was therefore natural that most postcommunist CEE governments launched complex transitional programs soon after they assumed power to facilitate a shift from centrally planned to market economies. Access to a market economy generated enormous hopes for prosperity and better living conditions. While the economic, social, and psychological costs of this transition have varied greatly, the economic restructuring has proved to be an extremely difficult process (Boh, 1993). Its price has been declines in production, employment, and real incomes, frequently accompanied with rises in unemployment and inflation—two phenomena that several generations of Central and Eastern Europeans had never experienced.

One feature that has characterized the post–1989 CEE era is the drama of self-discovery. Step by step, these countries have been discovering that they are left with crippled societies, with the remnants of dogmatism, the inertia of state-centralist practices, the egalitarian-uniformist psychology, the proliferation of mediocrity, and the absence of a long-range horizon for social action and motivation for people. Not surprisingly, issues bearing on population reproduction have emerged as important topics of debate. Welfare policies are at a crossroad: the former, "socialist" model is criticized and rejected in a wholesale way, but the future of social policies still seems to be open and largely dependent on economic prosperity. "Socialist" abortion laws are almost universally under attack in campaigns by the previously silenced Catholic Church, which asserts spiritual and moral values centered on the family and the traditional role of women. The only exception, at least for the time being, is Romania, where women were previously brutalized by the prohibition of abortion. In most other postcommunist countries, restrictive legislative measures to reduce abortion have been proposed, although it was only in Poland and in Hungary that new laws limiting the availability of abortion for nonmedical reasons were enacted. More specific information is provided in the chapters to follow. Detailed reviews of restrictive abortion policies in Poland and Hungary may be found in Fuszara (1993); Zielinska (1993); David and Titkow (1994); Gal (1994); and Kulczycki (1995). In Poland the campaign to criminalize abortion was accompanied by efforts to restrict sexuality education and access to modern contraceptives. In Hungary the wording of the new law was more restrictive than its implementation.

A countervailing abortion-rights pressure stems from the emergence of national family planning associations, various pro-choice organizations, and the new autonomous women's movements (Einhorn, 1993; Funk and Mueller, 1993; Feischmidt et al., 1997). Regrettably, many policy issues that should ideally be seen as social or economic topics, represent very delicate political concerns. Public debates about abortion are seldom linked to sexuality, women's autonomy, or the private sphere. More typically, they involve notions such as anticommunist morality, nationhood, demographic decline, and the defense of humanity. Abortion thus exemplifies one of the highly symbolic themes in relation to which much broader political arguments are discussed.

From the perspective of ordinary people, the effects of recent political changes appear to be somewhat less dramatic. In countries like the former German Democratic Republic, the Czech and the Slovak Republics, and Hungary, the supplies of modern contraceptives are now, mainly thanks to the privatization of pharmaceutical markets, incomparably better than they were during the 1970s and 1980s. But, there are still many spheres of reproductive life where individuals do not have much choice. Inertia is most evident in the area of health services, where it seems extremely difficult to shift professional attitudes away from curative to preventive health care in general, and from abortion to contraception in particular (Visser et al., 1994). Sexually oriented journalism has not greatly affected the tradition of public as well as private silence in sexual matters. Moreover, inasmuch as children are the manifestation of the survival of their parents' value systems, they may also carry the reproductive culture of previous generations. Behavioral stereotypes which came into being in the CEE countries over the last 40 years represent a bizarre combination of precommunist values, Marxist teachings, and the survival techniques that grew out of a culture of permanent scarcity. They helped individuals to endure the hardship of communism. There is no reason to expect that what proved efficient so many times in the past will simply disappear in the future.

A great deal of the divergence in the evolution of individual postcommunist states reflects variations in the depth and breadth of the transformation and the attitude of the people. Intercountry differences in socioeconomic conditions and policy strategies play an instrumental part in shaping the social as well as demographic developments of particular populations (Macura, 1995). Recent changes in abortion levels in the Czech and the Slovak Republics, two countries in which statistical recording of abortion is considered to be of very good quality, are illustrated in Chapter 6. The substantial declines in abortion rates since 1989 have been attributed mainly to a shift away from paternalistic welfare policies, to the introduction of fees for procedures performed for nonmedical reasons, and to the liberalization of markets and the subsequent improvements in the supply of modern, Western-made contraceptives (Pavlik, 1994; Tirpak and Adamica, 1995; Kucera and Fialova, 1996). There is little doubt that apart from these policy-induced factors the declines in abortion rates have been by-products of other circumstances as well.

These include alterations in family formation patterns and gender roles, and the emergence of a wide range of new, previously unavailable or even unthinkable lifestyle choices for younger generations. However, it is worth noting that during 1990–94 the decline in abortion paralleled a decline in overall fertility. Both trends seem to have a common underlying cause: the advent of a new socioeconomic climate that has greatly accentuated the need for economic calculation as a means of individual survival and well-being. Reacting to this new environment, large segments of the population apparently have abandoned the old reproductive habits and pragmatically adopted wait-and-see forms of behavior, including a more cautious approach to sexual risk taking. Nonetheless, judged by West European standards, abortion levels remain high. In 1995, around 30 percent of identifiable pregnancies in both the Czech and the Slovak Republics resulted in induced abortion, a proportion higher than anywhere in the West. In other words, it seems that the reasons for and the circumstances surrounding Czech and Slovak couples' recourse to abortion as a means of limiting family size have brought about a widespread societal pattern that is likely to fade away only gradually. Thus, like the communist interlude itself, the postcommunist recovery from this aspect of the region's legacy promises to be a historically unprecedented and regionally diverse process.

There is still not enough attention paid by CEE decision makers, medical professionals, and social scientists to abortion as a sociocultural phenomenon. Much still has to be accomplished to change people's inherited passive attitudes into active ones; and thus to reduce abortion to a backup method for rare failures of normally effective contraception (which many would agree is the only role it should have in a developed society). Bringing the issue out in the open and forcing societies at large to face the problem, instead of hiding it, has been only the first necessary step in moving toward more efficient fertility regulation. Although the emerging demographic trends of the 1990s suggest that rationality and responsibility for individual behavior is increasing, the transition from abortion to contraception remains an incomplete process in the CEE countries.

chapter 3

The Woman Question

Henry P. David and Joanna Skilogianis

INTRODUCTION

Interest in gender inequality has a long history. Often defined as a set of social and cultural practices that differentially influences the lives of women and men in nearly every society (Presser, 1997; Riley, 1997), concern with women's rights and gender inequality was also an early tenet of Marxist-Leninist movements. Equal rights for women were imposed from above in the Soviet Union in the 1920s and after 1945 in the CEE countries (Tobias, 1997).

Within the context of assessing the current position of CEE women, this chapter traces the historical development of socialist views on the "woman question," a phrase introduced in the nineteenth century writings of the German Social Democratic Party theoretician, August Bebel (1971{1875}), and subsequently used by others (e.g., Lenin, 1951; Zetkin, 1958). Since specific trends are reviewed in the separate country chapters, this overview provides a broad perspective that emphasizes developments before and after the October 1917 Revolution in the Former Soviet Union, considers the divergences between rhetoric and the realities of daily life in the formerly socialist countries, and examines the effects that the political and economic transitions are having on such issues as women's rights, reproductive health, and feminism. Preparation of the chapter benefitted from the suggestions of Nadezhda Azhgikhina, Elena Ballaeva, Harriet Presser, and Anastasia Posadskaya-Vanderbeck.

AN EARLY SOCIALIST PERSPECTIVE

In many parts of Europe, the late nineteenth century witnessed growing claims by women for legal rights (Malysheva, 1992). When initially published in 1875,

Bebel's book, *Women in the Past, Present, and Future* (later renamed *Woman and Socialism*), represented the first attempt to study the "woman question" from a socialist perspective. Although revised and republished in numerous editions and translated into 15 languages including English (Bebel, 1971), Bebel's book never assumed the theoretical significance attained by Friedrich Engels's *The Origin of the Family, Private Property, and the State*, originally published in 1884 and widely translated (Engels, 1972). Both authors held that the oppression of the proletariat in general and of women in particular was rooted in the institution of private property and the economic isolation of the nuclear family. As private property could be abolished only by a socialist revolution, the emancipation of women was seen as an element in the proletarian struggle against capitalism.

Both Engels and Marx were concerned with how wealth was created and distributed in society. They postulated that there could be little justice as long as one class worked and another appropriated the fruits of that labor. Within this largely economic concept, the relationship between the genders was perceived as part of the process of production. Engels noted that the first division of labor between man and woman was for the reproduction of the species. He held that the male-dominated family had existed in some form since the beginning of recorded history, usually operating as a production unit to secure its members' existence and the future of their children (Chao, 1977).

In an 1885 letter Engels wrote "It is my conviction that the real equality of women and men can come true only when the exploitation of either by capital has been abolished and private housework has been transformed into a public industry" (Marx and Engels, 1934: 462). It was anticipated that the nineteenth-century Industrial Revolution would mark the beginning of the final historical conflict between propertied and propertyless classes. The result would be a society where the class producing material commodities would, for the first time, also own the means of production. Women were in the special position of being exploited as unpaid workers in the home and as wage laborers outside. Even in the propertied bourgeois family, the wife was perceived as wholly owned by her husband. "Equal rights for men and women" or "equality of all citizens regardless of sex" became a key component of Marxist ideology (Marx, 1936; Lenin 1966a). Women's participation in the labor force was seen as the key to their emancipation (Hartmann, 1981).

Considering the traditional commitment of Marxist movements to women's emancipation and the total volume of writings by Marxists, the amount of theorizing devoted to the "woman question" was sparse. Engels's book was perceived as having laid the foundations for a socioeconomic theory of women's oppression and a political strategy for emancipation (Sacks, 1975).

Bebel's writings were important for his sensitivity about women as human beings. He realized that education and employment were important for achieving economic independence and for reducing frustration associated with closed career paths. He also recognized that many women were as concerned about their own sexual needs as with sexual exploitation by men. Not merely a bundle of maternal instincts, women yearned to be complete and useful members of society outside the realm of the family.

For early Marxists, the woman question was never the feminist question. (Hartmann, 1981). Bebel, Engels, and Marx analyzed the woman question in terms of women's relationship to the economic and political system rather than in regard to gender inequality or male dominance. For Engels, women's labor force participation was the key to emancipation. His primary focus was on class relations, not on the philosophical differences between women's and men's experiences under capitalism. Marxism dominated feminism.

Building on Bebel and Engels, Klara Zetkin (1958) prepared a report on the woman question for the 1896 Congress of the German Social Democratic Party that laid the organizational foundation for a socialist women's movement. She drew a distinction between bourgeois feminism (a few privileged women struggling against the men of their class) and socialist feminism (women struggling together with men of the working class against the capitalist class as a whole).

The first major Russian work on the woman question was written in 1900 by Nahezhda Krupskaya, the wife of Lenin. Drawing on her experience among St. Petersburg workers in the 1890s, she wrote about perpetually pregnant working women without maternity benefits whose children either died or were nursed by strangers and raised on the streets. Under socialism, society would be responsible for the care and education of all children. Krupskaya acknowledged existing male prejudices and criticized workers who argued that politics should be a male preserve (Stites, 1978).

Little else was written by Russian Marxists until 1908 when Aleksandra Kollontai published *The Social Basis of the Woman Question*, a powerful appeal for a separate organization of women, a view not widely supported by the Communist Party at that time or later (e.g., Kollontai, 1971a; Stites, 1978). In 1913 Kollontai was among the few Russian social theorists who expanded on Engels's views on personal relationships and called for greater experimentation and the liberalization of the restrictive abortion legislation (e.g., Kollontai, 1971a, 1971b, 1972, 1977; Clements, 1979; Heitlinger, 1979; Farnsworth, 1980). However, for many Marxists, feminism was less important than class conflict and deemed divisive of the working class (Hartmann, 1981).

THE BOLSHEVIK REVOLUTION

The situation evolving in the Soviet Union after the October 1917 Revolution was not ideally suited for the implementation of goals established by socialist women's organizations. Party leaders confronted what they considered to be more urgent priorities in making the Russian Revolution secure. Propaganda work among women was deemed of secondary importance by male communists (Zetkin, 1929). Memoirs of the early 1920s suggest that Lenin was impatient with attempts to introduce radical new sexual notions into Communist Party policy, adhering to the view that the revolution demanded more discipline and less personal freedom. In two letters to Inessa Armand, Lenin (1966b) characterized "women's demands for free love" as bourgeois, which should be subordinated to the political needs of the time. In a 1920 conversation recalled by Zetkin (1929), Lenin deplored women's preoccupation with sexual matters, insisting that revolutionaries had better things to do. He rejected the polemics of Kollontai, admonishing women workers that "equality before the law is not yet equality in life" (Lenin, 1963). His stand on sex and "free love" was perceived by others as existing well to the right of the overall trend of party policy and thought at the time (Heitlinger, 1979; Besemeres, 1980). Feminism was considered nonrevolutionary and sometimes even as counterrevolutionary (Zetkin as cited in Malysheva, 1992). It became a victim of the closed society in which Marxism was "sex-blind" (Hartmann, 1981).

The period just before and after Lenin's death saw the approval of some of the most far-reaching reform legislation ever passed on sexual, marital, and family matters. While Lenin considered free access to abortion on request as a right to be made available to the working class, he also deemed a high incidence of pregnancy termination a roadblock to building a new socialist society.

With the rise of Stalin, traditional gender roles were reasserted, restrictions on abortion tightened, and Kollontai's works no longer published. An inverse relationship began to be postulated between sexuality and the class struggle, between the heroic "building of socialism" and the "unhealthy" preoccupation with private pleasure. The people of the Soviet Union were urged not to indulge in socially unproductive sexuality but to invest their energies in fulfilling five-year plans (Geiger, 1968; Goldman, 1993).

Innovative Marxist thought on the woman question came to a halt. Besemeres (1980) suggests that family and population policy promulgated during Stalin's time represented a retreat from Engels, and presumably from Marx, but probably not from Lenin's private views.

RHETORIC AND REALITY

The Socialist Era

As noted by Heitlinger (1979), most early socialist writings on the woman question were based on Engels's hypothesis that the emancipation of women would result from the abolition of private property (along with class rule and exploitation), the productive employment of women (giving them a public social role and economic independence), and the socialization of private domestic work and child care (which would provide time for self-cultivation and public life). In the socialist era much more attention was given to the first two principles than to the third. In the time of Stalin, for example, publications tended to equate formal equality with day-to-day practical equality, implying that the high level of women's labor force participation was tantamount to complete liberation. However, questions of sharing housework with male partners and/or providing services to ease daily household burdens were largely ignored amidst the priorities given to increasing stocks of capital goods.

In the years after World War II the Soviet model of socialism was adopted by the diverse CEE countries. The situation of women differed in each country according to its cultural traditions and aspirations and as influenced by its economic development. The numerous advances said to have been achieved by women, their supposed equality before the law, and the struggle to attain equality in economic and family life have been well described in several publications (e.g., Scott, 1974; Sokolowska, 1977; Heitlinger, 1979; Funk and Mueller, 1993; Rueschemeyer, 1994a, and others) and are noted in the separate country chapters.

The reliberalization of abortion in the mid-1950s represented a significant advance for women's reproductive rights. While legal abortion provided a safe alternative to unsafe clandestine abortion, the new family policies did not transform motherhood into a blessing that could be readily combined with self-fulfillment in other areas of life (as Marxist theorists had assumed). By the 1960s the socialist countries had some of the highest abortion ratios and lowest birth rates in the world. These developments produced prolonged controversies between advocates of women's rights and policymakers who were more concerned about changes in population structure, future labor resources, and prospects for long-term national survival. The ambivalence with which the former socialist countries regarded abortion is apparent from the differences in legislation promulgated, ranging from availability on request to strict prohibition. At the same time, nationalist, demographic, and economic objectives fostered pronatalist incentives of varying monetary value.

In considering the official legal rights gained by women and the difficulties encountered in translating these rights into positions of importance and responsibility equal to men, the major barrier appears to be woman's biological role as the bearer of children. Few of the early Marxists perceived that socialism did not alter the reality that the working world had been structured largely by men for their convenience and that any inability to adapt to male standards would be considered a female shortcoming. Marx and Engels assumed that rational behavior would prevail once the capitalist traditions and influences had been drastically altered, that socialism would create ideal conditions for family life, and that couples would wish to have more children. It apparently did not occur to them that aspirations for a better quality of life would influence fertility behavior and motivate people to reduce family size.

While the CEE countries were ideologically committed to the goal of full equality for women, this commitment was secondary to the overriding economic concern of how to make fullest use of women's productive and reproductive capacities (Peers, 1985). Work was a duty, not a right. Low wages necessitated two incomes for family survival (see Chapter 2). Innovations in family policy and legislation introduced over the years were perceived as attempts to encourage joint parental responsibilities and foster equilibrium between women's employment and childbearing at maximum effectiveness, given individual and cultural circumstances (Scott, 1977; Wolchik, 1989). The availability of prolonged maternity benefits for women who elected to stay home for two or three years with their child emphasized gender role differences that weakened the actual position of women in the labor market. Women who returned to work after such an absence often found themselves several steps behind their male colleagues and seldom caught up. The frustrations they encountered were further complicated by the perception that gainful employment was a central component of a woman's social role. Women were educated with the expectation that they would work outside the home for a significant portion of their lives and that, with society's assistance, they would rear children as successfully as women who devoted full time to their families.

In actuality, socialism did not bring with it the desired expectations of equality for women. As Hartmann (1981) notes, societies can make the transition from capitalism to socialism yet still remain partriarchal. Despite creating a cultural milieu where economic involvement was assumed to be as central to a woman's life as to a man's, the CEE governments did not achieve the structural transformation of society required to permit equality in career outcomes for women who chose to bear children. While the expansion or improvement of laundry, cleaning, food preparation, and shopping efficiency emerged as openly acknowledged pronatalist measures, steps were never taken to make maternity consistent with women's increasingly heightened career expectations (Meyer, 1985). There was little cultural pressure on men to lift the double burden and absorb at least an equal share of household obligations.

The socialist experience with the woman question reveals the true depth and enormity of the problem (Scott, 1976, 1977). The abolition of private property did not resolve what Engels called the "conflict between the sexes." Belief in women's inferiority was older and more deeply ingrained than economic class struggles and could not be entirely resolved by improved systems of law, education, or employment. Further changes depended on transformations in social microstructures, in traditional gender roles, and in family relationships. Laws were not always enforced and social policies were seldom well implemented. The idealistic goals of the socialist system were unfulfilled in practice. While progress was achieved in official terms, women's daily burdens actually increased (Heitlinger, 1980).

Efforts to translate rhetoric into reality were met with varying success in the different CEE countries (Gal, 1994; Verdery, 1994). The idea that gender equality could be achieved merely by increasing women's participation in the labor force and allowing them to attain some form of economic independence was not enough to achieve the goals of a socialist state. Entry into the labor market meant for many women an assumption of a triple burden as wives, mothers, and workers. The majority of women worked in low status unskilled or semi-skilled jobs. When women advanced to traditionally male positions, such as physicians or professors, these professions became "feminized" and tended to lose status and remuneration (Peers, 1985).

Production and Reproduction

The ways in which the socialist countries handled the balance between labor production and reproduction differed considerably. In the former German Democratic Republic (see Chapter 7), for instance, women were able to achieve some degree of economic and political influence although they never attained the most prestigious positions in their fields. The state continued to improve health services for mothers and children as well as provide affordable daycare, flexible work hours for working mothers, and an education system that gave women opportunities for advancement. However, despite major legislative endeavors, these programs had only limited success. Other examples of partial achievements are the Czech Republic and Hungary (see Chapters 6 and 8). Women were granted special provisions with respect to work accommodations, health care during pregnancy and maternity, and equal treatment regarding family legislation and juridical practice. Here too, the effects were only limited. Women became more dependent on the largesse of the state, often abdicating personal responsibility for their well-being.

In Romania (see Chapter 10) the strict enforcement of rules for both labor production and reproduction added to women's societal obligations. Reproduction assumed such priority that stringent policies were implemented to prevent women from limiting their fertility by using either contraception or abortion. Reproductive capacity was viewed as an essential source of future labor.

In the 1980s *perestroika* marked a change in the USSR from totalitarianism toward a more open society. According to Gorbachev, one of the mistakes made earlier was the failure "to pay attention to women's specific rights and needs, . . . a paradoxical result of our sincere and politically justified desire to make women equal with men in everything" (Racioppi, 1995: 824). Gorbachev did not, however, move toward gender equality, preferring instead a "womanly mission" in civil society.

In sum, while the lives of many women improved, socialist societies fell short of their stated expectations and often created more burdensome circumstances. The ideal of achieving gender equality was never fully realized as patriarchal traditions continued to prevail and service infrastructures remained inadequate for women's household needs. In Hartmann's (1981) view, to attain its goals socialism must go beyond the traditional Marxist focus on class solidarity and work towards the resolution of gender conflict. To achieve gender equality, the struggle must be against both capitalism and patriarchy.

POSTSOCIALIST TRANSITIONS

The political and economic transitions from socialism to more democratic political systems and from centrally planned to market economies affected each of the CEE countries in different ways (e.g., Feischmidt et al., 1997). For many women, the transition process imposed negative consequences, including high unemployment and a lack of job security, a decrease in political representation and activity, a lower level of health care, a worsening of living conditions, a return to traditional patriarchal values with respect to women's roles, the continued "triple burden" for many women as workers, mothers, and wives, and, finally, the "feminization of poverty" (Goldberg and Kremen, 1990; Kligman, 1994). The change in political systems was marked by a retreat from women's rights and equality.

The decrease in political involvement among women was largely due to apathy resulting from the negative associations with such activities in the past. While women had about a 30 percent representation in national, regional, and local assemblies, such representation was often the result of quotas, giving only an appearance of equality. Women office holders were expected to support the party line, not to voice women's concerns (Corrin, 1994). With the replacement of the quota system by democratic elections, representation in Parliament declined to between 5 and 9 percent throughout the region. Male-dominated governments largely ignored efforts to integrate women into political life (UNICEF, 1994).

GENDER ISSUES AND FEMINISM

As Bebel wrote in 1875 and Presser (1997) confirmed more than a century later, gender issues are often ideologically motivated. They point to and stem from gender inequalities and differences on various dimensions, most often highlighting constraints imposed on women by men, usually with the support of prevailing social institutions.

A feminist perspective considers gender inequalities to be socially constructed to the advantage of men, who are comparatively more empowered than women, both within and outside the family (Tobias, 1997). Again, not unlike Bebel, Presser (1997) notes that the nature and extent of such inequalities depend on context, varying by race, ethnicity, and social class. There is an expectation that bringing these processes into the open will generate social change, reduce gender inequities, and enhance women's control over their lives (Dixon-Mueller, 1993).

Reproductive health and reproductive rights are often controversial concepts in the area of population policy and as priorities for feminists. They relate to sexuality, the status of women, and their empowerment. The vitality of these concepts was apparent at the 1994 United Nations International Conference on Population and Development (ICPD) in Cairo (McIntosh and Finkle, 1995; Germain and Kyte, 1995). The adopted Program of Action represented a dramatic shift from past conferences by focusing on reducing gender inequality, improving women's health, and guaranteeing women's reproductive rights (United Nations, 1995b). These new directions moved significantly beyond family planning programs while still endorsing the importance of providing needed information and services (United Nations, 1996).

Truly independent women's movements are poorly supported in the CEE countries. Some women reject the goal of equality, associating it with unpleasant memories of past pseudo-equality. As noted by the Russian feminist Olga Lipovskaya (1994: 224), "Total rejection of the socialist ideals of the past has seriously challenged the notion of gender equality." The situation is further confused by the diverse definitions of feminism and the public identification of feminism with the stereotype of a domineering masculine woman (often lesbian) who hates men. Many advocates of women's political movements tend to avoid the term in public discussions of gender equality.

While still confined to a small group of the intellectual elite, feminism is slowly gaining adherents, nourished by United Nations conferences on women's rights, the Network of East-West Women with its internet-on-line forum, gender studies centers established with foundation support in Moscow, Kharkov, and several other major cities, the Gender Studies Program of the Central European University in Budapest, and the Regional Women's Program initiated by the Soros Foundation in cooperation with its Open Society institutes (Barber, 1997). The focus is to give women choices, with infrastructural supports to exercise them rather than dictate what those choices should be.

CONCLUDING COMMENTS

In varying forms, all the formerly socialist countries oriented their social policies primarily toward labor force participation. Women were perceived as producers and reproducers. Building on nationalist traditions, the party promoted patriarchy in the guise of gender equality. As will be seen in the country chapters, ideological deception reflected demographic and labor force objectives. The struggle for the rights of women was supplanted by the demand for labor productivity (Malysheva, 1992). Despite their rhetoric, governments postponed or ignored actual problems of gender equality.

The woman question remains unresolved in the CEE countries. However, the gradual rise of modern feminism with its commitment to gender equality and female empowerment offers women a renewed opportunity to work with policymakers to reconsider the woman question and break free from traditional patriarchal bonds. Parliamentary factions are seeking gender expertise and more nongovernmental organizations are drawn into political dialogue as "women's rights" are increasingly recognized as "human rights."

part II

COUNTRY REPORTS

chapter 4

Albania

Amilda Dymi and Pamela Pine

EDITOR'S INTRODUCTION

Because of its unique history, language, and self-imposed political isolation, Albania long was the most closed and least developed CEE country. Until the twentieth century, Albanian history was largely one of invasions and occupations, from the Roman Empire to Nazi Germany. During Ottoman Rule, begun in 1478, more than 60 percent of the population was converted to Islam whose legal strictures determined women's rights in society (Tarifa, 1994). It was not until 1912 that Albania proclaimed its independence.

Between the end of World War II and 1990, Albania endured one of the severest autocratic regimes in Europe. A combination of strict central planning, political isolation, and totalitarian control over every aspect of life blocked economic and social development (Marmullaku, 1975; Prifti, 1978; Andrews and Ovalioglu, 1994; Vickers and Pettifer, 1997). Very little was known about this small, remote European country whose population is still predominantly Muslim and whose language is not directly related to any other Indo-European tongue (Zickel and Iwaskiw, 1994; Gjonca et al., 1997).

Among all the European socialist countries, Albania was the last to fall as great political changes swept across the continent. In 1990 the communist regime began to disintegrate. At that time very few Albanians had international contacts, fewer than a dozen elevators functioned in Tirana and not a single traffic light existed in the country (Hockstader, 1997a). After two years of chaos, the first democratic elections, held on 22 March 1992, initiated political and socioeconomic progress. However, after the May 1996 elections, the most criticized in postcommunist Europe, Albania once again became a one-party state (Spolar, 1996b). The increasingly autocratic Berisha government strictly controlled all media while the collapse of pyramid schemes caused the loss of life savings for thousands of Albanians (Hockstader, 1997c). In the summer of 1997 a Socialist Party-dominated government took office. At the end of 1998 it was trying to cope with the massive influx of refugees from Kosovo and steadily worsening living standards (Leskaj, 1998).

This chapter examines current trends in historical perspective, focusing on cultural traditions that influence population policy, women's rights, abortion and contraception, sexual mores, and future perspectives. Particular emphasis is given to the evolution of the status of women. Documentation is provided where available. Official statistics produced during the communist era were not always accurate (David, 1992a; Senturia, 1996).—H.P.D.

HISTORICAL TRENDS

Geography, Demographics, and History

Bordered in the north by Serbia and Montenegro, on the east by Macedonia, on the south by Greece, and on the west by the Adriatic sea, Albania is a small mountainous country. The capital is Tirana. Some 98 percent of Albania's approximately 3.2 million people are ethnically Albanian. Albanians and their Indo-European language are descendants of the ancient Illyrians who lived in the Balkans as early as the second millennium B.C. Minority groups consist of ethnic Greeks, Vlachs, Macedonians, Montenegrins, Roma Gypsies, and a few others (Pollo and Pluto, 1981).

Major religious influences include Islam, Albanian Orthodoxy, and Catholicism. Massive conversions to Islam during the nearly five centuries of Ottoman Turk occupation changed the religious structure of the Albanian people. Estimates based on the 1938 census suggest that Albanians are approximately 70 percent Moslem, 20 percent Orthodox, and 10 percent Roman Catholic. With the banning of religion in 1967, legitimized by the 1976 Constitution, Albania became the world's first officially declared atheist state and remained so until 1990.

Albanian history includes a variety of political regimes. After more then 500 years of Turkish rule, Albania declared its independence in 1912 when the Turks were defeated in the First Balkan War. In the peace treaty negotiated in London, the Great Powers, serving as arbiters, gave territories with large Albanian populations to neighboring Montenegro, Serbia, and Greece.

After World War I, Albania became a republic. Its first president, a mountain chief, proclaimed himself King Zog I in 1928. He fled in 1939 when Italian troops invaded. During the German occupation, from 1941 to 1944, communist partisans controlled the mountainous hinterland. At the end of World War II, Enver Hoxha, seized power on behalf of the Communist Party.

Post–World War II Albania

Modeling his regime on Stalinism, Hoxha maintained complete control over the country until 1985. After breaking ties, first with the former Yugoslav Republic in

1949, then with the Soviet Union in 1961, and finally with the People's Republic of China in 1978, Albania's isolation from both East and West exacerbated its political isolation, economic stagnation, and poverty. To ensure security, more than 900,000 small concrete bunkers were built along the borders, the coast, and throughout the country at a cost the country could ill afford (Andrews and Ovalioglu, 1994).

In the process of isolating the nation, and playing on the people's fear of possible foreign invasion, Hoxha created the poorest country in Europe. Albania's dysfunctional centralized economy could not take advantage of its considerable natural resources and was unable to achieve self-sufficiency. A strict authoritarian state kept the population under control.

The Transition Process

After Hoxha's death in 1985, his hand-picked successor, Ramiz Alia, eased the situation somewhat by reestablishing diplomatic relations with the United States and the Soviet Union. However, his limited reformist approach, was not sufficient to stem the public drive for radical change. Thousands of Albanians, primarily from the larger cities, started an unprecedented mass exodus on 2 July 1990 by invading foreign embassies. This was followed by a huge migration over the borders to Greece and Yugoslavia and to Italy via dangerous sea routes. Other unrest followed. Continuing student protests, supported by prominent intellectuals and thousands of Tirana citizens, forced Alia to legalize opposition political parties.

The first pluralistic elections, held on 31 March 1991, did not bring the long-expected change. The Labor Party (the renamed Communist Party) won a two-thirds majority. Although the opposition Democratic Party won a majority in Tirana and in other large cities, the communists won in the countryside. The Labor Party's name was changed to the Socialist Party and a liberal free market-oriented policy was proclaimed but the economic situation continued to deteriorate.

On 22 March 1992, with a proportional electoral system in place, the opposition Democratic Party won 90 of the 100 electoral zones and nearly two-thirds of the 140 Assembly seats. The remaining seats went to the Socialist Party, the Social Democrats, and others. President Alia formally resigned and the assembly elected a heart surgeon, Sali Berisha, as the new president. In the subsequent three years, public order was restored, the inflation rate declined to less than one percent per annum, and the standard of living improved. While unemployment remained high (around 40 percent), a steady flow of remittances from Albanians working abroad contributed greatly to economic stabilization. Political events that occurred after the disputed 1996 elections were noted in the Introduction.

POPULATION POLICY

Pronatalism was one of the main concerns of the communist regime (Todhe, 1994). It was widely accepted because it reflected a traditional positive attitude toward childbearing (Senturia, 1996, 1997). Albania reported the youngest age structure in Europe with nearly 35 percent of the population below age 15. In the years 1950–80 the natural increase of the population was said to be about 2.4 percent, declining to 1.9 in the 1980s. As shown in Table 4.1 and Figure 4.1, birth rates reached 44.5 per 1,000 population in 1955, gradually declining to 25.3 in 1986, 23.7 in 1992, 21.9 in 1994 and 20.8 in 1996. In 1990 the birth rate for rural areas was 27.2 per 1.000 population, reaching 36.8 in some areas (Dumani, 1993). The total fertility rate declined from 6.85 in 1960 to 3.03 in 1988 and 2.80 in 1991, the highest in Europe at the time It subsequently decreased to 1.98 in 1994 before rising again to 2.32 in 1995 (Albania/MOHE, 1997). This trend was accompanied by declining infant and maternal mortality rates (Kallajxhiu and Gliozheni, 1994). Infant mortality dropped from 41.2 in 1993 to 30.0 per 1,000 live births in 1995 while maternal mortality declined from 40.6 per 100,000 live births in 1994 to 28.5 in 1995 (Albania/MOHE, 1997).

Almost all demographic information was considered a state secret until 1991 (Sjoberg, 1991). In that year a historical time series was published in the statistical yearbook (Albania/Institute of Statistics, 1991) and a Population

Figure 4.1
Births and Legally Induced Abortions: Albania, 1980–96

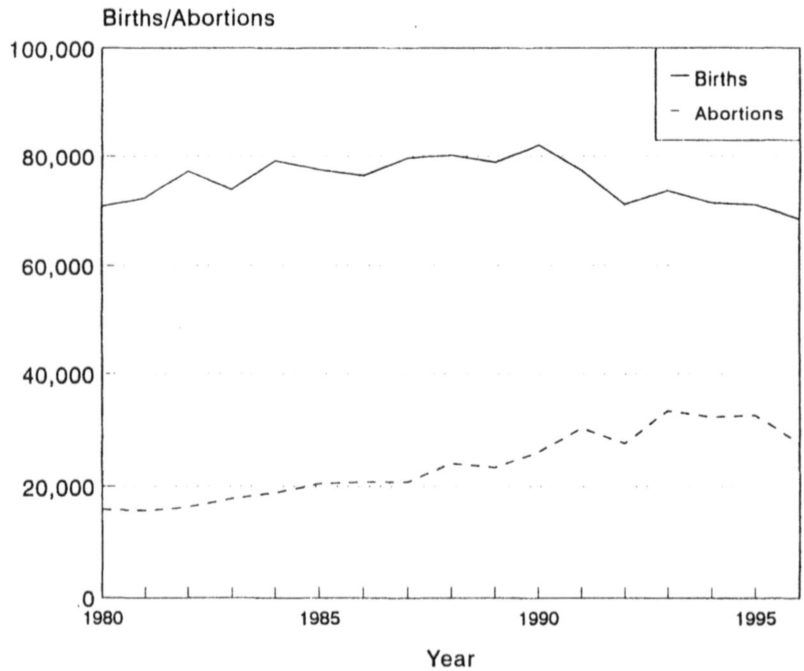

Sources: See data from Table 4.1.

Table 4.1
Birth, Abortion, and Fertility Rates and Ratios: Albania, 1950–96

Year	Live Births	Total Registered Abortions	Birth Rate	Abortion Rate[a]	Abortion Ratio[b]	Total Fertility Rate
1950	47,300	na	38.9	na	na	na
1955	61,300	na	44.5	na	na	na
1960	69,700	na	43.4	na	na	6.85
1965	65,700	na	35.2	na	na	na
1970	69,500	na	32.5	na	na	5.16
1975	70,700	11,400	29.4	na	161	na
1976	70,100	na	28.7	na	na	na
1977	73,100	na	29.3	na	na	na
1978	70,100	na	27.5	na	na	na
1979	72,100	na	27.5	na	na	na
1980	70,700	15,900	26.5	na	225	3.61
1981	72,200	15,700	26.5	na	217	3.46
1982	77,200	16,300	27.8	na	211	3.58
1983	73,800	17,800	26.0	na	241	3.28
1984	79,200	18,900	27.3	na	239	3.40
1985	77,500	20,500	26.2	na	263	3.26
1986	76,400	20,800	25.3	na	272	3.07
1987	79,700	20,800	25.9	na	261	3.16
1988	80,200	24,100	25.6	na	300	3.03
1989	78,900	23,400	24.7	30.9	297	2.96
1990	82,100	26,100	25.2	28.4	318	3.03
1991	77,400	30,400	23.8	39.9	393	2.80
1992	71,100	27,700	23.7	37.0	390	2.38
1993	73,600	33,400	22.4	45.0	454	2.08
1994	71,400	32,300	21.9	43.1	452	1.98
1995	71,000	32,600	22.7	42.7	459	2.32
1996	68,400	27,700	20.8	35.8	405	na

[a]Per 1,000 women age 15–44. [b]Per 1,000 live births.

Sources: UNICEF/TDB (1998), Dumani (1993), Todhe (1994), U.N. (1995a), COE (1998), Albania/MOHE (1997), and USBOC (1998).

Studies Unit was established in the Faculty of Economics of the University of Tirana. Clandestine abortion was reported to be widespread and associated with high fertility, high levels of maternal mortality, and scant knowledge of contraception (Dumani, 1993). A population policy, implemented in 1991, focused on maternal and child health; demographic data collection and analysis; and women and development. Regulations concerning contraception and abortion were altered, legalizing the use of contraception for family planning. With international assistance, medical equipment and supplies were imported and training for physicians improved. However, to understand the issues of women's health, including abortion and contraception, it is essential to review the evolution of the position of women in Albania.

THE WOMAN QUESTION

Ancient Traditions and Influences

The traditional law is drawn from the Kanun, derived from the arcane, patriarchal Code of Leke Dukagjini (republished in 1989). Together with the Albanian language, it was outlawed until the late 1870s (Camay, 1989; Tarifa, 1994; Dragoti, 1996). Its role in the development of civil society was largely negative and damaging to women.

The code defines a woman as "a sack, made to endure as long as she lives in her husband's house," and prescribes among her duties the obligation "to submit to her husband's domination." Women were expected to defer to their husbands and remain chaste (Whitaker, 1981). A husband had the right "to beat and bind his wife when she scorns his words and orders" (cited in Tarifa, 1994). By tradition, a bullet was included in a daughter's dowry that transferred the parents' right to punish their daughter by giving the husband the right to kill her if she dishonored the family. Parents often committed their daughters to marriage at birth, for which the groom's family was expected to pay a certain price to the bride's father.

While the Kanun compares women to priests (God's servants), banning and proclaiming it shameful for men to torture women physically or to touch those who are not their wives, women's lives were controlled by others. They were considered men's property from birth (Prifti, 1978). Virtually imprisoned in their homes, Albanian women had little understanding of the outside world. The average woman seldom strayed beyond the confines of the village of her birth or where she was taken by the husband her parents had chosen. Women's relations were restricted to members of their extended families. In the mountainous regions of Northern Albania, no woman could participate in a public meeting. Her opinion was not solicited in the family and education was often beyond her reach (Tarifa, 1994).

The Beginning of the Modern Era

In the mid- and late-1800s the French revolution and its progressive ideas encouraged a handful of Albanian scholars to start a cultural-political and nationalist movement, known as the Albanian Renaissance. It emphasized the importance and uniqueness of the Albanian identity and tradition by stressing its differences from the Eastern Arab culture and religion. In 1887 Renaissance leaders opened the first (gender segregated) Albanian language elementary school, taught by two Western-educated Albanian women. In 1891 the first Albanian girls were admitted. However, 80 percent of Albania's population and more than 90 percent of women remained illiterate until the end of World War II (Tarifa, 1994).

Invaders encountered difficulties in attempts to change traditional Albanian behavior. In rural Northern Albania, where only a minority converted to Islam and the presence of the Kanun was stronger, women never wore the veil and were not encouraged to do so. In 1928, four years after coming to power, King Zog, who was familiar with Western customs and had married an Austro-Hungarian princess, issued a special decree banning the Turkish-imposed tradition of the Arab veil in Albanian public places. However, Albania's rural character continued to contribute to the prevalence of a very traditional patriarchal society in which women had a submissive status and were among the least advantaged in terms of social and economic development (Biberaj, 1990).

World War II and the Communist Era

The advances in education and the massive employment of women that occurred after World War II and during the 50 years of communist dictatorship changed the status and role of women. The Communist Party supported the long held concept that the family is sacred. It accepted traditional values and added love as an element for the foundation of solid families. At the same time, it attempted to rearrange these values to form a new order of importance: politics ranked above all else, even the family. Applying totalitarian ideology, Hoxha's regime encouraged and rewarded family members who denounced other family members for their political beliefs.

Education also was affected. In 1946 elementary education became compulsory; in 1947 a literacy program was instituted for all men and women under age 40, and in 1952 compulsory schooling was extended to seven years. By 1955 illiteracy had been eradicated among those under age 40. Also in the 1950s, Albanian women gained access to higher education. By the early 1990s, women accounted for more than 60 percent of students in general secondary schools, slightly more women than men earned university degrees, and 38 percent of employed women had higher education with a further 47 percent having completed vocational secondary education (Tarifa, 1994).

Communist women spoke out against the humiliating tradition that considered women "short in mind but long in hair," and which banned them from speaking at the (men's) village assembly. In these open meetings, convened primarily by party activists in the northern part of the country where the influence of the Kanun was still very strong, fathers agreed to void engagements that had been arranged at birth, and to allow their teenage daughters to participate in the building of ferries or roads, a program organized by the state. For the first time in Albanian history these "youth actions," (which continued through the 1980s) permitted the unsupervised commingling of young men and women. They worked together and were able to socialize openly. It opened a new era for Albanian women.

As women gradually became an important part of the labor force their status was declared legally equal to men. If some women purposefully stayed home, party activists would knock at their door to ask why, stressing that their work was needed. By the 1970s, women constituted more than 45 percent of the Albanian labor force. To persuade women to build society and also meet their family responsibilities, the party encouraged high birth rates and rewarded those who gave birth to 12 or more children with the title of "Heroic Mother." The only economic incentive was a one-time payment at the birth of a baby and a small stipend for unmarried mothers (both abolished after 1991). As will be seen, pronatalism in Albania relied more on negative features, such as the lack of modern contraceptives and abortion services.

Despite the formally proclaimed gender equality, gender disparities persisted. Few real political or economic gains were made by and for women during this time. Women were given only token placement in highly ranked party or government positions. Families with only two or three children could not survive with only one parental salary. Both parents had to work to provide even the basics, creating a new lifestyle for Albanian families and placing additional burdens on women.

Life changed inside the home, often for the worse. Women had the multidimensional burden of working, caring for the children and in-laws, and cleaning and washing by hand, in most cases without help from their husbands. Though women had assumed nontraditional roles, men had not been taught to change their belief in their "privileged" status and authority.

As economic conditions continued to worsen, the family's dependence on women became greater than ever. This brought some societal change, at least in the cities. Women began to participate more in decision making. In practice, they often assumed the functions of the head of the household, although men continued to control the finances. As the economic situation grew worse again in the 1970s, as basic foodstuffs became more scarce, and as long lines of shoppers spent hours in line for rationed quantities, the Communist Party-supported Women's Union organized debates about the importance of cooperation in the distribution of household duties and the improvement of children's and families' lives. Most men, retired parents, and sometimes children, began to offer at least some assistance to the wife in the home. The need to cooperate facilitated the process of change and helped rewrite the rules of communication within urban families.

Education

As education became mandatory for girls, a popular preference arose for better educated brides. A university degree reduced the importance of the dowry and increased the chances for a secure job and a good salary. More and more women were encouraged by their families in their goal to attain a better education and delay marriage. In 1950 nearly half of the women married at age 19 or younger. By the 1990s, 56 percent of women married at age 20 to 24 years; only one-fifth of women aged 19 years or younger were married (Albania/MOHE, 1993a).

Gradually, as the levels of education increased for girls and women, their power in society and particularly within the family increased, though this did not occur without growing pains. The tendency to have smaller sized families with two or even one child became stronger in the cities while the number of children remained fairly high in the villages. Women reached a new level of independence by continuing their education and obtaining better paid jobs.

By the 1990s most women were far ahead of their 1940s counterparts who were largely illiterate. As of 1994 female enrollment at the primary and secondary levels was 98 percent in urban areas and 73 percent in rural areas, only slightly less than male enrollment. In 1988 half of the student body of Tirana University and other Albanian institutes of higher learning were female, reflecting the extent of acceptance of higher education for women (Albania/MOHE, 1993a).

Partner Relations

By the 1980s the changes in women's socioeconomic status in the cities further challenged these women's sense of self, their behavior, and their relationships with their partners and their families. Unhealthy marriages were and still are difficult to dismantle. The only acceptable reasons for divorce are adultery, bribery, or continuous/regular physical abuse. To obtain a divorce for spiritual or sexual unfulfillment is very rare. Women who dare pronounce their marriage unfulfilling are virtually committing social suicide. The concept of marital rape does not exist. Most men deem their sexual pleasure a marital right. Forced sexual relations are often considered by women to be a part of the marital package or bad luck that they have to endure.

With rising educational levels and socioeconomic changes, women have begun to expect more from their marriages. The number of failed marriages is increasing. In 1960 the divorce rate was 0.5 per 1,000 inhabitants, or 2.9 divorces for every 1,000 families. By 1990 the divorce rate had risen, particularly for city dwellers, to 0.8 per 1,000 inhabitants or 6.2 per 1,000 families in urban areas and 2.1 in rural areas (Albania/MOHE, 1993a). Later data are not available.

Socioeconomic Effects

The new era brought women into the mainstream and made the women's movement a firm part of socialist life. Women became entitled to paid maternity leave for nine months, daycare for their children, and free health services, including free drugs for children until the age of one. Nevertheless, strong patriarchal traditions and the very poor economic base for women contributed to the continuation of harsh conditions for women, especially those living in the hilly and mountainous areas of the country. These women fetched water for their homes in containers carried on their backs and walked as much as an hour and a half each way to do so. The daily workload for many young mothers had become so hard and exhausting at times that some physicians told of situations where babies had suffocated as a result of their mothers falling asleep while breast-feeding.

Although they were in a better situation compared to their rural comrades, women workers in industry (80 percent of whom are now unemployed) also suffered from difficult living conditions. Their day often started at 5 A.M. to have time to get the children ready for kindergarten or school. This was followed by eight hours at work, preparation of the family's food for the next day, and cleaning, washing, and ironing until long into the night. For those who worked late shifts, the schedule merely reversed itself. Brides often had to coordinate their schedules with their in-laws with whom they usually lived (and still live). The generation gap created problems. Some husbands' mothers still do not believe in having their sons help with the household chores. Small, cramped apartments and the lack of sufficient funds contributed to the development of highly stressful environments (Senturia, 1997).

The Influence of Women's Groups

The first mass women's organization, the Antifascist Women's Union, was founded in 1943 during World War II. In 1946 it evolved into the Women's Union of Albania, an arm of the Communist Party. It was primarily a political instrument but also focused on the development of new and more emancipated traditions for women: women should speak out; they should go to school; they should help and encourage their children to study; they should fight for higher standards of hygiene; and they should dare to compete with men in different areas of science. However, the Women's Union was only marginally successful in creating a genuine women's movement, or in closing the gap between political rhetoric and actual participation in political institutions. It dissolved after the introduction of the multiparty system in 1992.

In recent years a number of new women's organizations have been organized. Most continue to be affiliated with political parties and are struggling to define their roles (Carlson, 1994). As of late 1998, they have little influence in the rural areas, where most women live, or in the trade unions. The new women's organizations advocated greater consideration of reproductive rights, family planning, and social welfare issues by the public, the political parties, and the government. However, women lack political power. In 1993 only six women were in the 140-member Parliament. Women's concerns do not appear to be a priority on the state's agenda.

Women and the Law

Although shortcomings still exist in terms of women's rights (e.g., there are no laws banning discrimination concerning marital status or sexual harassment in the work place), women won equality before the law during the communist era and their status has not changed greatly since that time. Generally, the laws have offered some protection. The 1976 Constitution banned forced marriage, pregnancy has been a mitigating circumstance in court, and divorced women are protected in a number of ways. They automatically gain custody of children up to 15 years of age. When there is limited dwelling space in the home the ex-husband is often forced to leave. All personal belongings are split between the couple, unless a mutual agreement is reached to the contrary. Fathers have to pay monthly support no greater than one-fourth of their income until the child reaches 18 years of age (the age when children complete high school), an obligation of which they are relieved only if imprisoned.

Other support exists with regard to women's bodily rights. Domestic physical violence was and is punishable by law. The mandates on women's reproductive rights are changing. Family planning and abortion are now allowed under administrative mandate. Still, a true gender perspective is lacking in the programs and strategies established by the various ministries (Leskaj, 1998).

ABORTION AND CONTRACEPTION

Abortion

In contrast to other CEE countries (except Romania), abortion was banned, except for 30 strictly defined medical reasons that in each instance had to be approved by a district medical committee. In practice, abortions were sometimes permitted on all but economic or social grounds. The Penal Code of 1977 punished repeat offenders or those performing an abortion that resulted in the women's death or a serious health problem with up to eight years imprisonment. Otherwise, the punishment was reeducation through work or detention for up to two years. A woman performing an abortion herself without assistance was punished by a social reprimand or by reeducation through work (Albania, 1977, 1988; United Nations, 1992). With lack of information about contraception, especially in the villages, abortion was the traditional way of coping with unplanned pregnancies. Over the years, there was a steady increase in the number of "provoked" (self-induced) and "therapeutic" (approved) abortions, which were often performed on a subjective basis for an increasing number of reasons according to who knew whom. Clinicians estimated that the figures for self-induced and "therapeutic" abortions were perhaps as high as 50 percent of all deliveries (Todhe, 1994).

In 1989 the Ministry of Health broadened the list of medical indications for pregnancy terminations, subsequently expanding them to over 200 in early 1991. Government officials suggested that approved abortions constituted less than 10 percent of all deliveries (UNFPA, 1991). As shown in Table 4.1, there were 82,100 reported births and 26,100 reported abortions in 1990. "Provoked" abortions with complications requiring subsequent care in medical facilities were two to three times higher than approved abortions in 1989 and 1990 (Dumani, 1993). The actual number may have been substantially higher than shown in Table 4.1 because many unrecorded clandestine procedures did not require subsequent hospital care. Some gynecologists estimated in early 1991 that two of every three requests for abortion were accepted and that "therapeutic" abortion constituted 180 per 1,000 pregnancies (Todhe, 1994). During their reproductive years, some women may have induced a dozen or more abortions (Pine, 1991).

Abortions were most frequent in the larger cities. In Tirana, the official estimate was a ratio of 900 abortions per 1,000 live births (Dumani, 1993). A survey conducted at Tirana's Maternity Hospital in 1987 showed that 50 percent of abortions were sought for economic reasons, 20 percent for "having too many children," 12 percent for housing problems, 12 percent for "too frequent pregnancies," and 6 percent for not being married (Tare, 1992, 1993).

On 15 June 1991, by a Ministry of Health Directive, Albania revoked its law permitting abortion only on specifically defined grounds. The new directive made abortion available on request up to 12 weeks, for social reasons up to 15 weeks, and on medical grounds without a time limit. Various other factors removed barriers to obtaining an abortion. While a married woman required her husband's approval, this requirement was not always strictly enforced. (Theoretically, an unmarried woman should have her family's approval.) The Council of Ministers approved the Directive by Decision No. 226 of 27 May 1992 (David, 1994a). In 1992 the cost of a first-trimester abortion was 150 leks, approximately equal to two days' wages (*The Times*, (London) 1992). Abortions from 12 to 21 weeks cost 500 leks (Leshi, 1994).

In November 1995 Parliament approved the abortion law, making abortion on request legal for the firstl 12 weeks; after that time it may still be allowed for health or social reasons (Leskay, 1996). As shown in Table 4.1, the total number of registered legal abortions increased from 23,400 in 1989 and 30,400 in 1991 and to a peak of 33,400 in 1993 before declining to 27,700 in 1996 . During these years the reported legal abortion rate rose from 30.9 per 1,000 women 15–44 years of age in 1989 to 42.7 in 1995 before dropping to 35.8 in 1996. The reported legal abortion ratio increased from 297 per 1,000 live births in 1989 to 459 in 1995, decreasing to 405 in 1996. Many abortions continue to be unreported (WHO/EURO, 1997).

As shown in Table 4.2, the proportion of maternal deaths due to abortion was 51 per 100,000 live births in 1989. After liberalization, the proportion decreased to 39 per 100,000 live births in 1991, to 21 in 1993, and to 12 in 1995 (Rich, 1992; Todhe, 1994; Albania/MOHE, 1997). The infant mortality rate of 30.7 per 1,000 live births in 1994 continues to be among the highest in Europe (COE, 1998).

In 1991 the Ministry of Health reported that 33 percent of all abortions were spontaneous, 40 percent were induced for "therapeutic" reasons, and 26 percent were classified as "criminal" (self-induced) clandestine procedures (Carlson,

Table 4.2
Maternal Mortality: Albania, 1985–96

Year	Total Maternal Deaths	Maternal Deaths Due to Abortion	% Maternal Deaths Due to Abortion
1985	44	16	36
1986	32	20	62
1987	35	12	34
1988	40	21	52
1989	39	20	51
1990	31	15	48
1991	23	9	39
1992	24	5	21
1993	14	3	21
1994	29	6	21
1995	24	3	12
1996	19	na	na

Sources: UNICEF/TDB (1998), Rich (1992), Dumani (1993), Todhe (1994), U.N. (1995a), and Albania/MOHE (1997).

1994). By 1992 the number of provoked procedures had declined considerably (Sahatci, 1993). However, the abortion ratio at the Tirana Maternity Hospital rose from 889 per 1,000 live births in 1989 to 1,127 in 1993. An increase in abortions among teenagers and unmarried women was noted (Todhe, 1994). Meanwhile, the total number of live births in Albania declined from 82,100 in 1990 to 73,600 in 1993 and 68,400 in 1996 (Table 4.1).

Abortion service conditions remain substandard. Use of anesthesia, initially inadequate due to lack of supplies, has improved but is not always available (Mullaj, 1994). Sterilization of equipment can be a problem. Use of vacuum aspirators is not yet widespread. In 1992 they were available only at the University Clinic in Tirana (Sahatci, 1993). Contraceptive counseling is not provided in all clinics and privacy remains rare in public facilities (Leshi, 1994). Frequently, two abortions are performed at the same time in the same operating room (WHO/EURO, 1997).

Great differences exist between women with different levels of education and parity. Abortion frequency increases with parity and decreases with higher socioeconomic levels (Todhe, 1994). Of abortions performed in 1991, approximately 9 percent were for women with a university degree; 36 percent for women with a high middle school education; and 55 percent for women with an elementary school education. The lowest number of abortions were performed for childless couples.

Contraception

Between 1944 and 1992 family planning could not be openly discussed. Except for condoms, the concept of contraception was little known. The importation, sale, and distribution of modern contraceptives was prohibited. In 1981 condoms disappeared from pharmacies (UN, 1987). Medical training did not include information about sexuality or modern contraception. Tubal sterilization was performed only in rare cases for medical reasons when the woman had many children and the husband agreed. With the lack of knowledge among both the public and the medical profession, *coitus interruptus* remained the most practiced method of fertility regulation (Todhe, 1994).

Though many women have heard about modern family planning methods and appear to be strongly in favor of their availability, most know little about what is actually available or how the methods work. There are few data on knowledge, attitudes, and contraceptive practice. In a 1992 survey among well-educated, third-year social science and humanities students at the University of Tirana, 51 percent reported that they have only limited or no knowledge of any fertility regulating method. Among students of rural origin, 65 percent reported limited or no knowledge (Dumani, 1993).

Family planning was officially approved on 27 May 1992 as a fundamental right of all individuals (Albania, 1992). On Albania's First National Family Planning Day in 1992, 300,000 people marched in the streets of Tirana with banners calling for greater availability of contraceptives. While contraceptives remain scarce due to an undeveloped importation and distribution infrastructure, conditions are improving. Various forms of modern contraception are now available with a prescription from general practitioners, including pills, IUDs, and injectibles. In Tirana, condoms can be purchased on street corners, in kiosks, or in drug stores. Since 1992 family planning centers are operational in maternity hospitals and prenatal outpatient clinics (Kallajxhiu and Gliozheni, 1994) but motivation to use contraceptives remains low, in part due to fears about possible side effects and low acceptance of condoms by men. Withdrawal, backed by abortion, remains the most commonly practiced method of fertility regulation (WHO/EURO, 1997).

SEX AND SOCIETY

Sexual mores have been affected by recent societal changes. In larger cities, like Tirana, premarital sex is considered normal for couples in established relationships. Most city dwellers no longer consider a pregnancy shameful if it ends in marriage. According to a 1993 poll of 300 highly educated women of different ages, professions, and social status living in Tirana, organized by Reflections, an Albanian nongovernmental association concerned with women's issues, 55 percent of women interviewed had premarital sex; 75 percent of these women had premarital sex with partners other than their current husband (Albania/MOHE, 1993b).

Traditionally, women are not supposed to express pleasure during sex as men believe it reflects loose morals and appear to be threatened by it. It is only among the younger (20 to 30 years old), urban, and better-educated generation that this pattern is changing. Attitudes about sexual mores remain generally conservative. Homosexuality continues to be a punishable offense.

Adolescents

Albania's adolescents have been quick to embrace the new freedoms but have little or no information to help them cope with issues of sexuality. Age at sexual debut has declined and the incidence of teenage pregnancy and sexually transmitted diseases has increased (Reporteri, 1993; Cooper, 1994). Conversations about sexuality remain taboo within the family. Few young people are encouraged to seek advice from physicians although physicians claim that considerable interest is expressed (Albania/MOHE, 1993b).

Sexuality Education

Until recently, there was little sexuality education. Discussing sex was traditionally considered immoral. Available information was brief and superficial. There were descriptions of the maturation process in secondary school physiology textbooks, but no efforts to promote responsible reproductive behavior.

In 1990 several high school teachers and two female physicians organized a panel on "How to better know ourselves." The first of its kind, the panel addressed a mixed audience of nearly 600 parents and teenage students. However, when one of the physicians illustrated her presentation with enlarged drawings of the female and male sexual organs there was obvious embarrassment among the other panelists and the audience.

To strengthen resources for teenagers, students at Tirana University established a Student Organization for the Propagation of Sexual Education (SOPSE) in 1993. They became peer educators for the dissemination of information on sexuality, contraception, and AIDS (Iliriani and Asllani, 1995).

Albanian Family Planning Association

The Albanian Family Planning Association, established as an independent nongovernmental organization in 1993, works in close collaboration with local institutions, the Ministry of Health, the Ministry of Education, and the International Planned Parenthood Federation (Leskaj, 1996). It has its own family planning clinic in Tirana, offering contraceptives and related services free of charge. In cooperation with UNFPA, training courses were developed for physicians and midwives. Family planning has been introduced into the curricula of the Medical Faculty and School of Midwifery. Sexuality education courses are conducted for teachers in secondary schools (Sahatci, 1993). While the only TV channel is silent on family planning issues, public education is promoted in cooperation with radio stations and print media. In November 1997, the Albanian FPA was admitted to membership in the International Planned Parenthood Federation. High on its agenda is a stronger political commitment to a more comprehensive approach to population and development (Leskaj, 1998).

FUTURE PERSPECTIVES

While more than 45 years of dictatorship, followed by political chaos, have retarded social processes, Albanians continue to pursue democracy and a workable free market economy. This trend was affirmed in November 1994 when Albanian voters refused to ratify a new constitution that would have placed more direct power in the hands of the president (*The Economist*, 1994c). They did so again in 1997 by voting for a stronger Parliament.

Much of what was previously unacceptable in Albanian society has become part of an evolving culture. Women are free to choose between work or remaining at home, making a decision based on personal circumstances rather than official edicts. At the same time, women are less involved in political life, considering it part of the discredited communist past. As one result, women are underrepresented in Parliament. Career opportunities also continue to be limited by job discrimination and the lack of maternity leave, child care, and laws protecting women's rights.

To reduce reliance on abortion, continuing effort is being devoted to implementing sexuality education at early ages in all parts of the country. Mass media are utilized for informing young couples about access to modern methods of fertility regulation, and educating the public that abortion is always associated with risk when performed by untrained persons. Since family planning is perceived as a woman's issue, strengthening the position of women in Albanian society is a priority for the future.

chapter 5

Bulgaria

Dimiter Vassilev

EDITOR'S INTRODUCTION

Situated in the eastern Balkans, Bulgaria shares borders with Romania, Turkey, Greece, the Federal Republic of Yugoslavia, and Macedonia. After close to five centuries of Turkish rule, the country was liberated by Russia in 1878 and achieved full independence in 1908. In subsequent years Russia encouraged close cultural relations, fostered by the common Cyrillic alphabet and similarity of languages. While Bulgaria sided with Imperial Germany in World War I and with Nazi Germany in World War II, it refused to declare war on the Soviet Union when Hitler invaded it in 1941. In August 1944 Bulgaria withdrew from the war and in September Russian troops marched into Bulgaria unopposed. The Bulgarian army was placed under Soviet command and took an active part in the defeat of the German forces. By September 1944, a communist political system was established.

On 16 March 1945 George Dimitrov, former Secretary-General of the Communist International, returned to Bulgaria after many years in the Soviet Union. In a referendum on 8 September 1946 the monarchy was abolished. Bulgaria was proclaimed a republic led by a communist-dominated government. Dimitrov was elected prime minister in October 1946, Soviet troops left by year end 1947, and the Communist Party became the sole party. In 1954 Tudor Zhikov became the country's dictator and the Soviet Union's most loyal ally for the next 35 years.

Glasnost in Moscow was mirrored in Sofia. The collapse of the communist dictatorship in 1989–90 brought political liberty but did not radically change Bulgaria's economic institutions (Morris, 1997). Initially under "reformed" communists and then under various weak coalitions, Bulgaria was the economic laggard of the region. By 1990 "postcommunists," calling themselves the Bulgarian Socialist Party (BSP), were Eastern Europe's first to return to power in democratic elections. The BSP moved into opposition after the 1991 elections but regained its leading position in December 1993 on attaining a majority in the National Assembly.

After five years of political squabbling and corruption, little had been accomplished except to enrich former party officials who had become businessmen. The Bulgarian lev plunged against the dollar, grain shortages made bread scarce, and electricity and gas prices skyrocketed (Spolar, 1996c). Price controls on bread, milk, cheese, and oil, introduced on 1 October 1996, were ineffective because most grain had been exported. Bulgarians faced a slumping currency, eroding confidence in the banking system, and diminishing living conditions (Beck, 1996, *The Economist*, 1996a, 1997a). In November 1996, in the seventh election in seven years, the reform-minded candidate of the Union of Democratic Forces (UDF), Peter Stoyanov (1998), defeated the incumbent BSP president (Griffiths, 1996). The government resigned in December as the inflation rate reached over 300 percent, the value of state pensions dipped to two dollars a month, and churches opened soup kitchens for the elderly (Hockstader, 1997b).

The election of 19 April 1997 produced a UDF majority. The new government, led by Prime Minister Ivan Kostov, an economist, initiated a policy of financial discipline. Subsequently, the economy stabilized, the inflation rate fell to under 10 percent, and confidence in the banking system returned (*The Economist*, 1998c).

Within a historical context this chapter reviews demographic patterns and the interaction of changes in abortion legislation, pronatalist policies, and contraceptive practices. There follows a discussion of the development of the Bulgarian Family Planning Association, the woman question, and of sex and society, including family size intentions, sexuality education, and trends in adolescent sexual behavior. A commentary on future perspectives concludes the presentation. The chapter benefited from the helpful comments of Elwood Carlson.—H.P.D.

DEMOGRAPHIC TRENDS

Bulgarians constitute about 86 percent of the population of approximately 8.4 million people. About 10 percent are of Turkish origin and 4 percent are Gypsies (WHO/EURO, 1993; Obregon, 1995). Bulgaria's agrarian structure was based mainly on small peasant land ownership, which fostered a decline in marital fertility by the end of the 1930s (League of Nations, 1940; Coale, 1965). Until the mid-1940s three-quarters of the population lived in villages (Meyerfeldt, 1996). The subsequent collectivization of farms stimulated the creation of an agroindustrial economy and rural-urban migration (Carter and Zagar, 1977; Hoffman and Hatchett, 1977; Taaffe, 1977). There was a further exodus to the urban centers after 1989 when all residence restrictions were abolished. Large numbers of people emigrated to neighboring Turkey and Western Europe. Over 350,000 Bulgarians of Turkish ancestry left following coercive efforts by the communist regime to force assimilation, including the change from Islamic names to Christian/Slavic ones. By

1992, over 67 percent of the population lived in urban centers and towns, and 33 percent in smaller rural villages (WHO/EURO, 1993).

Bulgaria's demographic transition began approximately 25 to 30 years after that in Western and Central Europe (Van der Walle, 1980; Vassilev, 1986). While delayed marriage played a central role in the Western European fertility decline, Bulgaria began its fertility transition with the opposite pattern of early and nearly universal marriage (Sklar, 1974). Though overall fertility seemed to remain at high levels during a great part of the nation's demographic history, marital fertility declined over a long period (McIntyre, 1980), leading to very low overall fertility levels by the end of the 1930s and particularly after World War II (David and McIntyre, 1981).

Natality

In 1881, shortly after liberation from the Turks, the crude birth rate was over 35.1 per 1,000 population (David and McIntyre, 1981). At the end of World War II in 1945, the birth rate was 24.1 per 1,000 population. As shown in Table 5.1 and Figure 5.1, the birth rate gradually decreased from 25.2 in 1950 to 14.9 in 1966. The largest drop in the birth rate occurred between 1956 and 1966, when it decreased from 19.5 to 14.9. Rapid urbanization, forced agricultural collectivization, and industrialization contributed to this decline. From 1967 to 1974, when the government offered economic incentives to stimulate additional childbearing and made access to abortions more difficult, the birth rate rose from 15.0 in 1967 to a high of 17.2 in 1974 (David, 1982; Bulgaria, 1994). For a time, as younger rural couples moved to urban centers, the urban birth rate was higher than the rural birth rate (Bulgaria/CSO, 1988). In subsequent decades the birth rate continued to decrease. By 1997, it had plummeted to 7.7 (COE, 1998). Having a child became a "luxury" for many young couples (Meyerfeldt, 1996).

As also shown in Table 5.1, the total fertility rate (TFR) dropped from 2.31 per woman in 1960 to 2.03 in 1967, rose to 2.28 in 1969 and 2.30 in 1974, and then declined to 1.81 in 1990 and 1.09 in 1997. The brief rise may have been influenced by the 1967 law on maternal leave permitting grandmothers to care for their grandchildren and thus encouraging younger women to have children at earlier ages (Filipov, 1992). Early marriage and early childbearing were the norm. In 1982 mean age at first marriage for women was 21.3 years, the lowest in Europe. However, by 1997 it had risen to 23.1 years (COE, 1998).

Marriage and Divorce

Marriage rates have steadily decreased over the last half century. The number of marriages declined from 73,100 in 1970 to 44,800 in 1992 when the marriage rate reached a low of 5.3 per 1,000 population, dropping further to 4.2 in 1997 (Bulgaria, 1994; COE, 1998). Cohabitation prior to marriage became a dominant mode of behavior (Vassilev and Rushkarska, 1990). As of 1997, nearly all wives claimed they had been living with their future husband before the wedding. In 1992, 18.5

Table 5.1
Birth, Abortion, and Fertility Rates and Ratios: Bulgaria, 1950–97

Year	Live Births	Legally Induced Abortions[f]	Other Abortions	Total Registered Abortions	Legal Abortion Rate[g]	Legal Abortion Ratio[h]	Birth Rate	Total Fertility Rate
1950	182,600	na	na	na	na	na	25.2	na
1955	151,000	2,200	16,900	19,100	1.2	15	20.1	2.26
1956[a]	147,900	2,000	38,000	40,000	1.2	14	19.5	2.23
1957	141,000	31,700	14,500	46,200	18.2	225	18.4	2.12
1958	138,300	38,100	17,400	55,500	21.8	276	17.9	2.08
1960	140,100	51,000	23,900	74,900	29.0	364	17.8	2.31
1965	125,800	96,600	19,400	116,000	51.0	768	15.3	2.08
1966	123,000	101,500	19,600	121,100	53.5	825	14.9	2.02
1967	124,600	108,500	21,400	129,900	57.2	871	15.0	2.03
1968[b]	141,500	89,200	24,300	113,500	47.1	630	16.9	2.28
1969	143,100	106,900	22,800	129,700	56.4	747	17.0	2.28
1970	138,700	119,700	22,600	142,500	63.2	863	16.3	2.18
1973[c]	139,700	112,300	25,100	137,400	59.7	804	16.2	2.16
1974[d]	149,200	121,700	22,800	144,500	64.8	816	17.2	2.30
1975	144,700	120,400	23,100	143,500	64.3	832	16.6	2.24
1980	128,200	136,300	19,600	156,100	76.7	1,063	14.5	2.05

1981	124,400	132,700	19,600	152,300	74.9	1,067	14.0	2.01
1982	124,200	129,400	18,400	147,800	73.1	1,042	13.9	2.02
1983	123,000	116,000	18,200	134,200	65.7	943	13.8	2.00
1984	122,300	112,600	18,500	131,100	61.9	921	13.6	2.00
1985	119,000	112,200	19,800	132,300	64.4	943	13.3	1.95
1986	120,100	114,300	20,400	135,000	65.5	952	13.4	2.00
1987	116,700	113,000	20,800	134,100	64.7	968	13.0	1.95
1988	117,400	118,800	14,300	133,100	63.8	1,012	13.1	1.97
1989	112,300	111,000	21,000	132,000	59.2	988	12.5	1.90
1990[e]	105,200	127,600	17,000	144,600	67.8	1,213	11.7	1.81
1991	95,900	123,400	15,000	138,400	65.6	1,287	11.1	1.65
1992	89,100	118,900	14,000	132,900	67.5	1,334	10.5	1.54
1993	84,400	91,500	15,900	107,400	52.1	1,084	10.5	1.46
1994	79,400	87,500	10,100	97,600	50.0	1,102	10.0	1.37
1995	72,000	87,200	9,900	97,100	50.2	1,211	8.6	1.23
1996	72,200	89,000	9,500	98,500	51.3	1,233	8.6	1.24
1997	64,100	na	na	86,900	na	na	7.7	1.09

[a]Abortion available on request (April 1956). [b]Abortion restricted for childless women (January 1968). [c]Restrictions further tightened (April 1973). [d]Restrictions somewhat relaxed (February 1974). [e]First-trimester abortions available on request (February 1990). [f]See definitions in text. [g]Per 1,000 women age 15–44. [h]Per 1,000 live births.

Sources: David and McIntyre (1981), Henshaw and Morrow (1990), U.N. (1995a), Bulgaria/NSI (1997), USBOC (1998), UNICEF/TDB (1998), and COE (1998).

Figure 5.1
Births and Legally Induced Abortions: Bulgaria, 1955–96

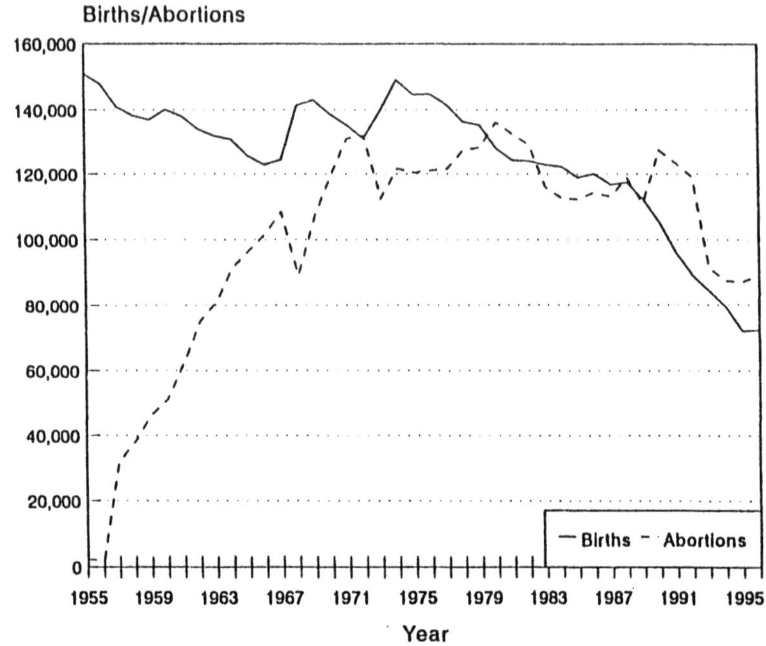

Abortion available on request, April 1956; abortion restricted, January 1968, April 1973; restrictions relaxed, February 1974; first-trimester abortions available on request, February 1990.

Sources: See data from Table 5.1.

percent of births were delivered by unmarried women, half of whom lived in consensual unions (Bulgaria, 1994). The proportion of out-of-wedlock births reached 30.1 percent by 1997 (COE, 1998).

Divorce rates have been consistently low, influenced in part by 1985 changes in the Family Code, which lengthened the time before a court would grant a divorce on the basis of mutual consent. Taxes levied on divorces were also increased. The divorce rate of 1.3 per 1,000 population in 1995 was the same as in 1990. It declined to 1.1 by 1997 (COE, 1998).

Mortality

Crude death rates have increased steadily since the mid-1960s, rising from 7.9 per 1,000 population in 1964 to 12.7 in 1992 and 14.7 in 1997 (COE, 1998). Infant mortality was very high at the end of World War II, reaching 144.5 per 1,000 live births in 1945. Over the next 45 years, with improvements in public health, the medical care system, and public education, the infant mortality rate (IMR) declined about 900 percent to 14.8 in 1990. However, this decline in IMR was based on the

COMECON definition of live births. A product of conception weighing less than 1,000 grams at parturition could not be counted as a live birth until it survived for seven days. If death occurred within that period, it was defined as "special abortion." This definition, resulting in a 15 to 20 percent undercount of infant mortality, has remained in effect (Carlson and Tsvetarsky, 1996). The IMR rose to 16.3 in 1994 and 17.5 in 1996, declining to 15.6 in 1997 (COE, 1998).

INTERACTION OF ABORTION AND PRONATALIST POLICIES

1936–68

Abortion for medically indicated reasons was first permitted in Bulgaria in 1936 for women suffering from serious diseases such as tuberculosis. After World War II, the Bulgarian National Assembly passed an amendment to the existing Penal Code (Paragraph 135), abolishing the punishment of service providers and women having abortions in an authorized medical institution. In April 1956 abortion was further liberalized by a special decree issued by the Ministry of Public Health and Social Welfare that allowed abortion on request (Bulgaria, 1956; Vassilev, 1994). Following receipt of a request for abortion at a women's consultation center, medical staff were instructed to try to persuade the woman not to terminate her pregnancy. If she persisted, the procedure had to be performed in an authorized hospital obstetrical unit. Abortion on request was available up to 16 weeks and up to 24 weeks in medically or socially indicated cases. Procedures were performed free of charge until 1960 when a tax of five leva (the equivalent of about seven U.S. dollars at the time), was introduced. Vacuum aspiration became available in 1963, gradually replacing sharp curettage.

The Bulgarian record system defined legal abortions as the sum of elective abortions plus those performed for medical reasons. "Other abortions" included women hospitalized for spontaneous and criminal abortions plus those otherwise unidentified. As noted in Table 5.1, liberalization resulted in a sharp increase in the number of legal abortions rising from 2,000 in 1956 to 31,700 in 1957. This increase included a transfer of previously illegal abortions, which decreased from 38,000 in 1956 to 14,500 in 1957. Hospital admissions for complications associated with illegal abortion declined (Starkalev et al., 1961; Stoimanov et al., 1965).

The change in legal status had only a limited effect on the already low birth rate. When the birth rate decreased further in the 1960s, a strongly pronatalist atmosphere emerged that would influence future abortion policy (Vassilev 1994).

1968–89

On 28 December 1967 a special decree of the Central Committee of the Bulgarian Communist Party (BCP) and the Council of Ministers was published to encourage an increase in the birth rate and reduce abortion. Effective 1 January 1968, family allowances were increased and restrictions placed on abortion for all married women having no children or only one child except in cases of serious medical indications (Bulgaria, 1967, 1968). The period of time in which abortions were allowed was reduced to 12 weeks. Abortion remained freely accessible only to women having three or more children (Vassilev, 1994). Unmarried women and widows with only one child were obliged to apply for an abortion to a special committee consisting of three persons, one of whom had to be a gynecologist. Unmarried women could be granted an abortion if they were under 16 years old or had been involved in rape or incest, or if "serious social indications" were determined by a woman's health center, provided the pregnancy was of no more than 10 weeks' duration at the time of the examination. Despite these guidelines, however, requests for procedures could not always be denied. Women with one or two children were entitled to an abortion if they persisted after a committee had tried to dissuade them. Abortions continued to be available on request without committee approval for women aged 45 years and older or for those who already had three or more children. Other requirements for an abortion included a six-month interval between consecutive abortions and an increase in the fee (of up to 30 leva, a sum equal to approximately one-third of the minimal monthly salary at that time).

Prior to 1968 family allowance and birth payments had been low but progressive. At the same time that abortion availability was administratively restricted in December 1967, the direction and intensity of the "positive" incentives to fertility were also changed. The family allowance structure was adjusted, with payments for third children greatly increased and payments for fourth and higher order births reduced by half. The lump sum birth payment was raised to a higher level. The new policy concentrated on second and third children. Payments were significant when measured in terms of prevailing average wage rates (Isenberg, 1963; McIntyre, 1975; David and McIntyre, 1981). The purpose was to stimulate Bulgarian families to have at least three children and to discourage (primarily Turkish and Gypsy) families from having more than three children.

Table 5.1 shows the changes that occurred after the 1968 restrictions were implemented. The birth rate increased from 15.0 in 1967 to 17.0 in 1969 and the number of legal abortions decreased briefly, from 108,500 in 1967 to 89,200 in 1968, only to increase again to 106,900 by 1969. Restrictions on abortion were relaxed somewhat in 1970, when abortion was allowed for all unmarried women and for married women with only one child (Kaser, 1976). It is difficult to separate the fertility effects of the increased financial incentives from the response to the decreased availability of legal abortion. The abandonment of the abortion restrictions in 1970 suggests that the fertility effect of the first pronatalist dosage was not very large.

The pronatalist measure had been introduced without having been discussed or voted upon in the National Assembly. When it became obvious that these measures were failing to achieve their objectives, further restrictive measures were announced in April 1973. Abortion was prohibited for all unmarried women, childless women, and women with only one child (Bulgaria, 1973). During this time, clandestine procedures posed a serious risk to women's health. Public reaction to the tighter abortion restrictions was sufficiently negative to force the Ministry of Public Health to make some slight modifications in February 1974. Exemption from the restrictions was granted for cases of rape or incest, if the woman was a minor or over 45 years old with at least one child, or for serious medical conditions (Bulgaria, 1974). The government did not fully return to the previous system of abortion on request that had prevailed from 1956 to 1968 (Stloukal, 1995). To obtain an abortion, the presence of an approved medical condition had to be demonstrated. Despite the limitations, more than half of all requests for abortion were approved by specially authorized committees. Gradually, medical indications for abortion became broader and more permissive as it was realized that all attempts to stimulate the birth rate over the longer term had failed.

At the same time that administrative restrictions on abortion were reimposed in April 1973, a revised Family Code was approved by the National Assembly. By the Decree on Encouragement of the Birth Rate, promulgated by the Central Committee of the Communist Party and the Council of Ministers, families were granted certain allowances and rights according to the number of their children (David and McIntyre, 1981). Upon giving birth, mothers received 100 leva (U.S.$113) for the first child; 250 leva (U.S.$283) for the second child; and 500 leva (U.S.$568) for the third child. If the family had twins, they would receive the maximum benefit. For fourth and higher order births, payments were equal to that received for the first child. Monthly child allowances were introduced and determined by birth order: 5 leva for the first child, 15 leva for the second, and 35 leva for the third. Any higher order births would receive 5 leva. Allowances were guaranteed for children up to 16 years of age. These monthly allowances were increased 20 years later because inflation had greatly reduced their value. However, their value had always been limited, the equivalent of only 5 to 10 percent of the minimum salary. They were more symbolic gestures than means of providing meaningful support for the family.

Pregnant women were entitled to 120 days maternity leave, paid at 100 percent of salary for the first child; 150 days for the second child; and 180 days for the third child. For higher order births, the leave was equal to that of the first child. Maternity leave was basically guaranteed for all pregnant women and could be extended for up to two years (three years in cases of multiple births) with payment equal to that of the minimum state salary. Additional incentives included: state assistance employment, annual leave, lower retirement/pension age, and transportation facilities. A broad system of daycare centers was developed. Industrial enterprises and cooperative farms were encouraged to create and support children's creches and nurseries.

The fertility effects of these program modifications are reflected in the birth rate increase from 16.2 in 1973 to 17.2 in 1974, shown in Table 5.1. As before, it is difficult to separate effects due to incentives to those resulting from the restrictions on access to legal abortion. Findings from sociopsychological studies suggest that the incentives were regarded as compensation for the costs of having additional children (Vassilev, 1982). More educated women had fewer children on average.

The sharp fertility rise from 1967 to1968 was concentrated in urban areas where the birth rate rose 1.9 points in that single year (Kostanick, 1977; McIntyre, 1980). Since the urban population was more likely to utilize induced abortion, reduced access to legal termination could be expected to have a greater impact on the urban birth rate, which reached its lowest level of 14.7 in 1965 (not shown) and generally rose thereafter. The rural rate continued to fall until 1968 and, after a one year reversal, resumed the decline during the next three years, rose slightly in 1973 and 1974 at the time of the renewed abortion restrictions, and then declined again during the next four years. Although there are age-structural factors that tend to depress rural fertility in Bulgaria, the broad program of pronatalist incentives introduced in the late 1960s played a role in the sudden development of this reverse differential. The types of inducements offered, including greatly extended paid maternity leave, improved child care facilities, tax advantages, and steeply progressive family allowances, had a more evident effect in the urban environment, where a greater proportion of mothers were in the labor force away from home and where the child care advantages of the extended family were less common (Bulgaria/CSO, 1988; David and McIntyre, 1981).

As shown in Table 5.1, the birth rate declined in 1981 while the abortion rate and ratio reached a peak. Another attempt to induce women to carry pregnancies to term was introduced in 1984 when every pregnant woman was required to have an interview with a representative of the Women's Organization. The fate of the pregnancy thus became a matter of public record. This measure did little to increase the birth rate. While official state population policy focused on pronatalist goals, echoing the motto "More children, more working power for the Fatherland!," the pronatalist policies implemented in the period 1945–89 largely failed to achieve their goals. Increases in the birth rate were small and temporary. By 1985 women with only a primary education had an average of 1.98 children, those with secondary education 1.57, and those with higher education 1.47 children (Bulgaria/CSO, 1988).

1990–97

The overthrow of the communist regime coincided with a relaxation of abortion legislation. In 1989 a new liberal abortion law (drafted by the Family Planning Association) was proposed. It was implemented by decree in February 1990 when the Minister of Public Health signed it the night before his dismissal from government (Bulgaria/MOH, 1990; United Nations, 1992). Abortion on request was legal again for all women regardless of age or citizenship provided that gestation was not more than 12 weeks and the procedure did not pose a danger to the woman's health. All hospitals and medical institutions authorized to perform abortions must admit the pregnant woman within three days of her request or inform her about contraindications. To keep the abortion confidential women have the right to select the place of operation.

The path to abortion includes diagnosis of pregnancy at the local (regional) women's consultation office, laboratory examinations, and final decision. All of this has to be accomplished within a period of 10 days or less if the pregnancy is close to the 12 weeks' term limit. The local/regional obstetrician is obliged to direct and refer the pregnant woman to the hospital she has chosen as a preferred venue for the operation. In case of medical contraindications, the local gynecologist has to treat the patient before sending her to the appropriate medical facility.

Terminations of pregnancy can be performed only in specialized obstetrical hospitals or in the obstetrical wards of general hospitals, not in an outpatient facility. Local or general anesthesia is obligatory, although the evidence is that many abortions are done without or with insufficient anesthesia. More than half the women leave the hospital the same day.

Medically indicated abortions have to be performed on request when the pregnant woman is suffering from a confirmed illness endangering her health or life (at the moment or in the future), or when risks to the fetus are evident. They can be performed up to 20 gestational weeks if the condition is included in the addendum to the decree. If the condition or the illness is not listed, the abortion is allowed as an exception. Termination of a pregnancy of more than 20 gestational weeks is permitted only if the woman's life is in danger or in case of severe morphological changes in the fetus, or if the fetus is severely harmed genetically ("genetic abortion").

Women requesting medical abortions are assessed by especially appointed medical committees authorized to grant the appropriate permission. For the safety of the patient, all questions connected with the interruption of a pregnancy of more than 20 gestational weeks are decided by special commissions established at the University Obstetrical and Gynecological Clinic and larger obstetrical hospitals. Throughout the decision making process and operation, a specialist competent to deal with the specific illness, for which the abortion is indicated, is required to be present.

The list of medical indications is long, including some obstetrical and gynecological indications. For example, a woman who has previously had a caesarian section followed by complications is allowed to decide, in the case of a subsequent pregnancy, whether to continue with the pregnancy or not. The same goes for any pregnancy likely to end in caesarian section. Genetically motivated abortion is the second largest group of medically indicated abortions. It includes a majority of pregnant women over 35 years of age. The concept behind the legislation has been to protect the offspring from some genetically determined damage caused by the age of the mother. At the opposite extreme, abortions for teenagers below the age of 16 are regarded as medically indicated. A new medical indication for abortion is if one or both parents has AIDS or is HIV positive (Vassilev, 1994).

If the commission denies the abortion request, this decision can be appealed within seven days to a special committee appointed by the Ministry of Public Health. The committee has to investigate the case and decide within one week. The decision is final and cannot be appealed.

The new legislation also provides contraceptive counseling and prescriptions for contraceptives. Fees were increased in line with the inflation rate (equivalent to about U.S.$10). Exemptions from fees exist for women under age 18, for those having abortions on medical grounds, and for pregnancies resulting from violence (WHO/EURO, 1993). The fee is considered a tax because it does not cover actual costs.

In summary, as Donkov (1991) and Stloukal (1995) have noted, Bulgarian abortion policy measures were anything but consistent, shifting from liberalization (1956–68) to officially strict but not always strictly enforced limitation (1968–73), to one of slight liberalization (1974–90), and back to full liberalization (1990 to the present). Bulgarian abortion legislation could be classified as among the most liberal. It protects the woman's right to abortion while allowing her to make choices with the safeguard of confidentiality.

A strong positive correlation exists between the age of the pregnant woman and the probability of pregnancy termination. In 1977 over 95 percent of women having abortions were married and more than half had two or more children (Vassilev, 1978). Women 30 years and older resorted to abortion as a fertility regulating method after marriage and delivery of the first child (41 percent of all abortions in 1989). A comparison of the fertility rates and abortion rates by age showed that abortion was practiced primarily between the births of the first and second child, less often to prevent having a third child (Vassilev, 1994). Once having fulfilled their reproductive goals, women terminate unwanted or accidental pregnancies. As of 1995 approximately half of all registered pregnancies were terminated. Abortions continue to exceed births (COE, 1997).

Abortions for medical reasons have varied in number and incidence depending on the abortion legislation. Restrictive changes in the law were followed by an increase in "medically indicated" and spontaneous abortions. When regulations were relaxed, women would choose legal abortions "on request," bypassing the longer bureaucratic procedure involved for medical abortions. However, when certain restrictions were introduced (such as the existing number of children a woman had, her marital status, the gestational age of the pregnancy, and so on.), pregnant women were forced to seek the intervention of medical committees to obtain abortions. This trend was most obvious after 1980 when restrictions were added for women having only one child. As a result, the proportion of medically indicated abortions rose significantly. Ultimately, the measures restricting abortion had little effect. The existing regulations left enough possibilities open for any pregnant woman to interrupt her pregnancy through reliance on medical indications that could be interpreted in diverse ways.

Abortion Rates and Ratios

The primary goal of liberalizing the abortion law in 1956 was to decrease health risks associated with clandestine abortions. This goal was attained soon after the legislation came into force. As shown in Table 5.1, the number of legally induced abortions rose steadily between 1957 and 1967 while the number of "other abortions" (spontaneous plus illegal) declined at first and then stabilized. The legal abortion rate increased from 18.2 per 1,000 women age 15 to 44 years in 1957 to 57.2 in 1967 while the abortion ratio jumped from 225 per 1,000 live births in 1957 to 871 in 1967. Within a context of traditional method failure and lack of ready access to modern contraceptives, abortion became the major means of fertility regulation. As shown in Table 5.1, the impact of political and economic dislocations in the early 1990s produced a sharp rise in the abortion ratio to 1,334 in 1992, matched by an abortion rate of 67.5. By 1996, the ratiio had dropped to 1,233 and the rate to 51.3 (COE, 1998).

Illegal Abortion

In Bulgaria, abortions are deemed illegal when performed by unqualified persons, whether medically trained or not, or when performed outside authorized hospitals. The registered number of such abortions has always been small (91 in 1990 and 69 in 1995) because they come to attention only when complications or fatalities occur. In the case of an abortion initiated or completed outside a medical institution, the law required registration of the data in the presence of witnesses such as a second physician or nurse. The report was subsequently signed by the patient or the witness.

After 1973, when the Ministry of Public Health severely prohibited or limited requests for abortions, the law required that all cases of spontaneous abortion in hospitals be regarded as suspect. The medical system had to fulfill its obligations by reporting spontaneous abortions as criminal. In addition, when a woman experienced a spontaneous abortion in a hospital, a physician had to be called as a witness and the director of the hospital had to be notified by a written report. Eventually, the strict enforcement of reporting spontaneous abortions proved ineffective and was deemed a witch-hunt.

Currently, if there is suspicion that an illegal abortion has been performed, the physician must inform the hospital administration, the nearest prosecutor, and the police within 24 hours. It is then up to the prosecutor's office and police to decide whether or not to investigate. Most medical personnel try to avoid this long procedure unless there is obvious evidence or there are life-threatening complications. An illegal service provider may incur a prison sentence up to three years or up to eight in case of fatal complications. The court rarely uses the maximum sentence, however, as it is generally accepted that abortions performed outside of hospitals are not necessarily unsafe or life-threatening because the service providers usually tend to be qualified gynecologists or midwives. Woman who have had illegal abortions are not punished (Vassilev, 1994).

Between 1956 and 1968, when abortion was legal and easily accessible, the charge for a clandestine procedure was 15–20 leva (three times the official charge), the equivalent of one-fifth of the minimal monthly salary. Between 1969 and 1980 the charges increased to an average of 50 leva for pregnancies up to 12 weeks and

500 leva for more advanced pregnancies. By 1991 the fee had increased even more due to inflation and a rapidly changing social environment. However, the cost of illegal abortions was never so high as to prevent pregnant women from seeking them. The factor regulating clandestine abortions was not money but safety. The cost for an illegal abortion was rarely more than the Bulgarian equivalent of U.S.$25 to U.S.$30.

The disfavor with which officials viewed abortion and contraception was reflected in the procedure justifying sick leave. Since abortion had to be listed as a reason for absence, it became public knowledge. Procedures performed at public hospitals could not be kept confidential. Measures to protect privacy or confidentiality were prohibited. The number of clandestine abortions varied with legal changes, the extent to which restrictions were enforced, and the rise and fall in "spontaneous" or medically indicated abortions.

As shown in Table 5.1, in the period 1980 through 1989, the number of legal abortions declined from 136,300 to 111,000, while the number of medically indicated abortions registered in the category of legal abortions rose by 86 percent from 3,850 to 7,150 (not shown). However, the number of reported illegal abortions (within the category of "other abortions") increased only 20 percent, rising from 180 in 1980 to 218 in 1989 (Vassilev, 1994). The medical profession cooperated in ensuring that women were not deprived of their rights.

Commentary

The continuing dominance of abortion as the major method of fertility regulation is the result of several converging factors. With its close ties to the former Soviet Union, the Bulgarian Government was relying on Soviet social and medical policy favoring abortion over imported "Western" contraceptives. While it was official policy to make pills available, their adoption was discouraged (Stefanov and Naumov, 1974). Side effects of modern contraceptives were purposely exaggerated in the media (Vassilev et al., 1974).

With women marrying at an early age and quickly having one or two children, fertility regulation became a salient concern for the rest of their sexually active adult life. Without access to effective contraception, they relied on abortion to cope with unwanted pregnancies. Culturally, there were few social, religious, or ethical constraints. Having an abortion did not lower a woman's status or reputation. The experience with legal abortion was good, seldom resulting in complications or fatalities. Since the law left the decision entirely in the hands of the woman, there was no need to request permission from a partner/spouse. Large segments of the medical profession, feeling comfortable with abortion, long resisted innovations in family planning. And, the Ministry of Public Health made most decisions regarding abortions, often without consultation with specialists. In all the years of the communist era, the subject was never brought to the attention of the National Assembly or publicly discussed.

There is little support for a pro-life movement. Recent surveys of medical students indicate that more than two-thirds are prepared to perform abortions on request (Marinova, 1997).

CONTRACEPTION

Despite early promises at the time of abortion legalization, effective contraceptives were largely unavailable in Bulgaria. Pronatalist ideology long prevented implementation of Paragraph 22 of the Instructions of 28 March 1973 (Bulgaria, 1973), which stated that women having abortions must be "educated in contraceptive practices suitable for their health status and to prevent repeat abortion." Modern contraceptives were not produced in the country. Pills were imported from the German Democratic Republic and Hungary and IUDs and condoms from Czechoslovakia, but the supply was erratic and unreliable, compounded by hard currency shortages (Blayo, 1991). A month's supply of pills initially cost less than a pack of cigarettes, compared to a U.S.$5 charge for abortion. IUDs were in short supply even at black market rates.

In 1973 Golemanov and Krustev (1973) noted that use of the most effective contraceptives was only "in its very beginnings," which they believed was responsible for the rise in abortions. *Coitus interruptus* was the most widely used method, mentioned as their preferred choice by 76 percent of 1,830 married Sofia women (Tzankov and Vassilev, 1976). Similar findings were noted by Stoimenov (1976). Menstrual regulation was not readily available. Female sterilization, regarded as a Nazi eugenic method, was unacceptable to the vast majority of the Bulgarian public and the medical profession. It was rarely used for family planning purposes. Vasectomy, rejected for psychological reasons, was never used or discussed. Small quantities of Copper T IUDs were imported intermittently from Finland in 1980. Research not attuned to the state's pronatalist policies did not receive financial support. Contacts with Western scientific institutions, producers, and researchers were strictly controlled.

After 1989 the Ministry of Health and other state institutions showed a fresh interest in family planning as a means of providing alternatives to reliance on abortion (Carlson, 1998). IUDs were designed, produced, and distributed. Cooperative production of condoms was arranged with a Singapore company. Pills could be obtained through pharmacies privatized in 1991. However, the cost of modern contraceptives gradually rose to international levels, out of reach for wider utilization. Within this context international assistance was requested and provided.

Family Planning Association

Initial steps for family planning began in the early 1930s when the first medical facility to assist pregnant women and newborns was opened in Sofia. At that time condoms were available but were recommended more for the prevention of venereal disease than for contraception. District women's health centers provided advice on sexual matters and contraception (Stoimenov, 1976). The Family Development Council (FDC) began its activities in 1973 as a semiautonomous organization under the umbrella of the Society of Gynecologists and Obstetricians. It worked with medical professionals, journalists, lawyers, and sociologists, and also tested some modern contraceptive methods.

In 1975, in an atmosphere of growing public support, the FDC became an autonomous organization, was renamed the Society for Planned Parenthood and Family Development (SPPFD), and was accepted into associate membership of the International Planned Parenthood Federation (IPPF). While ignored by the Communist Party, the SPPFD initiated collaboration with other organizations interested in problems related to fertility, sexuality, human reproduction, and sexuality education. With support from IPPF and the World Health Organization, the SPPFD generated ideas, produced policy papers, and resisted the attacks of pronatalist advocates on women's rights. The dominant Christian Orthodox Church and the Islamic religious system were not sufficiently powerful to oppose family planning. Full IPPF membership was attained in 1979.

The social and political tensions that followed the demise of the communist regime in 1989 impeded the development of new ideas and actions in the area of family planning. The Bulgarian Family Planning Association (BFPA) succeded the SPPFD, continuing the IPPF affiliation. Ten BFPA clinics opened between 1993 and 1996 in selected areas of the country. They are staffed by medical personnel who conduct examinations, prescribe, and dispense contraceptives. Supplies are obtained through IPPF and offered to clients at approximately 10 percent of market prices. Abortion services are not provided. In those areas where BFPA is not active, the government invited women's organizations to organize counseling centers and make referrals to obstetricians/gynecologists whose prescriptions (with such referrals) can be filled at local pharmacies at the same low subsidized prices as at the BFPA clinics (Carlson, 1998b).

Women in Transition Survey

In 1995, as the subsidized BFPA program was beginning, the Institute of Demography of the Bulgarian Academy of Sciences and the National Statistical Institute conducted a survey of a national probability sample of 8,642 men and women, known as the Women in Transition Survey (WTS). It included questions about knowledge, attitudes, and practice of contraceptive methods. Compared to findings from the Bulgarian segment of the 1976 World Fertility Survey, the proportion of women in first marriages and between the ages of 15 and 44, not using any contraception, decreased from 24 percent in 1976 to 14 percent in 1995. Similarly, the proportion of married women using modern methods increased from 6 percent in 1976 to 46 percent in 1995 and reliance on traditional methods dropped from 70 percent to 40 percent. Changes were particularly noticeable among 30 to 34 years old women whose use of modern methods rose from 6 percent to 54 percent

(Carlson and Omori,1998). For many women withdrawal backed by legal abortion remains the primary method of fertility regulation.

THE WOMAN QUESTION

Status of Women

The Islamic religion, though adopted only by a small minority of the Bulgarian population, greatly influenced gender roles and relations. Women were generally confined to their homes. The first women's association, called Milosurdie, was organized in 1857, just 20 years before liberation from the Turkish occupation (Kostova, 1994). It advocated educational advancement and cultural enrichment. Although women gradually gained access to some public institutions, the only broadly accepted spheres for women at that time were the family and the household (Panova et al., 1993).

Following national liberation in 1878, several women's groups, such as the Bulgarian Women's Union and the Association of University Women, were organized with the goal of achieving equality of men's and women's political, economic, and social rights as well as creating equality with respect to education, work, and income. However, once the new Bulgarian republic assumed responsibility for education, the organizations focused mainly on charitable and cultural activities. After the arrival of Russian troops in 1944, the government passed a law proclaiming the equality of men and women as part of its effort to implement communist ideology (Kostova, 1994).

With the establishment of the communist regime after World War II, the collectivization of agriculture, and the expansion of industry, a large proportion of women became industrial workers. They had equal rights on paper—political, civil, economic, social, and cultural—and were given various benefits such as monthly child allowances, paid maternity leave, and practically free daycare centers for the children, but remained unequal with respect to opportunities for advancement to top positions in public or political life. Women were faced with fulfilling three social roles simultaneously: that of mother and wife, productive worker (for the socialist state), and social activist (Petrova, 1993). By 1988 about 90 percent of women were in the labor force. The vast majority worked as kindergarten and primary school teachers, nurses, low level office workers, sales attendants, and manual workers. Less than 2 percent held management positions (Bulgaria/CSO, 1988).

The Bulgarian National Women's Union, founded in 1945, was an auxiliary of the Communist Party. It served primarily to perpetuate official party views, mobilizing support for the establishment of cooperatives and state enterprises (Petrova, 1993; Todorova, 1993; Kostova, 1994). In 1950 the union was absorbed by the Fatherland Front, an organization designed to advance the political goals of the Communist Party. By the late 1970s and early 1980s, the ideology shifted to more conservative views, emphasizing that women should return to their "authentic nature" by making their households and families a priority (Petrova, 1993). When a new women's organization, the Women's Democratic Union, was organized in March 1990, it advocated a strong commitment to new democratic ideals while at the same time placing emphasis on motherhood, home, and family.

In 1996 women's unemployment exceeded that of men. Presently, women's organizations are focusing their endeavors on cultural and economic issues, wary of drawing attention to women's problems for fear of increasing tensions and creating a possible backlash. There is only limited interest in political, social, and legal activities. Women's issues receive little attention in the media. It was noteworthy when the Union of Democratic Forces, elected in 1997, appointed a woman, Nedezhda Mikhailova, a poet and translator who has lived and worked in the United States, as the country's first female foreign minister.

SEX AND SOCIETY

Planning Family Size

Since the end of World War II, the two-child family has been the model of choice in various surveys (Vassilev, 1986). Whereas in the early decades of this century, 44 percent of children born were of fourth or higher order parity, 87 percent of births in 1987 were first or second children while only 5 percent were of fourth or higher order parity (Vassilev, 1989a).

Much of the research on family size intentions has been conducted by social gynecologists who have had both medical and demographic training. For example, in a study of aborting women in Sofia, it was noted that of those who had three children, only 5 percent wanted another child. The proportion wanting another child rose to 9 percent for those with two children and to 67 percent for those with only one child. In general, the women who had abortions had fewer children than they had initially planned on having (Vassilev, 1972). In a later study of working women, 75 percent expressed a desire to have two children but half had only one (Lagadinova, 1974). Other studies suggested that desired family size was 2.2 children while completed family size was 1.8 children (Stoimenov, 1976).

In another study of 1,830 married Sofia women in 1975, social, psychological and economic factors were found to be the major determinants of family size intentions, regardless of family background or the size of the family in which the woman was reared. There was an inverse correlation between a woman's educational attainment and mean number of children in the family. Housing was the single most important determinant of family size intentions. About 41 percent of newly married couples had no choice but to live with relatives. The couple very often was not allocated a separate apartment until the wife was over age 30, by which time there was much less interest in reproduction. Of nearly equal importance was communication between the partners and their perception of family harmony. There was some indication that younger husbands and those married for a longer time were more willing to help with household chores such as shopping and cleaning;

production workers seemed more prepared to do so than clerical and other service workers (Stoimenov et al., 1977).

Sexuality Education

The first steps toward organized sexuality education were initiated in the second half of the nineteenth century, directed primarily to preparation for marriage and the prevention of sexually transmitted diseases. In 1929 the Public Health Legislation Act was approved, containing, in Paragraph 53, the statement that potential marriage partners must have the chance for a free premarital consultation. Legislation enacted in 1936 required applicants for marriage licenses to present a certificate proving that they were free of venereal disease.

In 1967 the Communist Youth League published several documents emphasizing the need for sexuality education. Ten years later the president of the republic addressed an open letter to the youth organization accusing it of neglecting the needs of the younger generation regarding sexuality information. Shortly after this "signal" from the top, a book on contraception was produced (Vassilev, 1979, 1989b, 1989c).

The Red Cross developed some programs oriented especially toward adolescents, including some aspects of sexuality education. Human sexuality courses were introduced into the curricula of pedagogic institutions preparing future teachers. The first organized program of sexuality education was added to the school curriculum in 1985. For students in the fifth grade, the subject was included as part of lessons on natural history. For eighth graders, sexuality education became part of the biology lessons. The focus was on the anatomy and physiology of the reproductive system.

As noted by Bostandjiev (1992), the attitudes to sexuality education in the communist era were rooted in a tradition of patriarchal and authoritarian human relations. The prescribed task of the authorities was to repress adolescent sexuality and maintain institutional control over personal behavior. The justification for this approach was to reduce the high rates of abortion. Human sexuality was considered in moralistic and medicalized terms. Deviant behavior was subject to sanctions. Until 1989, the official communist press abstained from publishing material on sexuality except for pronatalist purposes and for government documents.

In 1988 a group of specialists, mostly school psychologists and psychotherapists, organized *Modus*, an independent professional organization focusing initially on sexuality education. Among its first publications were programs entitled "Prevention of AIDS Depends on You and Me" and "In the World of Intimacy" (Vassilev, 1988). Although well received by students, these initiatives were marginal and alien to the repressive character of the educational system.

After 1989 barriers to sexuality education began to crumble. The mass media disseminated better information on human sexuality although many journalists remained uninformed (Chernev and Stamencova, 1994). Special programs, TV series, and discussion groups focused increasingly on sexuality, parenthood, and family life. The Bulgarian Family Planning Association developed programs designed to avoid unwanted pregnancy and foster contraceptive responsibility. There were never any objections to sexuality education from the Christian Orthodox Church or from Islamic religious institutions as long as the manner in which the material was presented to adolescent girls conformed to Islamic traditions. Sexual phenomena were discussed in the context of biology, a compulsory school subject. Courses were introduced into medical school curricula. In March 1992 several *Modus* members formed the Council for Sexual and Family Health. It developed sexuality education programs for university and postgraduate students of pedagogy, psychology, and medicine as well as workshops and training courses for other specialists (Bostandjiev, 1992). Meanwhile, pornography flooded the country, much as it did in other formerly socialist societies.

Adolescents

The age of sexual debut among many male students is 14 years (more than 19 percent claim to have their first contacts at that age) and 19 years for female students (17 percent claim their first contact then). Studies indicate that 62 percent of boys and 45 percent of girls experience sexual debut prior to age 18. Only 7 percent of partners used a contraceptive method at that time. Another study showed that 61 percent of girls delivering out-of-wedlock children had become pregnant by their first partner (Bostandjiev, 1992).

The motivations for sexual debut demonstrate their accidental and unpredictable character. For the majority of girls' motivation for sex was love (41 percent), or liking (23 percent). Another 11 percent allowed sexual contact in order to "keep" the partner or to "make him happy." Other motives included curiosity (3 percent) and indifference (2 percent) (Vassilev and Rushkarska, 1990).

Almost 40 percent of women marry under age 20. In 5 percent of marriages, both partners were under 20 years of age (Bostandjiev and Vassilev, 1989). Bulgarian mothers were the youngest in Europe. The fertility rate per 1,000 women aged 15 to 19 years was the highest, reaching 84.1 in 1982 before declining to 69.9 in 1990 and 45.1 in 1997 (COE, 1998). By 1989 a longer interval was noted between first and second children (Vassilev, 1989a).

Abortions among women under 15 years of age and in the age group 15 to 19 years were relatively uncommon (9.9 percent of all abortions in 1989). Pregnant women in these age groups usually married and gave birth (Bulgaria, 1989). However, the proportion of those young women having abortions is slowly rising (Vassilev, 1994).

Sociological and psychosocial studies reflect a strong tendency to question and change moral and ethical principles inherited from the past (Institute of Youth, 1974; Bostandjiev and Vassilev, 1989; Sougareva, 1992). The acceptance of premarital sex is seen as normal by 80 percent of boys and 42 percent of girls in high school. However, the so called "double standard" still persists as a result of traditional influences. While male premarital sexual relations are freely tolerated, those of females are much less so.

The new preoccupation with "sexual freedom" has resulted in an increase in adolescent pregnancy, abortion, and delivery. In a six-year study of 1,088 pregnant girls, ages 13 to 16 years (average age 15.4 years), during 1988–93 at the Department of Obstetrics and Gynecology at the Kliment Orchidsky Medical University in Pleven, it was noted that 625 had deliveries while 463 chose abortion (Prozhanova and Tantchev, 1995). The ethnic composition was almost evenly divided between Bulgarian girls (52 percent) and girls from minorities (48 percent) such as Turkish and Gypsy. The Bulgarian girls were twice as likely to choose abortion than the ethnic girls who, in line with their traditions, generally preferred to carry to term. Among all the girls, 74 percent had received primary education but 17 percent had not completed primary schooling. In the group of young mothers, only 29 percent had an official marriage contract; the rest lived in union with their partners. About 79 percent of these pregnancies were desired. While 51 percent of the delivering girls had initiated partner relations up to one year prior to pregnancy, only 24 percent of the aborting girls had had relationships with their partners for as long a period of time. Of the total sample, 64 percent accepted their pregnancies while 35 percent attempted interruption, 9 percent with drugs and 27 percent with nonmedical methods. It appeared that 33 percent had been raped (Prozhanova and Tantchev, 1995).

FUTURE PERSPECTIVES

It is difficult to speculate about the future (late 1998) when Bulgaria is emerging from a period of political and economic chaos. About 25 percent of the population was unemployed with women most severely impacted. More than 80 percent of the population existed below or barely above the level of poverty. The situation was further destabilized by various protest movements, increased criminality, and a general feeling of insecurity. Modern contraceptives were economically out of reach when priority had to be given to food and shelter. In such a context legal abortion remained the major method of fertility regulation. In 1997 the birth rate declined to 7.7 per 1,000 population, the lowest rate ever recorded in Bulgaria.

Since the 1997 election of the Union of Democratic Forces financial stability has returned and the economic situation is improving. Greater access to modern contraceptives coupled with more public emphasis on family planning are reducing reliance on traditional methods backed by abortion. Total registered abortions are declining. In all, there are hopeful indications that Bulgarians are becoming more efficient in fertility regulation with less dependence on abortion.

chapter 6

Czech Republic and Slovak Republic

Vladimír Wynnyczuk and Radim Uzel

EDITOR'S INTRODUCTION

The Republic of Czechoslovakia was established in 1918 as one of the successor states to the multiethnic Austro-Hungarian Empire. The Czech lands consisted of Bohemia, Moravia, and Silesia, while Slovakia and Ruthenia constituted the Slovak component. The Czech lands traditionally were more closely associated with the Austrian monarchy while the Slovakian component had been a part of Hungary. Significant differences developed in population patterns, in economic development, and in religious affiliations.

As a researcher closely associated with Czech colleagues for more than 30 years, I recall the emotionally charged days of 1968 when Soviet tanks crushed the Prague Spring. It was not until 21 yars later that the "Velvet Revolution" brought an end to the communist era. It had been my privilege to initiate, organize, and develop the still continuing Prague Study of children born in 1961–1963 to women twice denied abortion for the same pregnancy and pair-matched controls.

In the postcommunist era the Czechs' pre–World War II identification with Western Europe reasserted itself, strengthened by a stable government, economic reforms, and low unemployment (Hoagland, 1996; Pavlik, 1996). The Slovak Republic experienced a slower transition to democracy (Hockstader, 1997d, *The Economist*, 1997b, 1997c). A pro-Western coalition came to power in October 1998 (*The Economist*, 1999).

This chapter's historical focus is on Czechoslovakia, followed by postpartition developments in the Czech Republic and the Slovak Republic. Demographic trends, family and population policies, the woman question, and fertility planning are discussed. There follows a review of surveys on contraceptive knowledge and practice, abortion trends, and a summary of the longitudinal Prague Study. The chapter concludes with a discussion of sex and society, including activities of the Czech and Slovak family planning associations, sexuality education, and a commentary on future perspectives.

The summary of the Prague Study was jointly prepared with Drs. Z. Dytrych, L. Kubicka, and Z. Matejcek. The constructive suggestions of Drs. Tomas Frejka, Jirina Kocourcova, and L. Stloukal are gratefully acknowledged. With apologies to our Czech and Slovak colleagues, diacritics were omitted from all citations. The names of Czechoslovakia, the Czech Republic, and the Slovak Republic were chosen to denote the territories discussed.—H.P.D.

HISTORICAL AND DEMOGRAPHIC TRENDS

After World War I Czechoslovakia was proclaimed an independent state in 1918. With the exception of the interlude during World War II, when Nazi Germany gave Slovakia formal independence, Czechs and Slovaks lived in a common state for most of the twentieth century. In 1948, three years after the end of World War II, a communist government assumed power, installing an authoritarian political system and a centrally planned economy.

After considerable political and economic experimentation in the early and mid-1960s, a series of reform measures, implemented during the 1970s, produced some improvement in material life. The more rapid growth in the Slovak Republic led to a steady reduction in the differentials of living standards with the Czech lands that had existed prior to 1918. Following the collapse of the communist regime in 1989, the country was peacefully divided on 1 January 1993 into two independent states: the Czech Republic with slightly more than ten million inhabitants and the Slovak Republic with slightly more than five million inhabitants.

Natality and Mortality

Demographic differences have always existed between the population living in what is now the Czech Republic and the one living in what is now the Slovak Republic. The Czech population exemplified a demographically more advanced pattern typified by lower fertility and mortality levels. The Slovak population retained a pattern of higher fertility and mortality rates.

Births, birth rates, total fertility rates, and other demographic data for Czechoslovakia as a whole are shown in Table 6.1 and Figures 6.1 and 6.2. In the Czech lands the crude birth rate had declined to around 36 per 1,000 population by the 1890s (Pavlik, 1964). After continued decline into the 1930s and higher rates during the war years, the birth rate reached a trough of 13.3 in 1960, from which it recovered to 19.1 in 1975 (Table 6.2). This Czech natality rate, together with the Slovak rate of 20.6 per 1,000 (Table 6.3), marked a peak from which there was a considerable decline in subsequent years.

Compared to most European countries, there are oddities in the demographic development of the Czech Republic. The first is the increase in births during World

Figure 6.1
Birth Rates: Czech and Slovak Republics, 1950–95

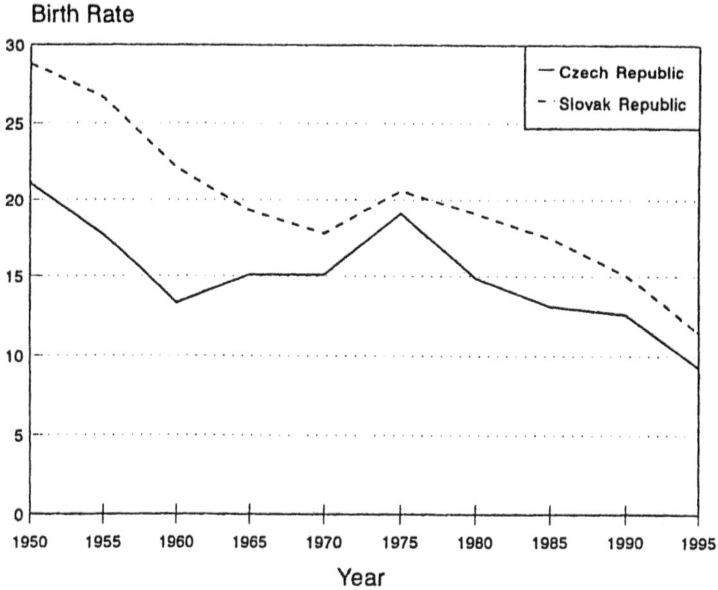

Sources: See data from Tables 6.2 and 6.3.

Figure 6.2
Total Fertility Rates: Czech and Slovak Republics, 1950–95

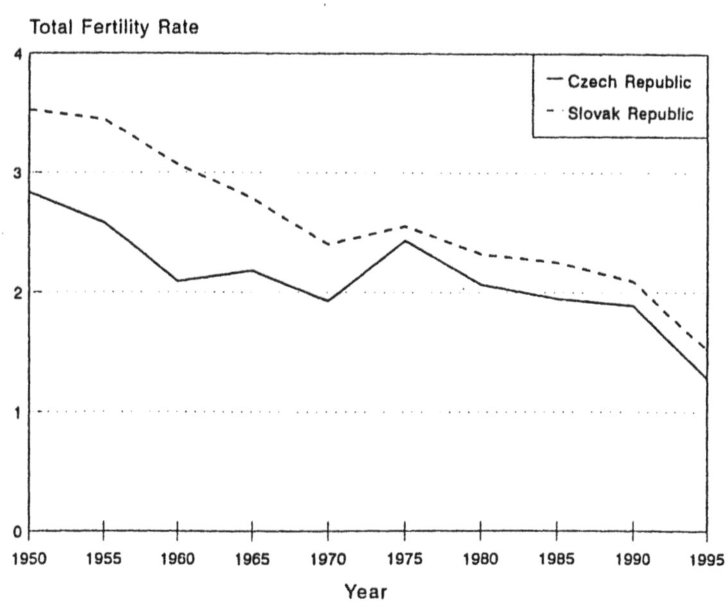

Sources: See data from Tables 6.2 and 6.3.

Table 6.1
Birth, Abortion, and Fertility Rates and Ratios: Czechoslovakia, 1950–91

Year	Live Births	Legally Induced Abortions	Birth Rate	Legal Abortion Rate[g]	Legal Abortion Ratio[h]	Total Fertility Rate
1950	288,100	na	23.3	na	na	3.04
1955	265,200	2,100	20.3	0.8	8	2.85
1957[a]	252,700	7,300	18.9	2.7	29	2.75
1958	235,000	61,400	17.4	22.8	261	2.57
1960	217,300	88,300	15.9	32.5	406	2.39
1961	218,400	94,300	15.8	34.0	432	2.38
1962[b]	217,500	89,800	15.7	31.5	413	2.34
1963	236,000	70,500	16.9	24.1	299	2.50
1965	231,700	79,600	16.4	26.3	344	2.37
1966[c]	222,600	90,300	15.6	29.6	406	2.32
1968[d]	213,800	99,900	14.9	32.4	467	2.01
1969	222,900	102,800	15.5	33.3	461	2.04
1970	228,500	99,800	15.9	32.3	437	2.07
1972	251,500	91,300	17.4	29.3	363	2.21

1973[e]	274,700	81,200	18.9	25.9	296	2.39
1975	289,400	81,700	19.6	25.9	282	2.46
1979	272,400	94,500	17.9	29.5	347	2.33
1980	248,900	100,200	16.3	31.0	403	2.15
1981	237,700	103,500	15.5	31.9	435	2.09
1982	234,400	107,600	15.2	33.0	459	2.10
1983	229,500	108,700	14.9	33.1	474	2.08
1984	227,800	113,800	14.7	34.5	500	2.07
1985	226,000	119,300	14.6	36.0	528	2.06
1986[f]	220,500	124,200	14.2	37.3	563	2.03
1987	214,900	159,300	13.8	47.5	741	2.00
1988	215,900	164,700	13.8	48.7	763	2.01
1989	208,500	160,300	13.3	47.0	769	1.95
1990	210,600	159,700	13.4	46.4	758	1.96
1991	208,000	na	13.3	na	na	1.93

[a]Abortion law liberalized (December 1957). [b]Stricter interpretation of abortion legislation (December 1962). [c]Abortion administration procedures relaxed (July 1966). [d]Pronatalist measures introduced (July 1968–April 1973). [e]Availability of abortion restricted (June 1973). [f]Abortion committees abolished (November 1986, effective 1 January 1987). [g]Per 1,000 women age 15–44. [h]Per 1,000 live births.

Sources: David and McIntyre (1981), U.N. (1995a), Srb (1995) and Stloukal (1995).

Table 6.2
Birth, Abortion, and Fertility Rates and Ratios: Czech Republic, 1950–97

Year	Live Births	Legally Induced Abortions	Birth Rate	Legal Abortion Rate[a]	Legal Abortion Ratio[b]	Total Fertility Rate
1950	188,300	na	21.1	na	na	2.83
1955	165,900	na	17.7	na	na	2.58
1958	141,800	49,000	14.8	26.2	346	2.28
1960	128,900	67,600	13.3	35.7	524	2.09
1961	131,000	70,100	13.7	36.6	535	2.11
1963	148,800	51,500	15.4	25.4	346	2.31
1965	147,400	58,600	15.1	28.0	398	2.18
1969	143,200	74,300	14.5	35.0	519	1.87
1970	147,900	71,900	15.1	34.2	486	1.93
1972	163,700	65,100	16.6	30.9	398	2.09
1973	181,800	55,900	18.3	26.5	308	2.31
1975	191,800	55,500	19.1	26.3	289	2.43
1979	172,100	64,500	16.7	30.3	375	2.27
1980	153,800	68,900	14.9	32.2	448	2.07

1981	144,400	71,600	14.0	33.4	496	2.00
1982	141,700	74,500	13.7	34.5	526	2.00
1983	137,400	75,000	13.3	34.6	546	1.97
1984	136,900	79,500	13.3	36.5	581	1.97
1985	135,900	83,000	13.1	37.9	611	1.95
1986	133,400	83,600	12.9	38.0	627	1.94
1987	130,900	107,700	12.7	49.6	823	1.91
1988	132,700	108,400	12.8	51.2	817	1.94
1989	128,400	107,400	12.4	49.9	836	1.87
1990	130,600	107,100	12.6	49.3	820	1.89
1991	129,400	103,100	12.5	47.1	797	1.86
1992	121,700	93,400	11.8	41.8	768	1.72
1993	121,000	69,400	11.7	31.3	574	1.67
1994	106,600	53,700	10.3	24.3	504	1.44
1995	96,100	48,300	9.3	22.0	503	1.28
1996	90,400	46,500	8.8	20.6	513	1.18
1997	90,700	43,300	8.8	19.3	477	1.17

[a]Per 1,000 women age 15–44. [b]Per 1,000 live births.

Sources: Rychtarikova (1995), U.N. (1995a), COE (1998), Czech Republic/CSO (1997), and Bartonova (1998).

Table 6.3
Birth, Abortion, and Fertility Rates and Ratios: Slovak Republic, 1950–97

Year	Live Births	Legally Induced Abortions	Birth Rate	Legal Abortion Rate[a]	Legal Abortion Ratio[b]	Total Fertility Rate
1950	99,700	na	28.8	na	na	3.54
1955	99,300	na	26.6	na	na	3.46
1958	93,300	12,400	23.9	15.1	133	3.24
1960	88,400	20,700	22.1	25.1	234	3.07
1961	87,400	24,200	20.8	28.2	277	2.96
1963	87,200	19,100	20.4	21.1	219	2.93
1965	84,300	21,000	19.3	22.5	249	2.78
1969	79,800	28,500	17.7	29.1	357	2.43
1970	80,700	27,900	17.8	28.2	346	2.40
1972	87,800	26,200	19.1	26.0	298	2.49
1973	93,000	25,300	20.0	24.8	272	2.55
1975	97,600	26,200	20.6	25.1	268	2.55
1979	100,200	30,000	20.3	27.8	299	2.43
1980	95,100	31,200	19.1	28.7	328	2.32
1981	93,300	32,000	18.6	29.1	343	2.28

1982	92,600	33,100	18.3	29.9	357	2.27
1983	92,100	33,600	18.1	30.2	365	2.27
1984	90,800	34,300	17.7	30.6	378	2.26
1985	90,200	36,300	17.5	32.2	402	2.25
1986	87,100	40,600	16.8	35.8	466	2.20
1987	84,000	49,700	16.1	43.4	592	2.15
1988	83,200	51,000	15.9	44.1	613	2.15
1989	80,100	48,600	15.2	41.6	607	2.08
1990	80,000	48,400	15.1	40.9	605	2.09
1991	78,600	45,900	14.9	38.4	584	2.05
1992	74,600	42,600	14.1	35.8	571	1.98
1993	73,300	38,800	13.8	31.8	529	1.92
1994	66,400	34,900	12.4	28.3	526	1.66
1995	61,400	29,400	11.5	21.0	479	1.52
1996	60,100	25,200	11.2	20.3	419	1.47
1997	59,100	22,300	11.0	18.0	377	1.43

[a]Per 1,000 women age 15–44. [b]Per 1,000 live births.

Sources: Czechoslovakia/FSO (1993), U.N. (1995a), Tirpak and Adamica (1995), COE (1998), Slovak Republic/SSO (1997), Stloukal (1998), and UNICEF/TDB (1998).

War II, when almost all European countries experienced a decrease in natality rates. It resulted, in part, from specific conditions in the German Protectorate (the name of the territory covering the largest part of the Czech Republic between 1939 and 1945). Men were at home and, by marrying, lessened their chances of being drafted for forced labor in Germany (Ross and Frejka, 1998). Pregnant married women received special food rations. Since universities were closed, more young people became parents while older couples compensated for previously postponed births (Srb,1993).

There were several reasons for rising crude birth rates in the late1960s and early 1970s. One was the large number of women of fertile age born during and just after World War II. Another was the pronatalist population measures introduced at that time. They included: extended paid maternity leave for 26 weeks; increased family allowances, especially for parents with two or three children; financial assistance (birth grants) to mothers after the birth of a child; a maternity allowance which enabled most mothers to stay at home and take care of a newly born child; and additional nurseries and kindergartens (Frejka, 1980). Still another reason was the 1973 restriction on access to abortion, as will be discussed later in the chapter.

Overall, birth rates decreased over time from 24.5 per 1,000 population in 1920 to 8.7 in 1997 in the Czech Republic, and from 32.6 in 1920 to 11.0 in 1997 in the Slovak Republic As shown in Table 6.2, the total fertility rate (TFR) in the Czech Republic declined from 2.43 per woman in 1975 to 1.17 in 1997. A similar decline occurred in the Slovak Republic from 2.55 in 1975 to 1.43 in 1997 (Pavlik, 1994; Srb, 1995; Stloukal, 1996a; Pavlik and Kucera, 1997; and COE, 1998).

Marriage Patterns

Between 1950 and 1990 the crude marriage rate fluctuated between 7.2 and 10.7 in the Czech Republic and between 6.9 and 11.3 in the Slovak Republic. This resulted in the highest percentage of marriages for both men and women in Europe. According to the 1991 Czech census, 95 percent of persons within the age group 45 to 49 years had been in at least one marriage (Rychtarikova, 1994).

The political system generated policies and a system of economic incentives that led to early marriages and the high proportions of people married. Social benefits were available primarily to married couples. Discriminative valuation of female work and the lack of part-time jobs for women of reproductive age made women economically dependent on their spouses. Being married and having one or more children were major conditions for applying and subsequently obtaining an apartment on a preferential basis. Moreover, unsatisfactory consumer services necessitated conjugal cooperation, requiring wife, husband, and children for shopping, maintaining a flat, keeping a financial equilibrium, and so on. It was far more difficult for single persons to cope.

Another result of the governmental policies was the low average age of brides and bridegrooms. At the end of the 1980s the Czech Republic was among the European countries with the highest percentage of brides and bridegrooms 18–19 years of age. In the mid-1990s the average in the Czech Republic was 21 years of age for women and 23–24 years of age for men. By 1997, the mean age of women at first marriage had risen to 23.3 years (COE, 1998).

The typically early entry into marriage during the communist era was the result, in part, of the prevailing material conditions of life. Salaries of persons below age 30 were generally low. Social benefits provided by the state were relatively significant for young families. Moreover, parents were commonly expected (and willing) to provide financial support for many years. Thus, there were few reasons for young people to postpone marriage until periods of greater maturity. A low standard of sexuality education also played an important role. Many unwanted pregnancies were eventually legitimated by marriages. More than 50 percent of first children in the Czech Republic were born less than nine months after marriage (Rychtarikova, 1995, 1996).

In the 1990s marriage and fertility levels declined among women under 25 years of age (Rychtarikova, 1996). The mean age of women giving birth climbed from 24.7 years in 1991 to 26.2 years in 1996 (Ross and Frejka, 1998). The rising level of unmarried cohabitation since the late 1980s has contributed to a rise in the proportion of nonmarital births from five percent in the 1970s to 17 percent in 1996 (Stloukal, 1997; Ross and Frejka, 1998).

FAMILY AND PUBLIC POLICIES

For purposes of discussion, the term *family policy* is defined in a social welfare sense, reflecting governmental financial support aimed at families with dependent children to decrease the differences in per capita income compared with childless families or families with economically independent children. *Population policy* is defined in a pronatalist sense, including governmental, mainly financial, support to couples as an incentive to have children (or more children) or to create conditions for couples to have children as early as possible and thus perhaps to have additional children in the future without a basic decrease in their living standards.

Family allowances were introduced as early as 1888 for certain target groups, such as state employees, when the country was part of the Austro-Hungarian empire. In 1926 the Czechoslovak Government introduced a law according to which families of civil servants were entitled to an allowance, such as financial contributions to educational expenses, for first and second children up to 18 years of age. However, the effect of this social measure, as well as of other policies implemented during the interwar period, was largely offset by very high unemployment in the late 1920s and 1930s.

Maternity benefits were improved in the period after 1945 when the Family Allowances Act was implemented. As previously noted, new family policy measures were introduced in the 1960s and the existing ones further improved. In general, state support was directed to all families with dependent children, regardless of the average income of the family.

Besides family allowances, the most important measure was paid maternity leave, which in 1962 equaled 90 percent of the normal salary. Originally fixed at 18 weeks by the National Insurance Act of 1948, paid maternity leave initially covered only civil servants. In the mid-1960s, when natality in Czechoslovakia, especially in the Czech lands, decreased, policymakers approved a number of pronatalist measures. All payments under the family allowance system were raised, with an especially large increase for the second child. A maternity leave program was introduced, allowing the mother to remain at home, without incurring loss of employment rights, until the child reached the age of one year.

Other measures were introduced in 1964, including rent reductions and early retirement for women, both differentiated by family size. Eligibility for retirement for women without children was 57 years, with one child 56 years, with two children 55 years, and with three or more children 54 years of age. (However, in 1995 the government changed the retirement age for women, raising it gradually from 55 to 60 by 2006.)

Since 1968 population policy was perceived as an integral part of state policy (e.g., Pavlik and Wynnyczuk, 1974; Havelka, 1974, 1978; Kucera, 1974; Vidlakova, 1976; Heitlinger, 1976, 1979, 1980, 1987, 1993; Pavlik, 1978; Frejka, 1980). The State Population Commission was established at the time of the abortion liberalization in 1957 as an advisory body to the government (Decree 918 of 1957). Its main task was to study why the birth rate was declining and to recommend policies that might reverse the trend. The commission represented the first link between researchers, theoreticians, and policymakers responsible for population-related social measures. In 1971 the State Population Commission was upgraded to a higher level Government Population Commission (Decree 185 of 1971), consisting of federal and national vice-ministers (from the Czech lands and Slovakia), plus representatives from trade unions, the Women's Council, and the media, along with university professors and heads of research institutes.

At the beginning of the 1970s, most pronatalist measures were improved. A lump-sum payment was given to mothers at the time of childbirth. This was 2,000 crowns, representing more than the average salary of 1,937 crowns a month at that time. Also, a household loan policy was established, permitting newly married couples to borrow up to 30,000 crowns to purchase a dwelling with low interest. The loan was reduced with the birth of each child: 2,000 crowns for the first child and 4,000 crowns for the second and subsequent additional children. By the late 1970s the Czech government was spending almost 4 percent of its annual budget on direct cash benefits and an additional 7 percent on services and subsidies in kind (Wolchik, 1991). The cash and tax benefits had a low value relative to the average wage and actual costs of raising children.

In the 1970s voluntary childlessness was viewed as selfish and an evasion of one's obligation to socialist society which "does not regard the motherhood of women as their private matter" (Mala, 1974). However, mothers were burdened by inadequate household services, a continuing housing shortage, and insufficient places in child care centers, all of which combined to increase the work load on women and reduced chances for equal career advancement and achievement (Musil, 1971). At the same time influential psychologists openly noted the poor quality of many child care facilities, manifested by high child care provider ratios, excessive regimentation, and impersonal staff attitudes toward the children, all of which had a detrimental effect on child development (e.g., Langmeier and Matejcek, 1975).

By 1978 daycare centers could accommodate only 7.6 percent of children under three years of age (*Demosta*, 1979). In 1976 kindergartens accepted about 65 percent of those between ages 3 and 6 (Paulickova, 1976). The special role of grandmothers was apparent from the 1970 housing statistics; 23 percent of households shared their living quarters, most frequently with a mother or mother-in-law.

The number of children qualifying for the highest total child allowance was gradually reduced from eight in 1947–53 to five in 1957–59, four in 1968–72, and three in 1979–90 (Kucera, 1994). The main political aim of the pronatalist incentives was to induce families to have more than two children, and to penalize childlessness (for example, through the tax system). The prevailing view among Czech demographers is that pronatalist policies implemented during the 1960s and 1970s were largely inadequate and did not motivate families to desire more children, except perhaps among the gypsies (Stloukal, 1996c). As in the German Democratic Republic and in Hungary, pronatalist measures produced shifts in the timing of childbearing. There were significant period fluctuations in fertility levels but not increases in the ultimate family size. However, without the pronatalist incentives, birth rates might have been even lower (Frejka, 1998).

THE WOMAN QUESTION

As recalled by Havelkova (1993), Charlotte Masaryk, the American-born wife of the first Czechoslovak president, T. G. Masaryk, shared with her husband the conviction that "solving the female question means solving the male question as well. Both concern sheer humanity." (Homolkova-Krickova, 1933: 212). While Charlotte Masaryk's statement in support of gender equality, made at the beginning of the twentieth century, did not reflect the realities of her time, historical sources document a progressive tendency in regard to the woman question both before and after the founding of Czechoslovakia in 1918 (Bohac, 1990; Siklova, 1997).

The proportion of economically active women of reproductive age in the total labor force rose from 37 percent in 1948 to 45 percent in 1979 (Tesarova, 1979, 1981). The proportion would have been 48 percent with the addition of women participating in the special maternity leave program (Solnickova, 1979), a figure exceeded only in the former Soviet Union and in the German Democratic Republic. The proportion of working women among all women similarly increased from 54 percent in 1950 to 71 percent in 1960, 85 percent in 1970, and 90 percent in 1989. On average, men earned 25 percent more than women. While total unemployment remains low, 55 percent of the unemployed in 1994 were women (Czech Committee for UNICEF, 1994).

Articles in Czech publications frequently referred to the woman question in terms of theoretical principles (e.g., Marx et al., 1973). Throughout much of the post–World War II communist era, a woman's role was consistently defined as a "unity of economic, material, and political functions" (Horna, 1978; Heitlinger, 1993: 95). The special treatment of women was embodied in protective labor legislation, health care provisions for pregnancy and maternity, short- and long-term maternity leave, and earlier eligibility for retirement (Wynnyczuk, 1969). Family legislation and juridical practice similarly reflected both equal and special treatment (Heitlinger, 1987). Such state intervention in women's wellbeing also increased state power over women's lives (Heitlinger, 1980).

As noted earlier, over the course of the communist regime important adjustments were made in response to the declining birth rate. Women were no longer expected to fulfill multiple responsibilities simultaneously (Wolchik, 1981a, 1981b). With the arrival of pronatalism in the mid-1960s, many measures originally adopted on egalitarian grounds in the 1950s were expanded to meet explicit demographic objectives (Heitlinger, 1979, 1993).

While women's access to education increased and sizeable groups of women entered the labor force, significant inequalities remained. Most women were employed in low priority areas that had lower than average wage structures. Women were seldom found in important managerial, academic, or political positions (Wolchik, 1994a). The proportions of women in the Communist Party and among its leaders were greatest in the period before and immediately after the establishment of communist power (Wolchik, 1994b). Legal equality was not synonymous with daily life reality.

After the collapse of the regime, the official women's organization, the Union of Czechoslovak Women, was hurt by its past subordination to the Communist Party and the negative stand taken by its leaders in the student demonstrations in November 1989. In 1993 the union separated into Czech and Slovak organizations that, among other goals, advocate free choice of the number and spacing of children couples wish to have, sexuality education, and free access to legal abortion. The Czech Union remains the largest and financially best endowed women's organization in the Czech Republic. Efforts to coordinate the actions of the multitude of diverse women's groups, organized since 1989, have not been very successful. Most have limited resources, operate largely in urban areas, and lack political influence. There are few high-ranking women officials in the Czech Republic. The July 1998 cabinet does not include a single woman minister. Women's representation in Parliament is less than 10 percent - in a country where the first woman was elected a member of Parliament in 1912 and suffrage was extended to women in 1919. In 1994 public opinion was so opposed to the theory and practice of the previous regime that many political leaders and ordinary citizens rejected women's equality as a goal, opposed reproductive freedom, and advocated a significant reduction in women's labor force participation (Heitlinger, 1993; Czech Committee for UNICEF, 1994).

FERTILITY PLANNING

Although sociodemographic research has a well established tradition in both Czech and Slovak Republics, an independent demographic research institute specifically oriented to studies in demography, including family planning and reproductive behavior, does not exist. In 1991 a Department of Demography and Geodemography was established at the Faculty of Science of Charles University in Prague. It conducts research studies, surveys, and field investigations. There is also a Demography Department at the Economic University in Prague. Several institutes in the Czech Republic perform statistical surveys and analyses. These include the Research Institute of Labour and Social Affairs, the Sociological Institute, the Czech Statistical Office, the Institute of Psychiatry, the Demographic Society of the Academy of Sciences, the Masaryk University in Brno, and several private enterprises.

Research Studies

Starting in 1956, following the revival of the social sciences, a number of investigations were conducted, primarily concentrating on the number of children wanted (planned), the ideal number of children, and the conditions under which a woman would be willing to have three children. The Czechoslovak Statistical Office (later the Federal Statistical Office) joined with the Ministry of Health in organizing an inquiry among 10,645 married women, aged 20 to 39 years, who visited medical facilities in 1956. Replies to the question of how many children a woman desired at the time of marriage yielded an average of 2.10 in the Czech Republic and 2.76 in Slovakia (Srb and Kucera, 1959). A subsequent study, conducted in 1959 by the State Population Commission, of 3,191 women aged 15 to 49 yielded somewhat higher findings but was not directly comparable to the previous study in terms of the women's ages (Pavlik and Wynnyczuk, 1974).

In a 1961 survey of married women more than half the respondents expressed an unwillingness to have three children (Prokopec, 1963). A 1963 poll among engaged couples showed that difficult housing conditions functioned as a brake for having more than one child (Prokopec, 1965, 1966; Pavlik and Wynnyczuk, 1974; Scott, 1974).

As part of the World Fertility Study (WFS) a study was conducted in 1977 of 3,041 married women between 18 and 45 years of age, all in their first marriage. On average, the women were married 11 years, were 33 years old, and had planned to have 2.27 children at the beginning of marriage. They actually had an average of 2.02 live births plus 0.35 induced abortions and 0.19 spontaneous abortions. There were 1.98 living children on average per woman. While 84 percent of all women declared that they had planned their family size before marriage, about 30 percent of births were said to have been unwanted (Srb, 1979; Haub, 1981).

Subsequent studies, conducted in 1986–87 by the Research Institute of Labour and Social Affairs (Haberlova et al., 1987) and in 1991 by the Federal Statistical

Office (Czechoslovakia/FSO, 1992) confirmed that the housing shortage discouraged most couples from having more than one or two children. This situation still persists. In sum, all the sample surveys from the period 1956–91 reflect a congruence between the planned and the ideal number of children of around two. This is a remarkable finding given the considerable sociopolitical changes that occurred during these years. However, it should be noted that differently selected samples of respondents answered differently posed questions, which raises some concern about representativeness and comparability of the survey results (Stloukal, 1996c).

FERTILITY-REGULATING BEHAVIOR

Contraception

Premarital consultation centers, established in the interwar period, were closed in the early 1950s. When reestablished in the late 1950s and early 1960s, information on family planning was not included in their work program (Heitlinger, 1979). The medical profession opposed contraception, partly for ideological reasons because modern contraceptives were perceived to be a product of Western-style neo-Malthusianism. International research was not publicized. Misinformation was widespread. Physicians' distrust was also a component of the prevailing philosophy of health care, which preferred curative measures to preventive ones (Stloukal, 1996c).

Instruction about contraceptive devices can only be provided in the gynecological departments of hospitals performing abortions. As directed by the Ministry of Health, every gynecologist has been obliged since 1958 to convey information on contraceptives and contraceptive practice to each woman requesting abortion. Obstetrical departments are also required to instruct women in contraceptive methods when they leave the hospital after a delivery. An oral contraceptive, *Antigest*, and the *Dana* IUD have been produced in Czechoslovakia since 1966. Both are available only by prescription; the IUD can only be inserted by gynecologists in hospitals. Condoms are produced for export and domestic use, and are sold in pharmacies and from vending machines. Sterilization was regulated in the mid-1960s. Available only for medical reasons, not for purposes of family planning, it requires the consent of the patient as well as medical authorization. Although there is no official record of the annual number of sterilization operations, their incidence was presumed to be statistically insignificant (Vidlakova, 1976).

In 1994 Slovakia enacted new legislation restricting sterilization to women who have at least three children or are over 35 years old. Vasectomy is allowed only if the wife cannot be sterilized for health reasons. Until 1994 these operations were provided free of charge (Kliment, 1994).

Contraceptive Practice

During 1965–66, before the pill and the intrauterine device became available, the State Population Commission conducted a survey of the sexual life of 293 young urban couples. Husbands and wives were interviewed separately in medical settings. Knowledge about contraception was very limited. Books, rather than families or schools, were the main sources of information. Most women indicated their dependence on the responsibility, experience, and skill of their partners both before and after marriage. The most frequent method of contraception was *coitus interruptus* (42 percent), rising to 73 percent when combined with the rhythm method (Prokopec, 1968; Heitlinger, 1976).

In 1973 an inquiry among 400 couples who married in 1968 in Ostrava, a North Moravian industrial city, showed that after five years of marriage, contraception was practiced by 57 percent of the couples. The trend was toward increased use of modern methods, mostly condoms (Wynnyczuk and Sracek, 1976). A similar study, conducted in 1976 among 400 couples married in 1970 and living in rural communities, showed that after five years of marriage roughly the same proportion of young married couples (57 percent) practiced contraception. However, 33 percent of village couples with three children did not practice contraception and a further 18 percent relied on *coitus interruptus* and/or the rhythm method (Wynnyczuk and Sracek, 1977).

The Czech component of the 1977 World Fertility Survey found that about one-third of the women interviewed in towns were fully informed about family planning compared with one in five in the rural areas. After adding women who responded that they were "partly informed," more than half of all the women surveyed indicated that they had little contraceptive knowledge. Withdrawal was the most frequently mentioned method followed by the IUD. The use of the pill, even compared with the IUD, was very low, mainly because of fears of side effects and the limited choice of different pills in both the Czech and Slovak regions toward the end of the 1970s (Srb, 1979; Haub, 1981). While condoms were always sold, contraceptives for women were free of charge until 1990 when a small fee was imposed (David, 1992a; Uzel and Wynnyczuk, 1994).

A 1991 study of contraceptive practices among a nationally representative sample of 1,072 Czech and Slovak women showed that 32 percent of women at risk of unwanted pregnancy depended on unreliable traditional methods of contraception, such as rhythm or withdrawal. About 12 percent did not practice any contraception. The condom was used by 29 percent; IUDs by 17 percent; and oral pills by 7 percent of the women. The low rate of pill use was ascribed to the general negative image of oral contraception (Uzel et al., 1992; Ketting et al., 1993). It was, in practice, easier and faster to complete a form and perform an abortion than to monitor women for years as contraceptive users (Buresova, 1991a).

More recently, gynecologists have been encouraged to write prescriptions by the family planning associations, pharmaceutical companies, and the government (Stloukal, 1998a). In 1993 a system of competing health insurance companies was introduced. These companies pay 20 crowns per month for domestically produced oral contraceptives, covering the total cost of Czech pills. However, if they choose, women can pay 50 to 110 crowns (U.S.$2-U.S.$4) per month for imported pills.

During 1993 the (U.S.) Centers for Disease Control (1995) in cooperation with the Czech Statistical Office, the Institute for the Care of Mother and Child, and Factum (a Czech public opinion firm), conducted a representative Czech Republic Reproductive Health Survey. Interviews were obtained at the respondents' homes by trained female interviewers. At the time, 69 percent of women who were married or living with a partner were using some type of fertility-regulating method. Of these, 22 percent depended on withdrawal, 17 percent on condoms, 15 percent on IUDs, 8 percent on pills, 4 percent on combinations of condoms and natural methods, and 3 percent had resorted to tubal ligation. No respondent reported that her partner had undergone a vasectomy. Thirty-one percent were not using contraception. About two-thirds of these women indicated as reasons the lack of current sexual activity, infertility, or wanting to become or being pregnant. Couples with two or three children were most likely to be using contraception, while only 29 percent of childless couples were doing so. Contraceptive use appeared to increase with educational level. Women unaffiliated with any religion were more likely to be using contraception than were Roman Catholics or members of other religions. There was virtually no difference in contraceptive prevalence between Catholics who regularly attended church and those who did not (Centers for Disease Control, 1995; Serbanescu, 1998).

During 1993–94 the Sexological Institute (1994) in partnership with DEMA conducted a sample survey of 1,719 inhabitants of the Czech Republic (862 men and 857 women) older than 15 years. One of the findings was that at sexual debut 57 percent of men and 64 percent of women did not use contraception. With permanent partners, 41 percent of men practiced withdrawal, 19 percent used a condom, 17 percent relied on the IUD, 16 percent on pills, and 10 percent on natural methods. Several couples used more than one method. Among female respondents, 38 percent preferred withdrawal, 23 percent IUDs, 22 percent condoms, 22 percent pills, and 12 percent natural methods. Again, some overlap was noted.

Insertion of locally produced IUDs is covered by the health insurance system. Foreign IUDs may be purchased at a pharmacy for about 250 crowns (U.S.$9). Delfen foam (Cilag) costs 160 crowns (U.S.$6) per package. Its costs, like condoms, are not reimbursed by insurance companies.

ABORTION

Historical Development

After 1918 the Republic of Czechoslovakia adopted an act prohibiting abortion from the Austro-Hungarian Empire Code of 1852. In 1920 a section was added allowing abortion for narrowly defined health reasons (Sracek, 1988). Although abortion was widely tolerated by the public, conservative parties and the Roman

Catholic Church obtained a 1936 legal interpretation restricting abortion to cases when pregnancy or childbearing endangered the life of the woman (Heitlinger, 1976). In 1950 induced abortion was declared a criminal offense. The pregnant woman could be punished by imprisonment of up to one year and whoever assisted her was liable for up to 10 years in prison. Exceptions were permitted only on strictly defined medical grounds. Restrictions were gradually relaxed when legalization for nonmedical reasons was increasingly discussed in the media (Scracek, 1988).

Parliament approved a new law (No. 68 of 1957) on 19 December 1957 (Czechoslovakia, 1959; Scracek, 1988). Its key phrase (Paragraph 3, Subsection 2) stated that "permission for termination of pregnancy may be granted on medical grounds or other reasons deserving special consideration." It also required that the woman be less than 12 weeks pregnant and should not have had an abortion in the preceding six months (Uzel, 1994). As expressed in the legislation, the major reason for liberalizing abortion was to reduce detrimental health effects associated with illegal termination by unqualified practitioners. While the new law did not change the earlier provision for punishment of illegal abortionists, self-induced abortion was no longer deemed a crime.

The "other reasons" for deserving "special consideration" were not specified in the 1957 legislation but were established in practice to include a broad range of economic and social grounds. The legislation was formally amended in Acts No. 126 of 1962, No. 94 of 1964, No. 54 of 1966, and Nos. 71 and 72 of 1973. To request an abortion the pregnant woman had to forward an application with stated reasons, either personally or through her physician, to the head of the gynecological department located in the district of her residence. If the woman did not withdraw her application after information on possible consequences of abortion had been provided, her request was placed on the agenda of the Abortion Committee, usually consisting of a gynecologist, a social worker, and a representative of the trade unions or the Council of Women. If the District Abortion Committee rejected the application, the pregnant woman could appeal to the Regional Abortion Committee. (A district is an administrative unit with about 100,000 residents; a region has more than one million inhabitants.) The composition of the Regional Abortion Committee was similar to that of the District Committee.

As shown in tables 6.1, 6.2, and 6.3, implementation of the liberalized law in 1958 produced a steep rise in registered abortions both in the Czech lands and in Slovakia. To some extent, legal abortions replaced illegal ones. Contraceptive practice may have deteriorated as reliance on abortion increased (Frejka, 1983). Following the Decree of 21 December 1962, abortion commissions temporarily tightened restrictions, particularly for married women experiencing their first pregnancies and requesting an abortion for less than rigorous social reasons. The fee charged for abortions on other than medical grounds, dropped in 1960, was reinstated. It ranged from 200 to 500 crowns, varying according to individual economic circumstances. From mid-1964 the range of this "partial charge" was extended to 800 crowns, then roughly equivalent to half a week's wages (Kucera,

1994). The proportion of approved abortion applications declined from approximately 90 percent of all applications in 1960–62 to 86 percent in 1963 but rose again to 89 percent in 1964. If the woman was assertive, her request was usually granted. About 70 percent of all abortions during this period were performed for social reasons (Stloukal, 1995).

The 1961 peak of 94,300 legal abortions was followed by a decline to around 70,500 in 1963 (Table 6.1). After considerable public discussion, restrictions were relaxed in 1966 and abortions again increased in number, rising to a new peak of 102,800 in 1969, when the ratio of legal abortions was 461 per 1,000 live births. The 1957 Abortion Act and the 1962 Amendment were further modified by order of the Ministry of Health of the Czech Socialist Republic (No. 71 of 1973) and the Ministry of Health of the Slovak Socialist Republic (No. 72 of 1973), both effective from June 29, 1973. As before, abortion requests were directed to district and regional abortion committees.

Under the revised administrative guidelines the committee's decision on the woman's request was made on medical and/or other specific grounds including (a) age over 40, (b) at least three living children, (c) pregnancy following rape or other criminal acts, (d) special difficulties in the case of a single woman, (e) recent death or serious illness of husband, (f) difficult housing situation or inadequate family income, or (g) disturbed family life. The last three criteria enabled childless married women and those with only one child to obtain an abortion (Vidlakova, 1976). An abortion was generally permitted up to 12 weeks from the last menstrual period, provided no other pregnancy termination was performed during the immediately preceding 12 months. The previous limitation of six months was retained for women who had four children.

As shown in Table 6.1, following the 1973 administrative restrictions, abortions declined from 91,300 in 1972 to 81,200 in 1973. Pregnancy terminations rose again in subsequent years, reaching 94,500 in 1979 and 159,300 in 1987 when abortion commissions were abolished. Terminations of unwanted pregnancies became available on written request of the woman, provided the length of pregnancy did not exceed 12 weeks and there were no contradictory health indications. Application had to be made to the hospital nearest to her permanent residence. Abortions peaked at 164,700 in 1988. The abortion ratio reached 769 per 1,000 live births in 1989. Figure 6.3 compares births and legally induced abortions in Czechoslovakia, 1953–90. Within the framework of an age-period-cohort analysis, Stloukal (1998b; 1998c) suggests that the decrease in fertility levels after 1958 may be viewed primarily as a response to the 1957 abortion liberalization whereas the 1986 abolition of abortion commissions appears to have had little impact on 1987–90 fertility levels.

As shown in Table 6.2, the abortion ratio in the Czech Republic peaked in 1989 at 836 per 1,000 live births, falling to 477 in 1997. The abortion rate peaked in 1988 at 51.2 per 1,000 women age 15–44, declining to 19.3 in 1997. Similar data are presented in Table 6.3 for the Slovak Republic, which had higher birth rates and

Figure 6.3
Births and Legally Induced Abortions: Czechoslovakia, 1953–90

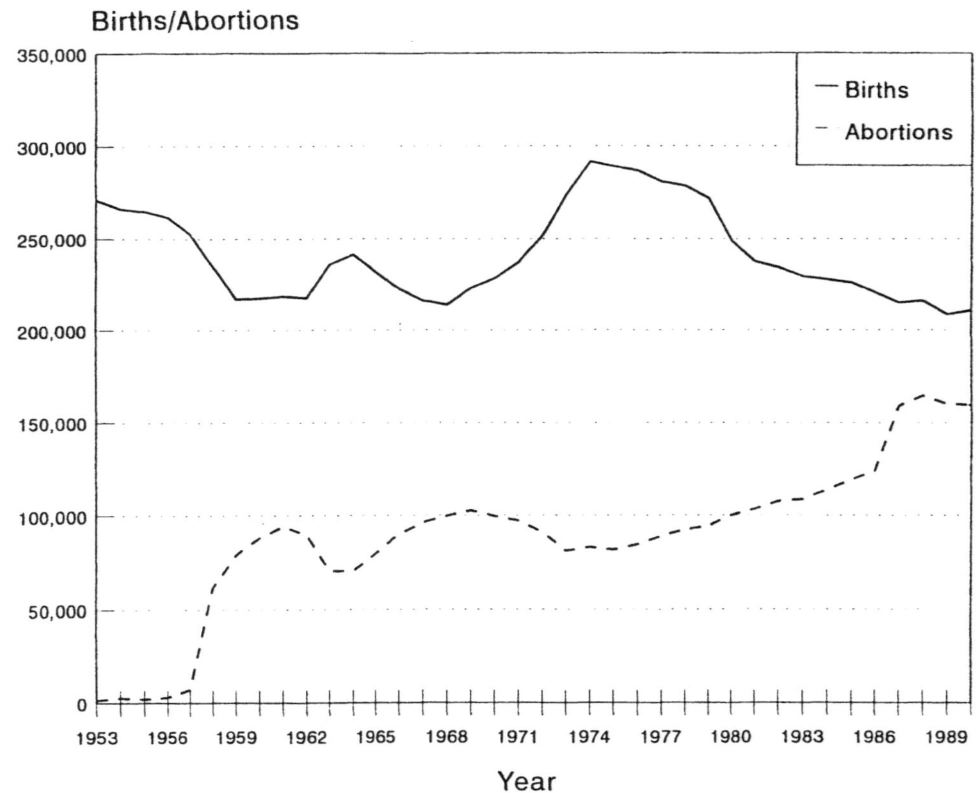

Abortion law liberalized in December 1957; pronatalist measures introduced July 1968–April 1973. Abortion committees abolished in November 1986, effective 1 January 1987.

Sources: See data from Table 6.1.

Figure 6.4
Legal Abortion Ratios*: Czech and Slovak Republics, 1960–95

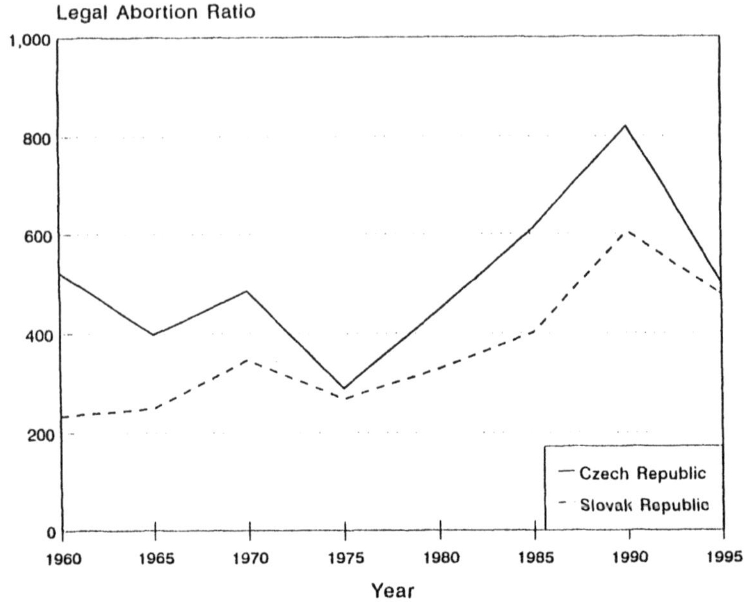

*Legal abortions per 1,000 live births.

Sources: See data from Tables 6.2 and 6.3.

lower abortion rates and ratios. Legal abortion ratios for the two republics are compared in Figure 6.4.

As also shown in Table 6.2, registered induced abortions decreased from 107,100 in 1990 to 43,300 in 1997 in the Czech Republic while the number of births declined from 130,600 to 90,700. The increasing cost of contraceptives has not had a negative influence. Sales are rising, contraceptive practices have improved, and the total number of pregnancies has dropped. The trend is moving from primary reliance on abortion to more effective contraception, encouraged by the information dissemination activities of the Czech Family Planning Association, pharmaceutical advertising, and a more rational approach to contraceptive risk taking. A similar trend can be observed in Table 6.3 and Figure 6.3 for Slovakia. Combined data are shown in Figure 6.2.

Abortion in Practice

Following the 1 January 1987 changes in the law (Hoderka, 1988), abortion continued to be easily available in the Czech and Slovak Republics for all women, regardless of their social position, provided they had not had a procedure in the immediately preceding six months. However, the requirement that the woman must request the procedure in a medical facility close to her permanent residence, employment, or school was somewhat disadvantageous in small towns and villages where people knew each other well.

To encourage early termination, no fee was charged for mini-abortions, performed up to the eighth week of pregnancy. A moderate charge was introduced in the early 1990s for mini-abortions performed for nonmedical reasons. In the Czech Republic, 82 percent of all 1993 abortions were mini-abortions. The 500 crown charge for women pregnant between eight and 12 weeks was equal to about one-fifth to one-sixth of a woman's average monthly salary. Sometimes it was difficult to ascertain the accurate length of pregnancy because ultrasonic diagnostic apparatus was not always available to physicians and information on the last menstrual period was often inaccurate.

In 1996 termination of pregnancy of six to eight weeks gestation ("mini-abortion") cost 1,200 to 2,800 crowns (U.S.$43–U.S.$100) in a hospital. Between nine and 12 weeks, the maximum price was 3,800 crowns ($136). Abortions for medical reasons continued to be free of charge. According to the Ministry of Labour and Social Affairs, the average monthly salary in the Czech Republic in 1994 was 6,870 crowns or about U.S.$277.

After 1987 abortions were usually performed by vacuum aspiration, followed by a curette examination. Mini-abortions were performed under local anesthesia and abortions from eighth to twelfth weeks gestation under general anaesthesia. Since the implementation of the abortion law in 1957 only two cases of deaths resulting from complications have been recorded. Other complications are not followed statistically. The provision of abortion services must be considered within the context of the Czech health system. Until 1990 physicians were state employed civil servants who performed abortions in hospital wards. They received a regular monthly salary that, for doctors, was often lower than the average salary of a qualified production worker. Considering these circumstances, medical practitioners were used to accepting abortions as a necessary evil. There were no reported cases of refusal to perform an abortion. Physicians who did not want to do abortions for moral or religious reasons usually chose other medical fields or worked outside hospitals. Since 1990 and the establishment of private medical practices, some abortions are provided in out-patient facilities.

Most Czech women considered abortion an acceptable method of fertility regulation. This attitude derived from several factors. Modern contraceptives (pills and IUDs) were not readily available when the abortion law was liberalized in 1957. Women became accustomed to solving the question of unwanted pregnancy by abortion. Married Czech and Slovak women talked openly about their abortions. Younger and unmarried women often kept the procedure secret, especially from their parents. With the previously noted improvement in contraceptive practice, there is now less reliance on abortion. By the mid-1990s abortion was used largely as a means of avoiding additional children by women who had already achieved their desired family size (Stloukal, 1996b).

Abortion Denied

Unique circumstances made it possible to organize, conduct, and continue for over 30 years the study of 220 children born in 1961–63 to Prague women whose request for termination of an unwanted pregnancy (UP) was twice denied (once on initial request and again on appeal) and 220 pair-matched controls born to women who accepted their pregnancy (AP) and did not request an abortion (e.g., David, Dytrych, Matejcek, and Schüller, 1988; David, 1992b; Kubicka et al., 1995). The matching criteria included age and sex of the child, school class, mother's age, number of children in the family, birth order, and socioeconomic status of the family as determined by its completeness and parental educational and economic levels.

The children were initially examined at age nine by a research team that did not then or later know which child belonged to which group. While differences between UP and AP children were not statistically significant, they were consistently in disfavor of the UP children. School, social, and family life were less satisfactory and less stable, especially for the "only" children.

In 1977, when the children were between 14 and 16 years of age, it was possible to locate 216 of the 220 UP children and 215 of the 220 AP children, achieving a 98 percent follow-up rate. Although intelligence test scores did not differentiate between UP and AP children, school performance continued to be poorer in the UP children. A significantly larger number of UP children did not continue their education to the secondary school level, but instead became apprentices or started jobs without prior vocational training. Many of the UP children described their mothers as inconsistent and variable in their emotional behavior toward them whereas the fathers were often perceived to be "warmer" than the mothers. Analysis of the findings from the Czech Child-Marriage-Family Scale suggested that the emotional gap between UP boys and their mothers widened and deteriorated over time whereas the relationship between the UP girls and their mothers remained about the same or improved.

A third follow-up wave was conducted in 1983–84 when the study participants were young adults aged 21 to 23 years. Data were also obtained from child care councils and from drug, alcohol, and crime registers. The findings showed a significantly greater problem proneness among UP than AP study participants. Compared to the AP controls, the UP young adults reported significantly less job satisfaction, more conflict with co-workers and supervisors, fewer and less satisfying relations with friends, and more disappointments in love. Among the married young adults in both groups, the UPs more often judged their marriages to be less happy and more often expressed the desire not to be married at all or not to be married to their present partner.

Another study was conducted in 1989 to assess partner choices among married UP and AP study participants. There were significantly more single and divorced women in the UP group than in the AP group. Significantly more UP than AP men had married, but there was no significant difference in their divorce rate. The partners of UP study participants were found markedly more often in the files of alcohol and drug treatment centers and on the crime register than the spouses of AP study participants. Compared to married AP women with children, married UP women with children more frequently reported that they were unprepared for their first pregnancy, felt "less happy" during pregnancy, and "felt like mothers" only after delivery (whereas many AP women experienced such a feeling in early pregnancy). For information on pregnancy, childbirth, and child care, UP women more often turned to the media and friends, whereas AP women turned more often to their mothers. They planned to stay at home with their child for a shorter period of time and returned to work earlier than AP women. The UP women also more frequently rated their marriages as "less happy" or "definitely unhappy."

A fourth follow-up was initiated in 1991 when the study group participants reached 30 years of age. For the first time, the siblings of UP and AP subjects were included. It was possible to establish contacts with and obtain data from 84 percent of all original UP subjects and 88 percent of AP subjects (i.e., more than 90 percent of those still available for the study), and 92 percent of UP siblings and 90 percent of AP siblings. UP subjects continued to manifest less favorable psychosocial development on average than their AP controls, although the differences narrowed. The greatest differences continued to be between the UP and AP only children. The sibling study was designed to distinguish which characteristics of the UP subjects were shared by their siblings and which were specific to those born of unwanted pregnancies. It was found that the UP females were more frequently emotionally disturbed than their AP female controls whereas no such difference was noted between the female siblings of the UP and AP subjects (Kubicka et al., 1995).

A fifth follow-up wave was conducted in 1997 when the subjects and their siblings were around 35 years of age. It was possible to reinterview 75 percent of the original 440 study participants plus 247 siblings. Initial findings reinforce the earlier observation of longer term detrimental effects of unwantedness in early pregnancy on later psychosocial development.

SEX AND SOCIETY

Family Planning Association

The Association for Family Planning and Sex Education (SPRSV) was organized in 1979 as a section of the Czechoslovak Medical Society. Its efforts to become a member of the International Planned Parenthood Federation (IPPF) were frustrated by disagreements between the Communist Party and the Federal Ministry of Interior. In 1989 SPRSV was reorganized into two independent family planning associations, one each for the Czech and Slovak Republics. With the removal of political barriers and governmental approval of their NGO status, the Czech and Slovak family planning associations attained full membership in the IPPF European Region in November 1994.

One of the major goals of the Czech Family Planning Association is to decrease the number of induced abortions. There is close cooperation with the mass media, explaining the importance of effective modern contraceptive methods (Uzel and Wynnyczuk, 1994). The priorities of the Czech FPA are: (1) to identify the target groups of women who do not use contraception for economic and other reasons, (2) to make sexuality education more effective, and (3) to persuade the public to move from abortion to better contraceptive practice.

Three modern contraceptive advisory centers have been established, two in Prague and one in Brno. Preparatory discussions have been initiated with the Ministry of Labour and Social Affairs and the Ministry of Health to add family planning clinics to the already existing "Counseling Centers for Family, Marriage, and Human Relations," established in every district (about 100,000 inhabitants) since the beginning of the 1970s.

The Czech FPA also tries very hard to increase the knowledge and information level of gynecologists, teachers, sociologists, psychologists, and others. A monthly *Bulletin* provides abstracts covering 12 Czech and 12 foreign journals. An annual national conference for specialists offers two to three workshops on new issues in the field. The need for well structured courses was documented by the findings of a 1992 survey of 155 Czech and Slovak gynecologists, showing that these specialists lacked knowledge about modern contraceptives (Visser, Uzel, Ketting, Bruyniks, and Oddens, 1993).

A major constraint for both the Czech and the Slovak FPA is the intensive effort to restrict reproductive and sexual rights by the increasingly powerful Roman Catholic Church and other pressure groups. They object strongly to sexuality education in the schools and exaggerate negative effects of contraceptive methods. Abortion is considered murder and physicians are blamed for pushing women into this decision (Kliment and Cupanik, 1997; Uzel, 1998).

The most vexing problem for the Czech and Slovak FPAs is how to assist the growing population of Romany Gypsies. It is estimated that their population is about 150,000 to 170,000 in the Czech Republic and 260,000 in the Slovak Republic (Pavlik, 1996). Their differing traditions influence their social and family life as well as their fertility-regulating behavior. The FPAs attempt to provide more sexuality education, better access to safe contraception, and improved services for reproductive health care (Uzel and Wynnyczuk, 1992).

Sexuality Education

After 1948 the inclusion of family life/sexuality education in the curriculum depended on the school principal who could invite a general practitioner to give some lectures. This type of education differed widely and was not very effective (Buresova, 1991a). Around 1970 "education for parenthood" emerged as a major theme of public discussion (Wynnyczuk, 1974, 1978). Experts (mostly physicians) writing for mass media, school textbooks, and personal advice columns generally subordinated sexuality education to parental education (Wynnyczuk, 1980). Factual clinical knowledge about the anatomy and physiology of male and female sexual organs, and the diverse methods of fertility regulation, was dispensed far more freely than information about nonprocreative sexual practices. Masturbation was played down. If discussed at all, it was usually only in the context of male sexuality (Heitlinger, 1987).

Responsibility for sexuality education, previously shared by diverse ministries, was coordinated by Government Resolution No. 137 of 1971 with instructions published by the Ministry of Education in 1974. Topics covered ranged from premarital to extramarital concerns (Scheuflerowa, 1977). One objective was to provide knowledge on sensitive topics to elementary school children before they encountered misinformation elsewhere. Special efforts were made to convey a sense of responsibility to adolescents and young couples, most of whom married before age 21 (Prokopec, 1968).

A late 1970s survey of young married couples found that over a quarter of the husbands and half of the wives considered their sexuality education to be insufficient or nonexistent (Kozakiewicz, 1980). Considerable differences were noted in the level of sexuality education between white-collar and blue-collar workers and their children, and between urban and rural areas. One problem was the reluctance of many women teachers to discuss sexual topics with adolescent boys. To strengthen sexuality education programs, a Family Planning and Parenthood Section was established in the Czechoslovak Academy of Sciences. The Sexological Society organized a Section on Parenthood Education and the Czechoslovak FPA organized biannual scientific meetings on problems of parenthood education (Buresova, 1991b). The Institute of Sexology, founded in 1921, participated in these endeavors, publishing lecture notes and textbooks, and attempting to improve teacher training. Since the 1990s these efforts have been greatly hampered by the Roman Catholic Church, which would prefer to abolish sexuality education in the schools.

Adolescents

A nationally representative 1994 survey (Weiss and Zverina, 1996) in the Czech Republic noted that age at sexual debut had declined in recent years but did not differ between men and women (18.1 versus 18.0 years on average). Most women (88 percent) had first intercourse with a stable partner whereas 34 percent of men began with a casual partner. A majority (57 percent of men and 64 percent of women) used no contraception at sexual debut. To obtain an abortion, women under age 16 need parental consent. Teenagers between 16 and 18 years of age do not require parental consent but their parents must be informed after the procedure (Law 66/1986, Instructions 75/1986; Buresova, 1991a).

FUTURE PERSPECTIVES

In the early 1990s intensive efforts to restrict access to legal abortion were launched by pro-life pressure groups, especially by the Roman Catholic Church. The pro-choice position was portrayed as a legacy of communist ideology, the abortion law as a product of the totalitarian regime, and termination of an unwanted pregnancy a result of 40 years of corruption of moral values. Antiabortion activists consider sexuality education a source of bad morals for society and call for the prohibition of information on all questions connected with sexual life. Presently, voices are more muted. There does not appear to be an immediate threat to legal abortion.

Most Czechs are nonbelievers. According to the 1991 census, 40 percent did not belong to any church; 16 percent did not answer the question about religious affiliation; 39 percent declared themselves to be Roman Catholics; and 5 percent claimed affiliation with a range of churches (Stloukal, 1996c). In the Slovak Republic, 29 percent claimed no religious affiliation, 17 percent did not respond, and 46 percent declared themselves to be Roman Catholic.

On 9 January 1991 the Federal Parliament adopted a Bill of Human Rights. After a long and controversial discussion, Parliament accepted a formulation that does not unambiguously guarantee a woman the right to freely and responsibly decide about the number of children she wants in accord with the international statements that Czechoslovakia signed. Instead, it agreed to a compromise formulation that "human life is worthy of protection even before birth."

A public opinion poll, conducted in March 1991 by the Federal Public Opinion Research Institute, found that 61 percent of people questioned agreed that a woman should have the right to choose by herself for or against abortion (in the Czech Republic 63 percent, in the Slovak Republic 56 percent). Twenty-three percent said that termination should be performed only on the basis of health and social conditions, 11 percent that it should be allowed only in cases where a woman's life is endangered, and 4 percent supported total prohibition. The greatest support for an absolute ban was in Bratislava, the capital of the Slovak Republic (13 percent). Forty-one percent of those questioned in the Slovak Republic stated that the partner's agreement should be required for the procedure, an opinion rarely expressed in the Czech Republic (Czechoslovakia/FSO, 1992).

In the 1993 Czech Reproductive Health Survey, an overwhelming majority of women (85 percent) of childbearing age expressed the view that women/couples should have the right to decide whether to have a pregnancy terminated by means of induced abortion for any reason they choose. Every segment of the population examined was similarly opposed to legal restrictions. Even among Catholic women who attended mass regularly, seven of every ten felt there should be no limitations on access to abortion. Among the 14 percent of respondents who thought that there should be some restrictions on access to abortion, an overwhelming majority felt that abortion was justified under at least some circumstances, particularly if a pregnancy endangered a woman's life or if there was risk of a fetal defect (Centers for Disease Control, 1995).

A 1995 opinion survey noted that among respondents who proclaimed themselves to be religious believers, 44 percent expressed the view that women are entitled to decide freely about abortion, 25 percent thought it was acceptable only for health reasons, and 23 percent only for life-threatening situations. Only 8 percent were opposed to abortion under any circumstances. The corresponding figures for nonbelievers were 73 percent, 21 percent, 4 percent, and 1 percent, respectively (Kalnicka, 1996).

Differences of opinion also affected the public stands of the associations of gynecologists-obstetricians. While the Czech Association unanimously adopted the pro-choice position, there was a controversial discussion among Slovak gynecologists at their national conference in February 1991. The leaders of the association, mostly older physicians representative of more restrictive attitudes toward abortion, opposed the younger generation of gynecologists who supported the views of their Czech colleagues. In an open vote the liberal attitude of the younger group prevailed. However, the leadership of the Slovak Gynecological Association continued in the hands of the minority that supports restrictive trends. Such a differentiation between pro-life and pro-choice attitudes is increasingly becoming a generational conflict in Slovakia, where members of older generations still occupy very influential posts in many institutions. Within that context, the Slovak Family Planning Association remains one of the few organizations defending the right of access to legal abortion services. It is also the only organized group advocating contraception as a means of reducing resort to abortion (Kliment, 1998).

The 1996 Human Development Report from the Czech Republic (Pavlik, 1996) sounds a note of optimism. A rise in the general state of health has increased life expectancy for the first time in thirty years. The population is stable. The political and economic transition process is stormy but continuing.

chapter 7

The Former German Democratic Republic

Jürgen Dorbritz and Jochen Fleischhacker

EDITOR'S INTRODUCTION

The German Democratic Republic (GDR) was established in 1949 following the collapse of negotiations among the powers then occupying Germany. After the fall of the Berlin Wall in 1989, the GDR ceased to exist in October 1990, when its states were incorporated into the Federal Republic of Germany (FRG) (Meier, 1997).

From the perspective of this book, the GDR is unique in several respects. During its 40 years' existence, the GDR implemented the most extensive pronatalist population policy in history while at the same time providing ready access to modern contraceptives and legal abortion free of charge. The GDR did not liberalize its abortion legislation until 1972, well after most other countries. in Central and Eastern Europe (CEE). This action followed the development of extensive sexuality education programs and the marketing of effective modern contraceptives. With the resumption of debates about pronatalist population policies in countries reporting low birth rates, it seems particularly appropriate to examine the GDR experience.

GDR residents experienced a more rapid and systemic transition than populations in other postsocialist countries. The situation was exceptional in the totality of the shift of economic and social order (Dorbritz, 1994b; Münz and Ulrich, 1994; Conrad et al., 1996; Adler, 1997). The erosion of the quality of life, hastened and exacerbated by sudden massive unemployment (especially among women), and loss of state-supported social and financial services, elicited emotional trauma, anxiety, and insecurity in large numbers of individuals and families. The period has been described as one of economic, cultural, and demographic "shock" or "catastrophe," resulting in a "birth strike" (Eberstadt 1994a, 1994b; Dorbritz, 1994b, 1995). This is reflected in the decline in birth rates (5.1 per 1,000 population in

1994), the lowest figure ever recorded in the territory of the former GDR or any naturally constituted population.

This chapter will focus on lessons learned from the GDR pronatalist population policy experience. In addition, reference will be made to historical trends, the woman question, sexuality education, and contraception and abortion, particularly during the Weimar and Hitler years, which set the stage for the sociopolitical evolution of GDR policies. The postreunification process and its effects on the new federal (former GDR) states will also be discussed in terms of marriages, births, and adolescent reproductive behavior.—H.P.D.

HISTORICAL TRENDS AND POPULATION POLICY

Prior to 1871, there was no German nation, only a collection of diverse states that shared a common ethnic and linguistic heritage (Heilig et al., 1990). The German Reich of 1871 encompassed about 40 percent more territory than the reunified Germany of 1994. Fundamental social and economic changes gradually transformed Germany from a predominantly agrarian state into a highly industrialized urban and largely Protestant society.

After losing World War I, the Weimar Republic struggled to establish a democratic state during a period of sharp fertility declines and economic uncertainties in Europe. With the advent of the Hitler regime in 1933, Nazi Germany made vigorous efforts to encourage marriage and childbearing (Finkle and McIntosh, 1980). Amidst mounting international tension, national power was perceived to be closely associated with population size. An intensive propaganda campaign was combined with financial incentives to convey the message that procreation was the duty of every loyal citizen (e.g., Frerks, 1938; Bleuel, 1973; Lenz, 1983). Higher fertility was promoted through interest-free loans with repayment reduced by 25 percent for every child born (Glass, 1940). However, it was never possible to demonstrate whether the subsequent upturn in fertility resulted from the pronatalist policies or the return of more favorable socioeconomic circumstances (Finkle and McIntosh, 1980).

The Nazi period cast a dark shadow over *Bevölkerungswissenschaft* (population science). In the 1920s and 1930s German demography was misused by the Nazis to lend ideological support to the rise of fascism. The concept of *Rassenhygiene* (racial hygiene or, more properly, biological heritage), was thought to determine the quality of human beings and their value for society. Natural selection of the strongest races was seen to be the driving force of history. Racially pure Germans were encouraged to have large families. Motherhood was touted as a patriotic duty and a woman's highest goal (David, Fleischhacker, and Höhn, 1988; Heilig et al., 1990; Bergmann, 1992; Ushborne, 1992). Access to contraception and abortion was prohibited or severely curtailed.

After World War II, population policy evolved gradually in the GDR, beginning with social welfare endeavors to improve the socioeconomic position of the family. It developed into a pronatalist policy designed to maintain women's high labor force participation and a high birth rate (Lungwitz, 1981; Wendt, 1993). At the same time, family planning was perceived as a social and societal responsibility, giving each couple the right to determine the number and timing of children it wished to have (Mecklinger, 1968; Mehlan and Geissler, 1973; David and McIntyre, 1981). Free availability of contraceptives and access to abortion on request without cost in the first trimester of pregnancy were perceived as an indicator of women's emancipation from compulsory pregnancy. Implementation of this policy was all the more noteworthy in view of the GDR's negative population growth (Institute for Sociology and Social Policy, 1984). Analyses of the evolving GDR population policies and the then-politically difficult position of demographic researchers stimulated considerable discussion in the postreunification period (e.g., Dorbritz, 1992a; Khalatbari, 1992; Speigner, 1992; Conrad et al., 1996).

THE WOMAN QUESTION

As noted in Chapter 3, the "woman question" was a phrase introduced in 1875 in the writings of the German Social Democratic Party theoretician August Bebel (1971 [1875]). Klara Zetkin's report to the 1896 Congress of the Social Democratic Party prepared the organizational foundation for a socialist women's movement (Zetkin, 1958). German feminists achieved the right to vote in 1918 as part of the post–World War I Weimar Constitution. By 1926, there were 32 women deputies in the *Reichstag*, a time when only three women served in the U.S. Congress (Steinem, 1980). After Hitler's rise to power in 1933, feminist organizations were disbanded. Women were banned from the *Reichstag*, from judgeships, and from other decision making positions. Married women were persuaded to stay at home and leave paid jobs to men. The emancipation that had been achieved was destroyed (Koonz, 1987). As women lost economic independence, they were taught to perceive themselves as destined to produce racially pure Germans, the highest value of motherhood (Beer, 1991; Klein-Schonnefeld, 1994).

Following the devastation of World War II, the number of persons leaving what had become the GDR rose year by year, reaching a total estimated at between 2.7 and 3 million individuals by 1961. One of the consequences of the ongoing shortages in the labor force was that women attained positions of considerable economic and political influence (Akkermann and Mehlan, 1978). Women's right to work and the principle of gender equality in pay were incorporated into the 1950 GDR Constitution. Laws were promulgated with the intent to reduce conflicts between career demands and motherhood without any explicit attempt to influence the total number of births. It was no longer an economic necessity to marry after the birth of a child. A woman could live alone with her child or join a community of like-minded persons.

Discussion of the power relationship between the genders was a taboo topic. Women remained largely excluded from the central decision making process in the economy, politics, and academia (Böhm, 1993). In general, the more prestigious the positions, the lower the percentage of women (Rueschemeyer, 1994b). Meanwhile, the average monthly income of women was about 25 to 30 percent lower than that of men (*Frauenreport '90*, 1990).

By 1975 the proportion of women in the labor force had risen to 49 percent. By the late 1980s, 91 percent of working-age women were engaged in paid employment. At the time, the GDR had the highest proportion of working women among industrialized countries. About one-third worked "short hours," combining work with child care (Ebert, 1981; Nickel, 1993).

In general, "women's work" in the GDR meant the pursuit of a career, paid employment, and a natural responsibility for children, family, and housework. Women not only had the right but the obligation to hold a job. The man's income alone was seldom sufficient to meet family needs. About 30 percent of women had no partners and were entirely dependent on their own resources. It was the ability to have children that defined the working GDR woman and mother, representing both production and reproduction (Nickel, 1993).

Improvement in the position of women was part of social family policy, embedded in Article 38 of the GDR Constitution. Over the course of two generations, a society of working wives and mothers had the benefit of low-cost daycare, flexible working hours, and an education system oriented to producing productive women (Atkinson, 1995). Overall, however, the interests of family and of women were subsumed to those of the state (Schneider, 1994).

PRONATALIST POLICIES

Evolution

During the early years of the GDR emphasis was on improving social conditions, especially for pregnant women and young children. As reflected in the 1949 Constitution, these were traditional social welfare policies designed to ease the burden of childbearing and childrearing but not specifically oriented to influencing reproductive behavior or demographic goals (Zielenska, 1987; David and Pierotti, 1993). Thus, in 1951, increasing birth grants were provided for second and third children with only a token amount for a fourth child. Additional support was granted to single mothers. In 1958 birth payments were sharply increased from 500 marks for the first child up to 1,000 marks for any fifth or subsequent child. In 1963 paid pregnancy and maternal leave was extended from 11 to 14 weeks. In 1967 monthly allowances were increased for parents with four or more children and in 1969 these payments were extended to three-child families (Vaskovics et al., 1994). As shown in Table 7.1, the total number of births stayed fairly stable around 290,000 from the mid-1950s to the mid-1960s, declined somewhat with the liberalization of abortion in 1965, and then dropped to 181,800 in 1975 after abortion on request had been implemented in 1972. Meanwhile, the crude birth rate declined from 17.0 per 1,000 population in 1960 to 10.8 in 1975.Trends in births and legally induced abortions are shown in Figure 7.1.

Figure 7.1
Births and Legally Induced Abortions: Former GDR, 1950–96

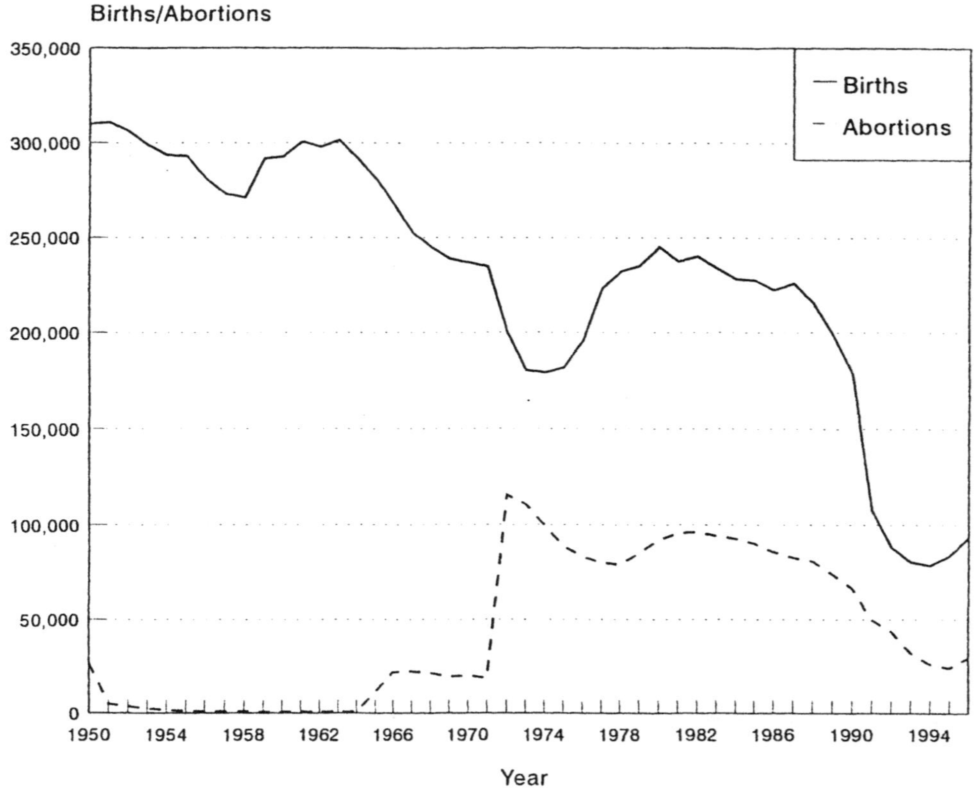

Abortion law liberalized in March 1965; abortion available on request in March 1972. Pronatalist incentives promulgated in May 1976.

Sources: See data from Table 7.1.

Table 7.1
Birth, Abortion, and Fertility Rates and Ratios: Former GDR, 1950–97

Year	Live Births	Legally Induced Abortions	Birth Rate	Legal Abortion Rate[f]	Legal Abortion Ratio[g]	Total Fertility Rate
1950[a]	309,900	26,400	16.5	6.4	85	na
1955	293,300	1,200	16.4	0.3	4	2.35
1960	293,000	800	17.0	0.2	3	2.33
1965[b]	281,100	11,400[e]	16.5	3.6	41[e]	2.48
1970	236,900	20,300[e]	13.9	6.2	86[e]	2.19
1971	234,900	18,700	13.8	5.2	80	2.13
1972[c]	200,400	115,600	11.8	33.1	577	1.79
1973	180,300	110,800	10.6	32.2	614	1.58
1974	179,100	99,800	10.6	28.8	557	1.54
1975	181,800	88,800	10.8	25.2	488	1.54
1976[d]	195,500	83,200	11.6	23.3	426	1.64
1977	223,200	80,100	13.3	22.5	359	1.85
1978	232,200	79,100	13.9	22.8	341	1.90
1979	235,200	85,100	14.0	23.9	362	1.89
1980	245,100	92,100	14.6	25.3	376	1.94
1981	237,500	95,600	14.2	26.3	402	1.85

1982	240,100	96,400	14.4	26.6	402	1.86
1983	233,800	94,100	14.0	26.0	402	1.79
1984	228,100	92,600	13.7	25.9	406	1.74
1985	227,600	90,300	13.7	25.4	397	1.73
1986	222,300	85,700	13.4	24.4	386	1.70
1987	226,000	82,700	13.6	23.7	366	1.74
1988	215,700	80,800	12.9	23.3	375	1.67
1989	198,900	73,900	12.0	22.0	372	1.57
1990	178,500	66,500	11.1	20.4	373	1.52
1991	107,800	49,800	6.8	15.5	462	0.98
1992	88,300	43,800	5.6	15.0	496	0.83
1993	80,500	32,200[h]	5.1	11.0	400[h]	0.78
1994	78,700	26,200[h]	5.1	9.0	333[h]	0.77
1995	83,800	24,100[h]	5.4	8.3	288[h]	0.84
1996	93,300	29,500[h]	6.0	10.1	316[h]	0.95
1997[i]	98,500	na	6.4	na	na	1.02

[a]Temporary liberal abortion regulations rescinded (September 1950). [b]Abortion law liberalized (March 1965). [c]Abortion available on request (March 1972). [d]Pronatalist incentives promulgated (May 1976). [e]Estimates based on data from Rostock area, representing 5 percent of GDR female population. [f]Per 1,000 women age 15–44. [g]Per 1,000 live births. [h]Not including East Berlin.

Sources: David and McIntyre (1981), GDR/CSO (1989), U.N. (1995a), FRG/FSO (1996), COE (1998), and Dorbritz (1998).

Four years after approving abortion on request in 1972, the GDR in 1976 embarked on a strong pronatalist policy whose objective was to encourage early marriage and early birth (David, 1982a, 1993). A grant of 1,000 marks, roughly equal to one month's salary for the average industrial worker, was paid at birth for every child. Monthly family allowances increased with each child and ranged from 2 percent of the average monthly wage for a one-child family to 29 percent for a family with six children. Newly married couples with both partners under age 26 and together earning less than one and a half times the average monthly wage were eligible for interest-free loans for obtaining and furnishing a home. The debt was reduced by 20 percent at the birth of the first child, 30 percent at the second, and 50 percent at the third (provided it was born within eight years of marriage).

In 1976 paid maternity leave was extended to six weeks before and 20 weeks after childbirth, with additional unpaid leave available until the child's first birthday. If no place was available in the local nursery school, the mother could take unpaid leave until the child's third birthday, with reemployment guaranteed. With the birth of the second and/or subsequent child the mother became eligible for paid educational leave until the last-born child reached one year of age, the so called "baby year." Working mothers with two or more children qualified for a fully paid reduced work week of 40 hours instead of the normal 44 hours. The pay for the four unworked hours was termed "mother salary." In 1984 conditions were further improved for families with three or more children, including the possibility of paid leave until the last-born child was 18 months old. In 1986 mothers became eligible for a year's paid leave after the birth of their first child (*Pro Familia Magazine*, 1987). This provision also applied to the father or another person, such as the grandmother, if they stayed at home to care for the child. Birth grants and monthly allowances continued to be improved. During pregnancy, women were not allowed to work overtime or at night. Mothers of preschool age children could refuse overtime or night work (Klein-Schonnefeld, 1994).

The network of state-subsidized nurseries, kindergartens, and child care support services was the most extensive of all the socialist countries. By 1989, 80 percent of children up to three years of age attended nurseries; 95 percent of three to six year olds were in nurseries, kindergartens, or daycare. Of school children in the first four grades, 81 percent participated in afterschool care. Babies as young as ten weeks were cared for from dawn to dusk if the mothers worked. Parental costs were minimal. Even the price of children's clothes was partially subsidized. No study was ever reported of the socialization effects on children of such early care outside the home. That it did not necessarily mean high-quality care was seldom mentioned.

Additional Sociopolitical Influences

The combination of a high value placed on children and extensive assumption of responsibility for social services by the state resulted in a pattern of marriage-divorce- remarriage and reproductive behavior that practically prevented childlessness. Despite its high divorce rate, the GDR remained a traditional society with a strong family orientation. Individuals continued to value marriage and children along with the windfall short term profits of child-related benefits (Dorbritz, 1992b). The state devalued marriage as a social institution and considered the family a symbol of a political culture without a public face, a conveyor of socialist values expected to produce conforming "socialist citizens" (Nickel, 1993; Peil, 1996).

Significant advances were made in the reduction of infant and maternal mortality. With improved prenatal and perinatal care, infant mortality was reduced from 72.2 per 1,000 live births in 1950 to 7.3 in 1990. Similarly, maternal mortality declined from 98 per 100,000 live births in 1960 to 12 in 1989. The high level of postnatal medical care plus the paid pregnancy leave induced many women to return to their jobs when the paid portion of their maternity level ended.

Demographic Effects

As shown in Table 7.1, the total number of births reached a trough at 179,100 in 1974, rose to 245,100 in 1980, and declined to 198,900 in 1989. After reunification, births dropped dramatically to 78,700 in 1994 before rising again to 98,500 in 1997 Meanwhile, the crude birth rate reached a low of 10.6 per 1,000 population in 1974, increased to 14.6 by 1980, and declined again to 12.0 by 1989. After reunification, it dropped precipitously to 5.1 in 1993 and 1994 before rising to 5.4 in 1995 and 6.4 in 1997 (Dorbritz, 1994c, 1998; COE, 1998). The TFR fell from 2.19 in 1970 to 1.64 in 1976. Strengthened pronatalist incentives produced a rise in the TFR to 1.94 by 1980. This seeming improvement tended to confirm the public's belief in the efficacy of pronatalist measures. However, the rise may well have been illusory. By 1989, the TFR had again declined to 1.57. It continued its decline to 0.77 in 1994 before rising to 1.02 in 1997.

Further analysis suggests that the seeming success of the pronatalist policies in the years 1976–80 derived in part from timing effects, that is, from a "make up" of births delayed from prior years and, to a lesser degree, from earlier than usually expected births by cohorts of younger women. After 1980, the trend of rising births reversed itself although pronatalist policies were further strengthened in 1984 and 1986 with the specific aim to promote the birth of a third child.

Whether the combined effects of earlier births and the make up of delayed births really contributed to an increase in the average number of children per woman can only be understood by analyzing the experience of generational cohorts of women. Women born in the 1950s, who presumably were most influenced by the pronatalist policies, halted the trend toward having fewer children. There seems to have been a stabilizing effect. Other factors may also have influenced the decision to have or not to have a child. However, when assessed against the goals of the population policy, the stabilization effect achieved was of little demographic consequence (Büttner and Lutz, 1990; Dorbritz, 1992b; Conrad et al., 1996).

An unexpected side effect of combining social and population policies was the high proportion of children born to unmarried mothers. This situation resulted in part from the high number of divorces (about 11.1 percent of the unmarried population over 18 years of age in 1991), plus the special privileges accorded to unmarried women, including preferences for nurseries and paid leave in case of a child's illness during the first three years of life. Under these conditions, births were often planned before the formal marriage. Most of these children were later "legitimated" by marriage. Between 1972 and 1990, the proportion of births to single mothers increased from 16 percent to 35 percent of all births. The upward trend continued to 42 percent in 1992 (Dorbritz, 1994c).

Several conclusions may be drawn from demographic studies of the GDR pronatalist policies:

1. Pronatalist measures have a certain attraction related to their financial incentive and time-determined privileges. If there is motivation for having children, families can be influenced to make up for delayed births or to have planned children earlier. This leads to the already noted timing effects. But, the state cannot buy childbearing. A pronatalist population policy cannot induce families to have more children than they want to have. Regardless of how the policy is designed, it can only reduce some family burdens, without, however, completely relieving the family of all burdens. That is one of the reasons why the GDR goal of "Replacement of the Parents' Generation" could not be realized.
2. A pronatalist population policy cannot alter general demographic trends. It can only influence short-term shifts around the trend. That is a second reason why the GDR population policy failed to attain its quantitative goals regardless of extensive financial incentives. Despite the considerable family assistance, the general trend of fewer births in nearly all of Europe could not be delayed or reversed in the GDR by a comprehensive population policy.
3. The GDR, like other countries, also saw a gradual rise in childlessness and decline in first marriages, coupled with an increasing number of divorces and an expansion of nonmarital lifestyles. These societal developments were not greatly considered in the evolution of the population policy that focused on the goal of combining motherhood with productive work.
4. The effects of pronatalist policies were limited. In some generational cohorts, the decline in the number of births was halted. An increase in the average number of children was, however, not reached. The desired third child did not appear often enough.

Although never specifically formulated, it is evident from theses published at the time by the Scientific Council for Social Policy and Demography that the objective of the GDR policy was to reach an average family size of two to three children (*Wissenschaftlicher Rat für Socialpolitik und Demographie*, 1979). This goal, to be achieved together with the continued high labor force participation of women, was an objective to which the government clung tenaciously despite trends pointing in the opposite direction. The policy of family promotion was placed into an ambivalent relationship between the opportunity for individual fertility regulation (ready access to free contraceptives and unrestricted accessibility to first-trimester abortion) and the sociopolitical goal to combine high birth rates with high female labor force participation. Schneider (1994) described these competing goals as the dilemma of the GDR family policy.

The pronatalist policy was directed with a "top down" approach, a policy for women but not by women (*Frauenreport '90*, 1990). It was as if the declining births and the decreasing total fertility rate represented a loss of status for a political system that wished to appear as "child-friendly." Moreover, it was widely believed that the "successful development of socialist society" should be manifest in a high birth rate environment. And, since, at the time, the FRG had the world's lowest birth rate, the GDR birth rate received particular political and societal attention. The pronatalist policy was further influenced by the continuing emigration, slowed by the closing of the state borders in August 1961, but never fully stopped. It increased the importance of women and supported the policy of combining motherhood and work.

The economic cost of the pronatalist policies was immense and accelerated the financial deterioration of the GDR. For example, in 1977, the cost of financing one kindergarten was about 116 marks per month without consideration of initial investment or maintenance charges (Winkler, 1980). In the 1980s, the state spent about 900 marks per month for all the subventions to which a family with two children was entitled. Total costs supporting a policy assuring "social security for all social levels in all stages of life" exceeded available economic resources. However, reducing the already approved subventions would have been equivalent to declaring political bankruptcy. Thus, no decision was made in the 1980s to choose between social policy and restructuring of the economy. Indeed, up to 1989, the concept of the unity between economic and social policy was propagated. It became one of the major factors leading to the economic collapse of the GDR and reunification with the FRG.

Lessons Learned

With the demise of the GDR and its 40 years of a multifaceted social policy, the question remains: What can be learned from its demographically oriented family and population policy? While there is evidence that specific social conditions have an influence on reproductive behavior, it is equally evident that policies designed to achieve increased family size are doomed to fail. The perceptions families have about their family life, especially decisions about the births of their children, can seldom be manipulated by the state. Thus, the goal of family policy can only be to improve the social environment in which the family lives. That means enhancing a family-friendly social atmosphere rather than financing specific stimuli designed to increase births. The emphasis on just one aspect of possible family policies—the attempt to combine improved working opportunities for women with better child care—could not surmount the reality that, with the complexity of modern societies, decisions about marriage and childbearing are motivated by multiple factors.

Thus, it can only be concluded that the great social experiment of "birth stimulation" was destined to fail. Schwarz (1992: 261) reached a similar opinion when he wrote: "our analysis must have a sobering effect for many demographers and policymakers. We do not claim that a pronatalist population policy, relying on instruments of family policy, is without effect. Its effects are limited even with the input of substantial means and strong ideological support." One reason was people's perception of their own lives, in which there was hardly room for more than one or two children and which could not be manipulated. Moreover, pronatalist policies were countered by the inability of a social welfare state to cover all the costs incurred.

Certain recommendations can be derived from the GDR experience:

1. Living with children entails financial costs, borne largely by the family, that contribute to socioeconomic differences when compared to families without children. Family policy should therefore strive to relieve the financial burdens on those who decide to live with children. Such financial relief should not be one-time stimuli oriented to the timing of a birth but rather should have a longer term supportive effect.
2. Decisions about childbearing are rarely made without alternative considerations. Most of all, there is pressure to decide between family and occupation/employment. More and more women want to combine both. The postreunification growth in childlessness suggests that women are increasingly making decisions against having children (Dorbritz and Schwarz, 1996). The social climate in the former GDR was such that there was pressure to bear at least one child.
3. Gender equality in the GDR was to be attained through equal opportunity in the workplace. This policy contributed to the emancipation of women but did little to relieve women of their multiple responsibilities in the family and at work. One of the goals of an effective family policy should be to try to reduce the traditional division of labor between the genders with more assumption of family responsibilities by men.
4. Special consideration should be given to alternative lifestyles, especially for those women who want to bear and raise a child outside of marriage. The disadvantages they incur should be compensated to some extent without demeaning the important concept of marriage and family.

In sum, the experience of the GDR population policy supports the hypothesis that public policies must be oriented to the diverse aspects of family life so as to foster congruence with private values (David, 1994b). Pursuit of a strict and narrow goal is inefficient, does not correspond to the complexities of modern societies, and limits individual choices to freely determine life and family form, the number of children, and the timing of their birth. And yet, history seems bound to repeat itself. The new federal state of Brandenburg announced in November 1994 a policy of paying parents U.S.$650 for every child or additional child they will have (Kinzer, 1994).

ABORTION AND CONTRACEPTION

Historical Trends

German abortion legislation was first codified in sections 181 and 182 of the 1851 Penal Code for Prussia, which formed the basis for the 1870 Penal Code of the North German Federation. On 15 May 1871 the latter code was incorporated into paragraphs 218–220 of the Penal Code for the German Reich. It mandated penalties up to five years imprisonment for the woman and the service provider for the crime of inducing an abortion. If there were mitigating circumstances, detention for a term of not less than six months could be substituted (David, Fleischhacker, and Höhn, 1988). The prohibition of abortion permitted no exceptions, causing considerable grief, deaths through self-inflicted or backstreet procedures, and the neglect of unwanted children (Zwerenz, 1980; Amendt and Schwarz, 1991; Klein-Schonnefeld, 1994).

The post–World War I struggle for the liberalization of abortion was part of a concurrent effort to improve sexuality education and make contraceptives more readily available to the general public, particularly to workers' families. The history of the birth control movement in Germany has been well described by others (e.g., Hodann, 1937; Glass, 1940; Mehlan, 1970; Lehfeldt, 1986; Von Soden, 1988). The beginnings occurred in 1905 in Dresden when Helene Stöcker founded the German League for Mother Protection and Sex Reform and established a counseling clinic. The first marriage counseling center was established in Dresden in 1911, shortly after the Fourth International Congress on Birth Control had been held there and one year before Margaret Sanger founded the American League for Birth Control.

In 1913 Theilhaber organized the Society for Sex Reform. In 1929 Hirschfeld founded the Institute for Sexology, whose Berlin clinic was the first to focus on sexual deviation. By the time Margaret Sanger visited Germany in 1927, contraceptive advisory centers were operating in most larger towns, with traveling services dispatched on a regular basis to less populated areas (Juchacz, 1929). By 1933 more than 15 organizations had merged into the National Association for Birth Control, with a combined membership exceeding 50,000. Over 100 centers gave contraceptive advice at little or no cost (Harmsen, 1936; Lehfeldt, 1986). Popular culture celebrated sexual liberation, and feminists fought to implement the equality of the sexes proclaimed in the Constitution (e.g., Stephenson, 1975; Bock, 1984). Meanwhile, the Nazi Party's views were summarized in Goebbels' (1929: 41, cited in Mosse [1968]) statement: "The mission of women is to be beautiful and to bring children into the world." The opposition of organized medicine was expressed in a 1928 resolution, which held that "contraception as a professional practice is considered undesirable" (Hodann, 1937: 205).

During the Weimar Republic representatives of the Communist Party and the Social Democratic Party introduced bills into the *Reichstag* for the legalization of abortion and argued for the rights of women. They noted that working women were seriously harmed by an excessive number of pregnancies occurring in quick succession and cited the dangers of illegal procedures induced by untrained persons (David, Fleischhacker, and Höhn, 1988). In 1926 the *Reichstag* voted to consolidate paragraphs 218–220 of the Penal Code into a single paragraph 218, reducing the severity of sentences and making abortion a misdemeanor when not performed for economic gain. In 1927 the Supreme Court ruled that abortion was not punishable if danger to life was not otherwise to be avoided. This ruling was broadly interpreted. Abortion services became available in major cities and ambulatory facilities in more sparsely settled communities.

In 1929 the Social Democratic Party and the Communist Party representatives jointly advocated abortion on request, if performed by a physician in the first three months of pregnancy. Nazi Party delegates to the *Reichstag* foreshadowed the future when they offered a bill designed to extend paragraph 218: "Whoever undertakes to artificially block the natural fertility of the German *Volk* to the detriment of the German nation, or promotes such endeavors by word, publication, picture or any other means, or who by mixing with members of the Jewish blood community or colored races contributes to the racial deterioration and decomposition of the German *Volk*, or threatens to contribute to such endeavors, will be punished with a penitentiary sentence for racial treason" (Wolf and Hammer, 1978: 281). Amid growing economic turmoil, political crises, and repeated dissolutions of the *Reichstag*, the draft revisions of the Penal Code were never voted into law.

The Hitler Era

On 30 January 1933 Adolf Hitler assumed office as Chancellor of the Third Reich. On 6 May 1933 the Nazis demolished Hirschfeld's Institute for Sexology in Berlin, closed all sex and marriage counseling centers, and destroyed research papers, books, and educational materials. Sex education associations were dissolved. Leading persons in the birth control movement were arrested, pushed into retirement, or "persuaded" to emigrate (Hodann, 1934; Schneck, 1985).

While the sale of contraceptives was allowed, advertisements were restricted in accord with a broad interpretation of the existing law prescribing a prison sentence up to one year and/or a fine for anyone publicly displaying, recommending, or distributing information about methods or objects for preventing venereal diseases, or if intended for "lecherous" purposes. The dispensing of contraceptives by private physicians was deemed a "gray area" (Lehfeldt, 1986) in view of Hitler's stated position that "the use of contraceptives means a violation of nature, a degradation of womanhood, motherhood, and love" (cited by Potts et al., 1977). Gradually, the production of contraceptives was prohibited, with the sole exception of condoms used for prophylactic purposes. Education about pregnancy prevention became an offense punishable by imprisonment and/or stiff fines. Leading gynecologists declared contraception to be injurious to health (Mehlan, 1988). In 1941 a Police Ordinance forbade the use and distribution of any device or substance "capable of preventing pregnancy" (Harmsen, 1950).

Abortion clinics were closed soon after Hitler's rise to power. Only specifically authorized physicians were permitted to perform abortions to save the life of the woman or for eugenic reasons. All abortions and miscarriages before the thirty-second week of pregnancy had to be registered. Punishment for illegal procedures became increasingly harsh. The number of convictions rose year by year. However, Jewish women were permitted to have an abortion on the grounds that paragraph 218 did not apply to the protection of Jewish embryos. After 1943 the death penalty could be imposed for inducing abortion. It was implemented in several cases involving mostly untrained providers (David, Fleischhacker, and Höhn, 1988).

Abortion in the GDR

After the end of World War II in 1945 the Allied Control Commission abolished the death penalty for abortion but reinstated the 1926 version of paragraph 218 along with the provision for therapeutic terminations in life-threatening situations. Faced with the rising tide of illegal abortions in the chaotic aftermath of the German collapse, the Soviet Military Administration in 1947 "temporarily" relaxed the prohibitions of Paragraph 218 in the Soviet Zone of Occupation. Termination of pregnancy was permitted on juridical, eugenic, and sociomedical grounds, broadly interpreted to include socioeconomic reasons (Mehlan, 1961, 1988). As shown in Table 7.1, the number of legal abortions rose to 26,400 in 1950. The number of deaths due to abortion complications was reduced by two-thirds but the desired reduction in illegal abortion was not attained (Mehlan, 1966, 1967).

With the mitigation of postwar social emergencies and "to increase the number of births," abortion was rerestricted by the newly established GDR on 27 September 1950 to terminations performed for eugenic and narrowly defined health reasons (GDR, 1950; Mehlan, 1988). The total number of induced interruptions dropped to 800 by 1960 and subsequently fluctuated between 700 and 800 per year until 1964. The proportion of legal abortions was estimated to have declined from one abortion per one live birth in 1947 to three or four abortions per 1,000 live births in 1954 (Mehlan, 1955).

With the reimposition of prohibitions on abortion, the birth rate rose slightly from 16.5 in 1950 to 17.0 in 1960 (Table 7.1). Representative inquiries reflected the gradually rising proportion of young women with one or no children coming to the attention of hospitals for complications associated with "secret" abortions. In the period 1960–63, legal abortions increased faster than births in the age groups under age 21, among the unmarried, among women with one child or none, and in the larger towns (Mehlan, 1966, 1967).

On 15 March 1965 the 1950 abortion legislation was administratively modified to again include juridical and sociomedical grounds similar to those promulgated on a temporary basis in 1947. Abortion was authorized by a district commission when, considering all the facts of the woman's "life situation," there existed a serious danger that her physical and mental health would be impaired by the pregnancy, delivery, or child care. Pregnant women deemed especially endangered were those under age 16 or over 40 years of age, those with five or more living children, those experiencing a rapid succession of pregnancies, victims of rape, or those with possible intrauterine damage to the embryo (GDR, 1965). The 1965 regulation was originally issued by the Ministry of Health as a secret instruction to public health service providers but became known to the general public through the protests of Protestant clergy (Schulz, 1971). While the Ministry of Health did not publish official abortion statistics for the years 1965–1970, estimates suggest around 21,000 legal abortions a year. These estimates are projections based on abortion figures collected in the Rostock District, which usually accounted for slightly more than 5 percent of legal interruptions in the GDR (Mehlan, 1974, GDR/ISH, 1979).

On 9 March 1972 the GDR removed abortion from the Penal Code and approved what was perhaps the most liberal CEE abortion statute. The law, which took effect immediately, declared in its preamble that "equality of the woman in education and occupation, as well as in marriage and the family, demands that the woman have the right to decide about the pregnancy and its resolution" (GDR, 1972). It then specified in the first paragraph that "in addition to the existing possibilities of contraception, a woman shall have the right to decide on the interruption of pregnancy on her own responsibility, so as to be able to control the number, timing, and spacing of births." This right extended through the first 12 weeks of pregnancy, after which abortion could be induced when a commission of medical specialists determined that continued pregnancy would endanger the woman's life or "where there are other motivating circumstances of a serious nature." A pregnancy could not be terminated if a period of less than six months had elapsed since the most recent previous termination unless a commission of medical specialists certified the existence of exceptional circumstances. Provisions were stipulated for appealing negative commission decisions. For women under age 18, written consent was required from parents or guardians (Mehlan, 1988).

Replacing many formerly illegal terminations, registered legal abortions rose sharply from 18,700 in 1971 to 115,600 in 1972 and then gradually subsided year by year to 79,100 in 1978, as shown in Table 7.1. The decline in induced abortion was probably the result of the growing success of a massive campaign to provide free distribution of modern contraceptives and to encourage women to use them, initiated concurrently with the liberal abortion legislation in 1972.

In subsequent years the number of abortions rose to a peak of 96,400 in 1982, after which they declined steadily to 73,900 in 1989. Meanwhile, the abortion rate decreased from 33 per 1,000 women aged 15 to 44 in 1973 to 22 in 1989. The abortion ratio per 1,000 live births also dropped from 577 in 1972 to 372 in 1989 (Dorbritz, 1994c). Resort to abortion prevailed primarily among young women under age 18 and among women over age 30 who already had the number of children they wanted (Roloff, 1997).

Introduction of free abortion legitimated the improved position of women in "socialist society." The decision to legalize induced abortion at a time of contentious public debate in the Federal Republic was driven by the political intent to demonstrate the superiority of the GDR, at least in the realm of abortion. An equally important reason was the recognition that with the steadily growing participation of women in the labor force and the deficit in service facilities, more and more multiparous women decided to terminate unwanted pregnancies illegally. Lack of knowledge among medical personnel increased the risk of complications and maternal mortality. The objective of the new law was to grant families favorable prospects for the development of the children they wanted to have (Mehlan, 1969, 1970). Research findings suggest that few unwanted children were born in the GDR (Henning, 1990).

Passage of the abortion legislation marked one of the few occasions when opposition votes were allowed to be registered. However, after abortion became a state-sanctioned method of fertility regulation, open public discussion was discouraged, whether on an ethical or medical basis. In the 1980s abortion data were no longer included in the *Statistical Yearbook.* The official silence reinforced women's perception that there was some moral shame about abortion and that having an abortion was a woman's fault because she had been irresponsible (Funk, 1993).

CONTRACEPTION IN THE GDR

Prohibitions against contraceptives were lifted at the end of World War II by the Soviet Military Administration. The first marriage and sexual advisory centers were reestablished in 1949 after the GDR became a state (Aresin and Aresin, 1969; Aresin, 1987; Neumann, 1987). In 1963 Mehlan founded the *Gesunde Familie* (Healthy Family) Working Group. In 1963 it became the Section of *Ehe und Familie* (Marriage and the Family) within the German Society for Health Protection. In 1956 Mehlan founded the Institute for Hygiene at the University of Rostock Medical School. In 1960 it convened the First International Conference on Abortion and Family Planning, followed by annual training seminars (Mehlan, 1988). After becoming an associate member of the International Planned Parenthood Federation in 1967, the section achieved full membership status in 1971. After reunification, *Ehe und Familie* merged with *Pro Familia*. Many advisory centers were forced to close when local authorities were unable to continue financial subventions (Mehlan, 1969, 1970, 1988; Neumann, 1991; Aresin, 1993).

In the early 1960s the GDR became the first socialist country to introduce medically prescribed contraceptives, such as oral pills and IUDs, and to supply them without cost (Neumann, 1990). The condom, which could be purchased at pharmacies, general stores, and from vending machines, was the only contraceptive for which a charge was made. Hormonal contraception predominated. Domestically produced oral contraceptives of varied dosages were available on prescription to all women over 16 years of age. By July 1973 about 40 percent of women in the fertile years were reported to be regular users (Hamand, 1989). A 1990 survey indicated that 94 percent of women 30 to 40 years of age used the pill over long periods of time and 52 percent did so permanently. Only 5 percent used condoms regularly. About 20 percent relied on an IUD. Natural methods had decreased in importance since the 1980s and were used infrequently (Oddens et al., 1993; Starke and Visser, 1994). Subsequent studies reported similar findings (Oddens et al., 1994).

In the 1980s there was a network of 220 government-financed marriage and sexual counseling centers throughout the GDR, with over 80 percent housed in local health facilities. The centers were typically directed by full- and/or part-time physicians (often gynecologists) with a staff of psychologists, lawyers, and social workers. In accord with the Ministry of Health guidelines (GDR, 1968), counseling was generally limited to medical aspects of marital conflicts and sexual disorders, infertility or abortion, and contraception (Aresin, 1977, 1979; Hamand, 1989). The center staff also assisted the schools in programs fostering the development of sexual ethics. Several popular books became well-regarded resources (e.g., Aresin and Müller-Hegemann, 1978; Neubert, 1967; Mehlan, 1970; Schnabl, 1970). Courses in family planning were mandatory in GDR medical schools, usually requiring about 30 hours; at Rostock, the training program consisted of 80 hours (Mehlan, 1969).

Sterilization was seldom used in the GDR. The very word evoked the past when, during the Nazi period, persons declared to be eugenically "unfit" on physical, mental, moral, and even political grounds were forcibly sterilized (Rothe, 1973). In 1946, immediately after World War II, the Soviet Military Administration annulled all eugenic legislation in the territory of the Soviet Occupation Zone. In 1954 the GDR Ministry of Health permitted sterilization on strictly defined medical grounds in cases of serious danger to the life and health of the woman and then only with her consent (Karcher, 1992).

A proposal for voluntary sterilization was developed in 1967 and discussed as a method of "irreversible contraception" in cases where reversible methods had proven ineffective and repeated pregnancies seriously endangered the life and health of the woman (Rothe, 1968). A directive was issued by the Ministry of Health on 21 April 1969 (GDR, 1969). Written application had to be made by the woman to the local district health officer. Vasectomies were discouraged as a method of family planning (Neumann, 1987).

SEXUALITY EDUCATION

Sexuality education, which had a well-established tradition in pre-Hitler Germany, was taken seriously in the GDR. In 1959 the GDR became the first CEE country to adopt sexuality education themes in the State Pedagogical Plan. Publishing houses produced a broad range of educational materials oriented to adults, adolescents, and children, as well as to parents, teachers, and medical personnel. The objective was to convey factual information and foster consideration of biological, psychological, and social aspects of human reproduction within the context of socialist sexual ethics (Grassel, 1976; Grassel and Bach, 1979; Bach, 1974, 1987, 1991; Hamand, 1989).

The responsibility of parents and teachers for sexuality education was confirmed in the GDR Family Code of 20 December 1965, and in the GDR Youth Law of 28

January 1974. Sexuality education in the schools was regarded as an integral part of education and was obligatory. Teaching syllabi were developed beginning with the first grade, becoming increasingly sophisticated and partner-oriented at every level through elementary and secondary school (e.g., Brückner, 1968; Knopp, 1980). Related books were written to assist parents when their children reached puberty (e.g., Neubert, 1969; Brückner, 1976). GDR television, radio, and youth-oriented periodicals were also enlisted in the effort to convey research results, respond to reader and viewer questions, and discuss the special needs of handicapped youth (e.g., Grassel and Radlbeck, 1975; Weinert, 1976). Surveys of contraceptive behavior and partner perceptions received increasing attention in the schools and in the media (Starke and Friedrich, 1984; Starke, 1987).

ADOLESCENT REPRODUCTIVE BEHAVIOR

Age of menarche declined in the 1970s to around 12.5 years. Sexual activity also began earlier. In 1975 one-third of 16-year-old and two-thirds of 17-year-old males and females had experienced sexual debut. While rural or urban residence or social level made little difference, education did. For example, college students reported a later sexual debut, often involving a longer term relationship and more regular contraceptive practice (Littmann and Szewczyk, 1977).

Surveys of adolescent sexual behavior were conducted by the Central Institute for Youth Research in Leipzig from its founding in 1966 to its closing on 31 December 1990 in fulfillment of reunification regulations (Starke and Visser, 1994). Three nationally representative sample surveys in 1972–74, 1980, and 1990 on partner relationships showed that young people entered into stable partner relationships at an early age in 1970 and 1980; by 1990 they were postponing binding commitments. In 1990 average age at sexual debut was 17 years. Women seemed to take the lead with 59 percent of female respondents and 47 percent of males experiencing sexual debut before age 17; by age 19 the levels were 80 percent and 60 percent, respectively. In 1990 more than 98 percent of the respondents were in favor of contraception and considered it a matter of mutual responsibility. While experiencing negative attitudes toward abortion as a means of fertility regulation, only 2 percent of respondents favored legislative restrictions. A tendency to postpone childbirth was apparent (Starke and Visser, 1994).

From age 14 a young woman could go to a gynecologist unaccompanied by a parent or guardian and obtain a prescription for the pill that would be filled by the pharmacy without cost. Since reunification, people have been obliged to pay the costs of contraception (Starke and Visser, 1994). In 1990, among 10 to 18 year olds, 67 percent agreed on a contraceptive method before sexual debut. This was a considerable improvement over the 51 percent who used contraception at sexual debut in earlier surveys.

The age-specific fertility rate of 15 to 19 year olds declined considerably between 1970 and 1976. Their proportion of births decreased from 18 percent of all live births in 1973 to 15 percent in 1976, while the proportion of pregnancy terminations increased from 12 percent to 13 percent of all legal abortions. In 1976 there were 750 abortions per 1,000 births among 15 to 19 year olds in urban centers compared to 420 abortions per 1,000 in rural regions. Among 15-to-17-year-old women, who accounted for 6 percent of all abortions in 1976, one of every two pregnancies was terminated. Among 18 to 19 year olds, who accounted for 7 percent of all abortions, only one of every six pregnancies was terminated (Mehlan and Akkermann, 1978; Leetz, 1977). In 1977 about 14 percent of all live births and 15 percent of all abortions occurred in women under 20 years of age (Mehlan, 1978). In 1991 the proportions were 6.9 percent and 8.5 percent, respectively (Dorbritz, 1994c).

POSTREUNIFICATION TRENDS

Demographic Shock

The intensity of the decline in fertility was surprising, even considering the net outmigration from Eastern to Western Germany in the period 1989–1992 (FRG/FSO, 1994). As shown in Table 7.1, in the two years between 1989 and 1991, the period of the fall of the Berlin Wall, monetary union, and reunification, the reported number of live births fell by over 45 percent. The downslide continued thereafter. By 1992, births amounted to 44 percent of the 1989 total, and birth rates were 55 percent lower than they had been three years earlier (Eberstadt, 1994c). The decline was most pronounced in the youngest age groups. Such an abrupt and precipitous decrease in fertility was unprecedented in any industrialized society during peacetime. Some stabilization became apparent after the number of births reached a low point of 78,700 before rising to 83,800 in 1995 and 98,500 in 1997, still less than half the 198,900 in 1989 (Dorbritz, 1997b). The crude birth rate remained at 5.1 per 1,000 women in 1993 and 1994 but rose to 5.4 in 1995 and 6.4 in 1997 (COE, 1998).

Between 1989 and 1993, the birth and marriage level decreased by more than 60 percent. The Total Fertility Rate fell from 1.57 to 0.8 (or less than 1 birth per women per lifetime) before rising again to 1.02 in 1997. Whereas in the 1980s about one-third of all marriages ended in divorce (1986: 39 percent), the Total Divorce Rate was only 8 percent in 1992 (Dorbritz, 1994a, 1994b, 1994c; 1995) as unemployed women were more economically dependent on their husbands. By 1996, the rate had risen again to 21 (COE, 1998).

Abortion/Contraception

When abortion threatened to become a major stumbling block in the reunification process, a temporary compromise was fashioned in Article 31 of the reunification agreement. It allowed the more liberal GDR legislation to remain in effect in the 5 new federal states and permitted women residing in the West to travel East while the more restrictive law continued to prevail in the 11 old federal states. It also stipulated that an all-German law was to be approved before the end of 1992. A June 1992 law permitted abortion on request in the first trimester but required obligatory preabortion counseling (Funk, 1993). The law was never implemented because a legal challenge was filed against it in the Federal Constitutional Court, resulting in a temporary injunction. The court ruled the law unconstitutional in June 1993 and issued temporary guidelines until another law could be approved.

The guidelines held that, except for specified medical reasons, abortion was incompatible with the constitutional duty of the state to protect life and thus illegal. However, neither physicians nor the woman would be prosecuted if the woman received counseling and if the abortion was performed during the first trimester of pregnancy. The court also stipulated that preabortion counseling must be oriented toward saving the life of the fetus and should be closely supervised. Physicians and hospitals have the right to deny abortion and health insurance companies are not obligated to pay for the procedure. The decision was perceived as a blow to women's rights, especially in the new federal states (von Baross, 1993). Women residing in the former GDR faced a new requirement of obligatory counseling, the reincorporation of the abortion law into the Penal Code, and the reintroduction of religious values. In June 1995 compromise legislation was approved, based on the 1993 Federal Constitutional Court ruling (*Pro Familia Magazine*, 1995).

In July 1996 the State of Bavaria passed legislation requiring a woman to explain why she did not want to carry a pregnancy to term. Moreover, physicians were not permitted to earn more than 25 percent of their total income of performing abortions. These moves were partially blocked by the Constitutional Court (*Choices*, 1996; *The Economist*, 1998b). Subsequently, the 294 Roman Catholic pregnancy counseling centers complied with Pope John Paul II's request to stop issuing the counseling certificates that women need to obtain an abortion without risk of legal prosecution (*This Week in Germany*, 1998; *The Economist*, 1998b). The more than 1,400 non-Catholic counseling centers remained unaffected.

While it might be thought that abortion played a major role in the birth dearth, official data suggest otherwise. As shown in Table 7.1, between 1989 and 1995, the number of reported induced abortions fell steadily from about 73,900 to 24,100. Even with the dramatic drop in births, the ratio of abortions declined from 372 per 1,000 live births in 1989 to 288 in 1995. Pregnancy avoidance, not pregnancy termination, appears to explain most of the change in the fertility level. Assuming that sexual activity has remained fairly stable and that underreporting of induced abortion is not significant, it seems reasonable to infer a substitution of more effective contraceptive practice with reliance on abortion as a backup procedure. This change in reproductive behavior is all the more remarkable when it is considered that contraceptives are no longer free of cost (except for women under age 20) and that legal abortion continues to be more readily available in the new federal states than in the old (Aresin, 1993). At a time when children are less wanted, a greater sense of responsibility among both partners seems to have led to more effective contraception. Whether the 1996 increase in reported abortions, accompanied by a substantial increase in births, reflects a reversal remains to be seen.

Births and Marriage

Another surprise is the persistence of the high percentage of out-of-wedlock fertility in the new federal states. Between 1989 and 1993, live births in the former GDR dropped by a reported 60 percent. But, out-of-wedlock births fell by 51 percent, whereas births within marital unions fell by 64 percent (Eberstadt, 1994c). This difference occurred despite the much greater economic and legal protective function of marriage in the old federal states that now applies to the territory of the former GDR. The close coupling of marriage and reproductive behavior, so characteristic of the FRG, has not yet occurred in the former GDR. The old patterns of behavior also persist when divorces are considered. The second year of marriage still has the highest divorce rate in the new federal states (GDR) whereas divorces crest in the fifth year of marriage in the old federal states (FRG). Thus, while earlier patterns of reproductive behavior have given way to new realities, there has as yet been little adjustment of values and lifestyles (Dorbritz, 1994b).

Causes of the decline in birth and marriage rates in the immediate postreunification period have been widely discussed (Dorbritz, 1994b, 1997a, 1997b). Most observers agree that the dramatic change in demographic behavior resulted from forced adjustment within an extremely short period of time to a vastly different economic and social system (e.g., Kolinsky, 1993; Fleischhacker, 1994; Ostner, 1994; Atkinson, 1995; Brähler and Richter, 1995; Marvin, 1995; Witte and Wagner, 1995; Adler, 1997).

The Woman Question Continued

Women were especially vulnerable. After reunification they quickly constituted nearly two-thirds of the unemployed (Aresin, 1993). Officially, 20.2 percent of able-bodied women were unemployed, compared with 11.8 percent of men (Atkinson, 1995). Mothers who still held jobs were hard hit by the 1990 closing of state-run daycare centers, kindergartens, and youth clubs. Only 40 percent of women who took maternity leave in 1991–92 returned to their jobs. Positions had disappeared for over a quarter of the 1,400 women surveyed by the Federal Ministry for the Family (*This Week in Germany*, 1995). In 16.4 percent of cases, the women reported their employers had gone out of business while they were on leave. Another 9.5 percent lost their jobs in absentia during mass layoffs. A slightly smaller portion of the young mothers gave notice while on leave or as their leave expired (*This Week in Germany*, 1995). In 1994, 77 percent of the long-term jobless were women (Atkinson, 1995).

Some women deemed having children a severe handicap to secure employment or career advancement. Numerous younger women became so desperate to make themselves acceptable to employers that they sought voluntary sterilization, which had become readily available and was covered by health insurance. For example, one hospital in Magdeburg reported 1,200 sterilizations in 1991 compared to 8 in

1989 (Fisher, 1992; Karcher, 1992). A "wave of sterilizations" was also noted in Dresden (cited in Adler, 1997).

FUTURE PERSPECTIVES

The sharp decline in total rates of fertility, marriage, and divorce, experienced in the former GDR, is not apparent in any of the other former socialist countries. This finding would seem to support Höpflinger's (1984, 1987) hypothesis that very rapid social shifts lead to individual insecurity and economic rationality, provoking a demographic shock. There is reluctance to marry and have children in an economically uncertain environment (e.g., Conrad et al., 1996; Adler, 1997). Under the new conditions determined by the market economy, having children represents a serious occupational, social, and financial risk that many women are reluctant to assume (e.g., Huinink, 1995; Beck-Gernsheim, 1997; Adler, 1997). An adjustment is gradually evolving, away from the fading traditional lifestyle of marriage and children to a greater recognition of other lifestyles, ranging from consensual unions to childless marriages. It is likely that there will be a longer term process of adaptation that may not end before the generations reared under the socialist conditions of the GDR move beyond childbearing age.

The increases in the 1995 and 1996 birth rates suggest that, as of early 1998, marriages and births are passing through a transformation phase (Dorbritz, 1997b). Family formation is characterized by parallel movements of postponement, catching up, and adaptation. Cohorts born before 1968 reduced family formation in 1990/91 and, with some exceptions, did not catch up again. With a few cohorts (1968–72) it appears that a continuation of the family formation process recommenced in 1995 and most probably continued to increase in 1996. Younger people born after 1972 at first reacted by postponing family formation but now appear to be following the West German age pattern.

Over time only a gradual renewal in the rate of births can be expected (despite the 1996 and 1997 increases in the number of births). This impression is strengthened by the 1990–96 surveys of "Life in East Germany." They reflect the loss of importance of marriage and partnership for the younger age groups, linked to dissatisfaction and unclear future perspectives (Dorbritz, 1997b). Within this context, a significant increase in marriage and birth rates appears unlikely in the near future (Conrad et al., 1996; Adler, 1997). Effective practice of modern contraception has lessened reliance on abortion since the early 1980s.

chapter 8

Hungary

Henry P. David

EDITOR'S INTRODUCTION

It is said to be 1,100 years since Arpad, a pagan chieftain from beyond the Urals, led his horsemen into the heart of Europe, swept into the Carpathian basin, and laid the foundations of the nation-state that became Hungary. The Hungarians are Magyars, a people whose language links them to Finns, Lapps, and certain Siberians. They settled the fertile plains on either side of the Danube in the heart of Central Europe in the ninth century. Most of present-day Slovakia, Romanian Transylvania, and much of the former Yugoslavia were integral parts of Hungary from the Middle Ages onward. The Ottoman Turks invaded the Balkan territories in the fourteenth century and occupied central Hungary in the sixteenth century. After the defeat of the Turks in 1686 and restoration of Habsburg rule, the Austro-Hungarian Monarchy was proclaimed in 1867. Under the empire, Hungary regained much of its previous territories, only to lose them and more after World War I in the Treaty of Trianon in 1920.

After World War I Hungary became a constitutional monarchy without a king, ruled by a regent, Admiral Horthy. Following an alliance with Fascist Italy and Nazi Germany, the former territories were briefly regained but lost once more after World War II. In 1948 communist leaders aligned Hungary with the Soviet bloc. It shared borders with Austria, Slovakia, Ukraine, Romania, Serbia, Croatia, and Slovenia. A popular uprising in 1956 was defeated by 200,000 Soviet troops as several hundred thousand Hungarians fled the country (Ash, 1996b). In 1968 Hungary became the first socialist country in Central and Eastern Europe (CEE) to relax central economic planning. In the subsequent decades, increasing flexibility marked the government's response to the deepening economic crisis. By the mid-1980s Hungary was only formally a communist state.

Reforms paved the way for political liberalization that was achieved in 1989 with the establishment of a multiparty system. The 1990 elections resulted in the ouster of the communist regime. However, the 1994 elections returned to power a socialist party led by reformed communists (Deak, 1994). In 1995 Hungary imposed an economic austerity program, devalued the currency, and reduced child allowances. As of year end 1997, Hungary had strong economic numbers with foreign investments and exports rising (Spolar, 1997). A right-wing government was elected in 1998.

Hungary was unique among the socialist countries in organizing a concerted campaign to foster more effective contraceptive practice, beginning in the mid-1960s. It resulted in a reduction in the average number of abortions per women of reproductive age from 2.8 in 1969 to 1.2 in 1989 (Blayo, 1992). While the availability of abortion was at times technically restricted, there was always a way for those seeking it vigorously. Particularly helpful in understanding the evolving Hungarian trends in fertility behavior are the longitudinal surveys of newly married women initiated under the auspices of the Central Statistical Office in the early 1950s and repeated in 1966 and again in 1974.

The chapter begins with a brief overview of Hungarian demographic trends, followed by a discussion of pronatalist population policies and changing abortion policies. Subsequent sections note contraceptive practices over time and studies of planned and actual fertility. The chapter concludes with a discussion of the woman question, sex and society, and the situation of adolescents.

The chapter was developed by the editor. It benefited greatly from consultation with Peter Jozan, Head, Population and Health Statistics Department, Hungarian Central Statistical Office, Ferenc Kamaras, a senior officer in the Fertility Statistics Section of the Hungarian Central Statistical Office, and Marietta Pongracz, a senior researcher at the Demographic Research Institute in Budapest. Gail Kligman reviewed an earlier draft. Diacritics were omitted from all Hungarian citations.—H.P.D.

DEMOGRAPHIC TRENDS

Hungarian demographic developments have the advantage of being unusually well documented. Continuous vital statistics are available for more than 100 years (e.g., Klinger, 1995). The early period of the demographic transition reflected a mixture of Western and Eastern European characteristics, with reductions in fertility occurring through delay and avoidance of marriage or through control of fertility within marriage (Andorka, 1979; Demeny, 1968; Coale, 1969). At the turn of the century the TFR was more than five children per woman (Ross and Frejka, 1998). After World War II Hungary became the first European country whose fertility declined below replacement level. Since 1980 the total fertility rate has remained below 2.0 despite concerted governmental efforts to influence reproductive behavior toward stemming the birth dearth and achieving at least population stabilization. In 1995 the total population was around 10 million. About 20 percent of the population lives in the capital city, Budapest. Some 7 percent of the population are gypsies.

The crude birth rate fell from 44 in 1880 to 35 in 1910, 25 in 1930, and 20 in 1940 (Klinger, 1977). As shown in Table 8.1, the birth rate peaked around 21.5 in 1953–55 when the 1878 law criminalizing abortion was strictly enforced. After the legalization of abortion in 1956, the birth rate declined to 12.9 in 1962, the lowest in Europe at that time. The broad range of pronatalist policies in the 1960s and 1970s led to somewhat higher birth rates. A subsequent peak of 18.4 was reached in 1975 after the implementation of abortion restrictions in 1974. Subsequently, the birth rate resumed an intermittent decline, sinking to 9.9 by 1997. With the total fertility rate at below replacement levels for all but three of the past 36 years and falling to 1.38 in 1997, the population decrease continues. With 14.3 percent of its population over age 65, Hungary currently has the "oldest" population of any of the formerly socialist CEE countries (Velkoff, 1993).

The inclination to marry has declined gradually from 9.9 per 1,000 population in 1975 to 4.6 in 1997 (COE, 1998). The remarriage rate decreased similarly.

The proportion of couples living in consensual union is rising, accompanied by an increase in the proportion of children born out-of-wedlock from 5.4 per 100 births in 1970 to 25.0 in 1997 (COE, 1998).

PRONATALIST POLICIES

Pronatalist measures began in 1924 when legislation introduced a 12-week maternity leave and established statutory protection for women during pregnancy (Kozlawska and Wojtyla, 1976). Following World War II several periods of population policy formulation evolved (Andorka and Vukovich, 1985; Vukovich, 1989; Klinger, 1984, 1991; Andorka, 1993; Kamaras, 1995). The first period began in 1953 when the government announced a decree on "developing the protection of mothers and children." It included regulations concerning the medical care of pregnant women, social security benefits for families, and the development of childcare institutions to improve conditions for child rearing and family living standards. The rationale was that the socialist system would automatically ensure optimal population trends synonymous with maximum population growth. The assumption was that government decisions designed to influence reproductive behavior could be implemented by administrative measures.

Policy considerations motivated the restrictive measures concerning access to abortion and contraceptives implemented in 1953. As shown in Table 8.1, the number of births increased temporarily; largely because of the severe penalties threatened for performing clandestine abortions (Hirschler, 1961). However, negative public reaction resulted in a relaxation of the restrictions in 1954, followed two years later by the sanctioning of abortion on request. The economic growth in the 1950s fostered improved living standards but did not result in increased fertility. On the contrary, certain societal changes, such as increasing female labor force participation, resulted in fertility declines.

Table 8.1
Birth, Abortion, and Fertility Rates and Ratios: Hungary, 1950–97

Year	Live Births	Legally Induced Abortions	Other Abortions	Total Registered Abortions	Birth Rate	Legal Abortion Rate[e]	Legal Abortion Ratio[f]	Total Fertility Rate
1950	195,600	1,700	34,300	36,000	20.9	0.8	9	2.62
1953[a]	206,900	2,700	40,000	42,700	21.6	1.3	13	2.79
1955	210,400	35,400	43,100	78,500	21.4	16.3	168	2.68
1956[b]	192,800	82,500	41,100	123,600	19.5	38.3	428	2.47
1959	153,300	152,400	35,300	187,700	15.2	72.1	994	2.08
1960	146,500	162,200	33,800	196,000	14.7	76.7	1,107	2.02
1965	133,000	180,300	33,700	214,000	13.1	80.1	1,356	1.81
1969	154,300	206,800	32,200	239,100	15.0	90.6	1,340	2.04
1970	151,800	192,300	29,800	222,100	14.7	83.5	1,267	1.97
1972	153,300	179,000	28,200	207,200	14.7	77.5	1,168	1.93
1973	156,200	169,700	28,200	197,900	15.0	73.5	1,086	1.95
1974[c]	186,300	102,000	30,700	132,700	17.8	44.3	548	2.30
1975	194,200	96,200	28,200	124,500	18.4	41.9	495	2.38
1978	168,200	83,500	21,700	105,200	15.8	37.0	496	2.08
1979	160,400	80,800	20,500	101,300	15.0	35.9	504	2.02
1980	148,700	80,900	20,000	100,900	13.9	36.3	544	1.92

1981	142,900	78,400	18,100	96,500	13.4	35.3	549	1.88
1982	133,600	78,700	17,500	96,200	12.5	35.5	589	1.78
1983	127,300	78,600	17,000	95,600	11.9	35.5	617	1.73
1984	125,400	82,200	17,900	100,100	11.8	37.0	656	1.73
1985	130,200	82,000	18,100	100,000	12.3	36.9	630	1.83
1986	128,200	83,600	18,300	101,900	12.2	37.7	652	1.83
1987	125,800	84,500	17,700	102,200	12.0	38.2	672	1.81
1988[d]	124,300	87,100	18,000	105,200	11.9	39.3	701	1.79
1989	123,300	90,500	17,700	108,200	11.9	40.8	734	1.78
1990	125,700	90,400	17,600	108,000	12.1	41.2	719	1.84
1991	127,200	89,900	17,100	107,000	12.3	40.7	707	1.86
1992	121,700	87,100	16,100	103,100	11.8	38.9	716	1.77
1993	117,000	75,300	15,900	91,100	11.4	33.7	644	1.69
1994	115,600	74,500	15,500	90,000	11.3	33.3	644	1.64
1995	112,100	77,000	15,300	92,200	11.0	34.5	687	1.57
1996	105,300	76,600	14,800	91,400	10.3	34.5	728	1.46
1997	100,300	74,600	15,300	89,900	9.9	33.9	744	1.38

[a]Medical boards established for abortions on medical and social indications (January 1953). [b]Abortion law liberalized (June 1956). [c]Restrictions on abortion imposed (January 1974). [d]Abortion committees abolished (December 1988). [e]Per 1,000 women age 15–44. [f]Per 1,000 live births.

Sources: David and McIntyre (1981), Henshaw and Morrow (1990), U.N. (1995a), Hungary/CSO (1997), and COE (1998).

In the second half of the 1960s a new population policy emerged, based on a long-term socioeconomic perspective. The demographic component was increasingly considered in societal planning. The aims were to ensure the reproduction of the population and to moderate the uneven age distribution.

Allowances for Mothers

Following the birth of their children, working mothers were entitled to a maternity leave of 24 weeks during which they received their total former wages. They could stay at home with their children until the child reached the age of two. The child care allowance made it possible for mothers to stay at home for a further year until the child reached the age of three. During this period they received a fixed monthly amount of 4,950–5,150 forints (U.S.$58–60) depending on the birth order of the child.

Additional measures were implemented in 1973. The amount of the maternity allowance was increased. Families and young couples received benefits to improve their housing conditions. Oral contraceptives were made available. The previously unrestricted access to abortion was somewhat limited. As will be discussed, abortions could be granted only for specified social or medical reasons.

A system of pregnancy allowances and child rearing allowances was introduced at the beginning of 1993. A mother became entitled to a pregnancy allowance from the fourth month of pregnancy until the birth of the child. The amount of the pregnancy allowance was equal to the family allowance, which varied from 2,750 to 3,950 forints per month (U.S.$32–46) depending on the birth order of the child.

A child rearing allowance could be given to families who cared for three or more young children in their own home. It could be paid until the youngest child was 8 years old. The monthly amount was equal to the minimum old age pension of 6,000 forints (about U.S.$70) in 1992. Time spent at home was regarded as employment and taken into consideration for the calculation of the old age pension.

Family Allowances

As of 1996 both two-parent families and single-parent families receive a family allowance for each of their children, independent of parental income. If they have one child, the allowance is paid until the child reaches six years of age; if there are two or more children it is paid until the children are 16 years old. In case of chronic disease and physical or mental handicap, family allowance payments are extended until the child is 20 years old.

The monthly amount of the family allowance per child increases with the birth order of the child. Compared to one-child families, the amount per child is 18 percent higher for families with two children and 36 percent for families with three and more children. Single parents who rear their child alone receive a higher family allowance than two-parent families.

Recent Trends

With the transition to a market economy and the pressures to reduce budget deficits came a reconsideration of pronatalist policies. Child care benefits were decreased in cash and in-kind. New legislation shortened the duration of benefits and based payments on means tests. The value of benefits was eroded, ranging between one-third to one-half of minimum living standards (Adamik, 1997).

Opinions on the effects of population policy in Hungary are divided (Hungary, 1994). In reviewing population policies between 1966 and the beginning of the 1980s, Klinger (1984) noted that the different measures influenced the shortening and lengthening of birth intervals, especially between first and second children. However, these policies did not significantly influence completed fertility. On average, the desired number of children hardly changed. More specific analysis is rendered difficult because population policies need to be considered within a context of cultural changes effecting the status of women, costs of educating children, motivations for childbearing, access to contraception and abortion, and related couple perceptions and values. There is considerable doubt whether population policies can influence couples to have an additional child, especially a third child (David and Pierotti, 1993).

ABORTION POLICIES

1878–1970

In the last 100 years Hungarian abortion policy varied from total prohibition to complete liberalization. Following the establishment of the Austro-Hungarian Empire in 1867, a law was enacted in 1878 declaring the interruption of pregnancy to be a crime except when performed for narrowly defined medical indications (Lammich and Nagy, 1988; Balogh and Lampe, 1994). As indicated in Table 8.1 and Figure 8.1, strong efforts to enforce existing laws against illegal abortion in 1952 and 1953 were followed by temporary baby booms. Also in 1953, medical boards were established to review and rule on requests for "therapeutic" abortions on selected social grounds as well as medical and eugenic indications (Hirschler, 1961). The subsequent increase in legal abortions probably represented a transfer from procedures that would have been unregistered before (Frejka, 1983).

Following liberalization of abortion in the Soviet Union in 1955, an Order of the Hungarian Council of Ministers was promulgated on 3 June 1956 "to increase woman's health protection and relax procedures relating to the interruption of pregnancy," virtually providing abortion on request of the pregnant woman (Hungary, 1956). Abortions could be performed by qualified physicians in hospitals or maternity homes (but not in outpatient clinics) after authorization by a local medical board consisting of three persons and meeting at least weekly (Klinger, 1974). The chairperson was a physician, designated by the local chief medical officer. The other two members were selected by the Executive Committee of the People's Council.

Figure 8.1
Births and Legally Induced Abortions: Hungary, 1950–97

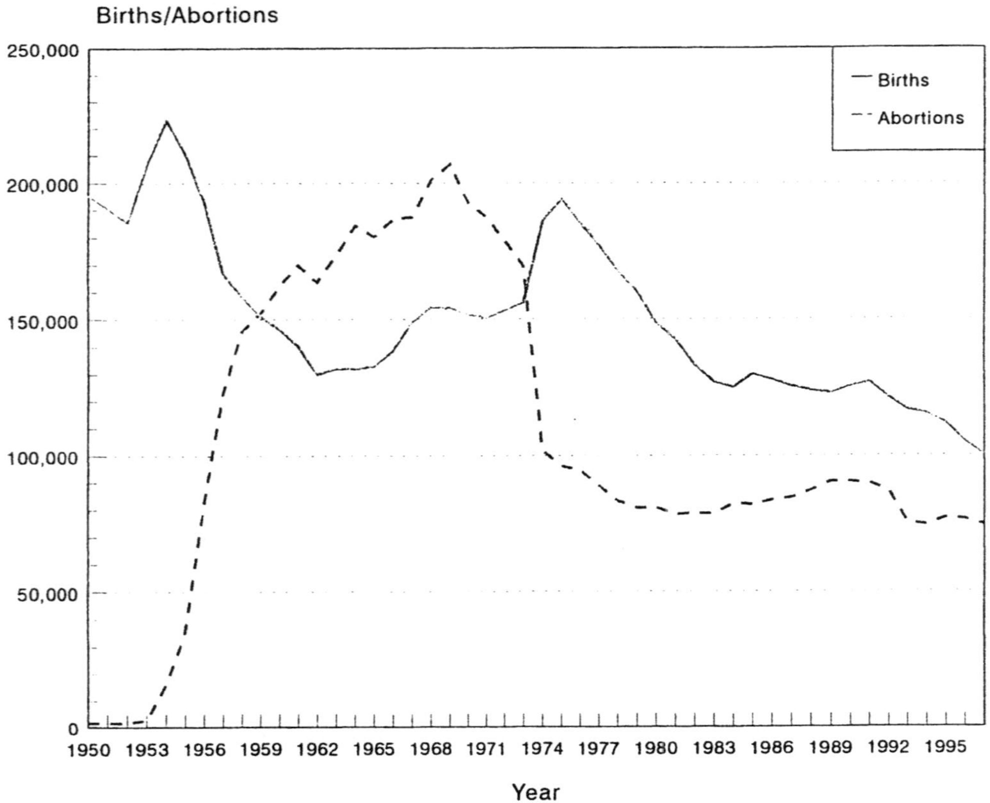

Abortions allowed on medical and social grounds in January 1953; abortion liberalized in June 1956. Restrictions on abortion imposed in January 1974.

Sources: See data from Table 8.1.

Typically, they were the head of the social affairs section of the People's Council and a woman proposed by the Trade Union Council. A gynecological examination was usually required by the board to determine the period of gestation (Hungary, 1956; Klinger, 1966, 1988).

Abortion was authorized until the thirteenth week of pregnancy, extended to 18 weeks for single women under age 20. In addition to health considerations, permission for termination was granted if justified by personal or family circumstances of "if the applicant insists on the interruption of pregnancy after the board has asked her to think it over." For terminations for nonmedical reasons, the applicant or her family was obliged to pay the cost of hospital treatment, about U.S.$13 at that time. Although abortions commonly required three days' hospitalization, social insurance benefits could not be applied to hospital costs or loss of wages until the fourth day (Maggs, 1977).

As shown in Table 8.1, the total number of abortions rose rapidly, becoming the main means of fertility regulation. By 1960 the number of legal abortions exceeded the number of births. A decade later abortions had reached their peak. By 1969 there were 1,340 abortions per 1,000 live births. Projections indicated that if terminations had stabilized at their then rate, a woman would have experienced 2.7 abortions during her reproductive years.

1970–92

Beginning in 1970 abortion showed a declining trend. This was partially the result of the introduction of domestically produced oral contraceptives and their increasing use under medical supervision. In October 1973, by decision of the Council of Ministers, the regulations controlling legal abortion were revised, effective 1 January 1974 (Hungary, 1973). Terminations on request remained available for single, divorced, separated, and widowed women and for married women who were over age 40, or had at least three living children, or had experienced three deliveries, or had two living children and had, in addition, undergone at least one "obstetrical event"—a term that included live births, still births, ectopic pregnancies, and induced abortions (which could be considered only after the second delivery). Married women who did not meet stated criteria could still request termination for medical, eugenic, juridical, and "serious social" reasons, including, for example, lack of adequate housing (defined as a separate dwelling unit), husband's military service, or husband's imprisonment (Hungary, 1973).

The October 1973 decision reflected a growing concern with demographic trends. In addition to promulgating restrictive regulations designed to lower the number of induced abortions, the Council of Ministers simultaneously introduced a range of measures aimed at ensuring population stabilization (about 2.1 children per family) while also promoting more effective contraceptive practice. Marriage licenses would be issued to couples under age 35 only upon presentation of proof that they had received contraceptive counseling and supplies from a physician. The network of family planning centers was greatly expanded, pronatalist incentives improved, and courses on sex education, family life, and contraception introduced at all levels of the educational system and military training.

The 1973 restrictions were announced well in advance of implementation, written in a manner that left considerable room for interpretation by medical authorities, and provided a procedure for appealing negative decisions. Regulations governing pregnancy termination remained largely as stated in the 1956 Ordinance. Although the fee for abortions performed for nonmedical reasons was increased to cover actual costs, it could be waived in cases of demonstrated need. Working women were eligible for two days' paid sick leave after abortion when a physician so recommended.

The change in abortion policy was initially perceived as so coercive that it precipitated a rare, and little publicized, moment of overt resistance from some social scientists and students. They petitioned Parliament and collected signatures against the regulations until they were threatened with job loss (Gal, 1994; Stloukal, 1995). Women's growing sophistication about the system, the loopholes in the regulations, and the abortion commission's benevolent laxness resulted in only a very small proportion of abortion requests failing to be approved, less than 3 percent in 1976 and fewer than 1 percent in 1978 (Jozan, 1978; Szalai, 1988).

The decline in numbers of legal abortions performed was considerable. As shown in Table 8.1, terminations decreased from 169,700 in 1973 to 102,000 in 1974, dropping further to 80,800 in 1979 and 78,400 in 1981. The abortion ratio fell from 1,168 per 1,000 live births in 1972 and 1,086 in 1973 to 496 in 1978 before resuming its rise as the number of births declined more rapidly than the number of abortions (see Figure 8.1) A similar decline can be seen in the abortion rate.

Access to legal abortion remained unrestricted. It was simplified in 1982 and in 1986 by ordinances designed to speed authorization (Hungary, 1982, 1986). By 1988 the obligation to consult a committee was abolished and abortion was made broadly available, particularly for women with two or more children (Hungary, 1988; Meszaros, 1990). As shown in Table 8.1, the numbers of legally induced abortions began to increase in 1984 and continued to do so until 1989. They then declined slowly until 1993 when there was a 15 percent drop from 1992. This was largely the result of Law No. 79 of 17 December 1992 on the protection of fetal life (Hungary, 1993; Balogh and Lampe, 1994).

1993–97

Since 3 January 1993 induced abortion is permitted in the case of threat to the health of the woman or of the fetus, a criminal act such as rape or incest, or when there is a grave crisis situation for the woman. Since the law leaves it up to the woman applying for induced abortion to decide whether she is or is not in such a crisis situation, the regulation represents no new restriction compared to the earlier practice.

The pregnancy can be interrupted until the twelfth week of gestation, and in certain medically defined cases, until weeks 20 and 24. The pregnant woman is obliged to give her petition for abortion personally to a staff member of the Family Welfare Service who informs and counsels her about financial assistance and adoption options. After an obligatory three-day waiting and reflecting time, the woman decides whether she wants to keep her pregnancy or, referring to a grave crisis situation, asks for its interruption. In the case of health-related reasons or a criminal act, the abortion is free of charge with costs paid by the Health Insurance Fund. For a crisis situation the fee is about 500 forints, approximately equal to a quarter of the net average monthly income. However, this fee can be reduced when indicated by the social situation of the petitioning woman (Batar, 1993).

As shown in Table 8.1, births continued to decline from 121,700 in 1992 and 117,000 in 1993 to 100,300 in 1997. Reported legal abortions decreased to 75,300 in 1993 and 74,500 in 1994 before increasing to 77,000 in 1995 and declining to 74,600 in 1997. With the continuing drop in births, the abortion ratio increased from 644 per 1,000 live births in 1993 to 744 in 1997. Trends in births and abortions from 1970 to 1997 are presented in Figure 8.1.

Since 1989 several pro-life organizations have emerged, usually with support from the Catholic Church. While vocal in their demands for a ban on abortion, they have attracted little mass support (Toth, 1993). Findings from a representative sample survey conducted by the Demographic Research Institute in late 1989 showed a majority favoring the right of each woman to decide on her own the fate of her pregnancy. Only 7 percent would support a complete ban on abortion and of those claiming to follow the teachings of their religion, only 17 percent supported prohibition (Pongracz, 1991).

ABORTION-RELATED DEMOGRAPHIC RESEARCH

Resources

Well established demographic studies have long provided a basis for research on reproductive behavior (e.g., Pongracz, 1976; Klinger, 1977, 1979, 1988, 1995). The reporting system for collecting statistics on obstetrical events from health facilities, devised in 1931 by Vilmos Tauffer, was greatly expanded after the 1956 abortion liberalization. As a result of special abortion and contraception surveys, plus a series of national family planning surveys conducted by the Central Statistical Office (CSO) in cooperation with the Ministry of Health, considerable information has become available. Included are data on abortion prevalence and related demographic and socioeconomic characteristics; health effects on women and their subsequently born children; and on fertility behavior including changes over time in family planning attitudes, contraceptive practice, and dependence on abortion. Although content, sample design, and survey techniques varied to a certain extent, the data derived from these studies permit a compilation of information on behavioral and attitudinal changes over time. The following paragraphs present selected findings from the CSO files.

Abortion Registration

Information on overall abortion trends is continuously available from the summary reports submitted by health institutions from 1956 to 1970 and, since 1971, from obligatory individual records on induced and spontaneous abortions, incorporated within the "accounting system" for the whole country and sent monthly to the CSO. These data facilitate analysis of the demographic, health, social, and other characteristics of all women having induced and spontaneous abortions, as well as continuous monitoring of total abortion levels.

Marital Status

Changes observed over the postwar decades reflect both changes in legislation and changes in attitudes toward abortion. At the end of the 1950s, 90 percent of women having an induced abortion were married; by the beginning of the 1970s this proportion fell to 80 percent. The restrictions enacted in 1974 affected mainly married women, among whom the number of abortions fell to the greatest extent while the number of births rose. Thus, in the middle of the 1970s the ratio of married females among those having had an induced abortion decreased to 70 percent. From the middle of the 1980s the ratio of unmarried women began to grow substantially, reaching 45 percent of all abortions by 1993. A partial explanation is the decline in marriages, which in 1996 was equal to 47 percent of the 1975 figure. Divorces have also been declining but at a slower rate (COE, 1998).

Age

Divergent trends were noted in terms of abortions and births by age. Whereas the age of women having first abortions decreased, the age of women having their first birth increased. In the 1960s the rise in abortions was most noticeable among 25–29-year-old women. In recent years the modal value of the increase of abortion shifted to the 20–24-year-old age group. Among women under age 20, willingness to give birth declined considerably in the past 10 years, while the rate of abortions increased by about 20 percent compared to the second half of the 1980s. The incidence of abortion was higher than it was in the mid-1970s (after abortion restrictions). All other age groups showed a decline in resort to abortion. In 1993 there were 900 abortions per 1,000 live births among women under age 20. Of 1,000 known pregnancies, 430 were terminated (Hungary/CSO, 1996).

Changes in marital status and abortion trends by age influenced observations on abortions by parity. Compared to the second half of the 1980s, the ratio of women without children rose considerably, while the proportion of women with two children fell significantly among the total number of women having an abortion.

The proportion of women with one child and those with three and more children did not change significantly. In 1993, 29 percent of women having an abortion were child free, an increase by 50 percent compared to the second half of the 1980s. About 21 percent had one child in 1993, the first year in which women without children exceeded the proportion of women with one child. One-third of the aborting women had two children, but this ratio was by 22 percent lower than in the second half of the 1980s.

Repeat Abortion

Surveys of psychosocial characteristics of repeat abortion were initiated in 1969 (Szabady and Klinger, 1970, 1972; Klinger and Szabady, 1978). More than half the respondents (57 percent) in the 1970 study said they would terminate a future unwanted pregnancy. The more prior abortions a woman had the less likely she was to tolerate another pregnancy. The decision to terminate was made jointly by two-thirds of married couples.

In the 1974 survey no differences were found in occupational groups between women having one or two abortions and women who experienced three or more terminations. Whereas 90 percent of the sample approved of abortion in situations of physical health risk for the pregnant woman, only 75 percent approved for cases of "bad" socioeconomic circumstances compared to 80 percent approval by women having had one or two abortions and 86 percent by women with three or more terminations (Hungary/CSO, 1975a).

Education

Mainly due to differences in contraceptive practice, women with less education had more children and also more abortions per 1,000 live births than women with a higher educational level. Differences declined over time. 1990 census data showed that educational level played a differentiating role mainly for 20–29-year-old women having abortions. For 30–34-year-old women the educational differences decreased considerably and for those over 35 years they became insignificant. Pregnancy interruptions were also more frequent among women under age 20 attending a grammar school or having passed a secondary school-leaving examination than among those who had completed eight forms. The situation was similar for women over 30 years. Those over 35 years who had completed university or a college were less willing to deliver an unwanted pregnancy. They terminated such pregnancies more frequently compared to women who completed eight grades or an apprentice school. Altogether these results reflected a difference of about 30 percent in the abortion rates between women with the lowest and those with the highest educational level.

CONTRACEPTIVE PRACTICE

Coitus interruptus and rhythm backed by legal or clandestine abortion long were the traditional methods of fertility regulation in Hungary. The condom was introduced at the beginning of the twentieth century and can now be purchased from pharmacies and vending machines. The diaphragm appeared in 1949 and chemical methods (e.g., spermicidal jelly and foam) arrived in 1954.

Oral contraceptives were introduced in 1967. The pills were available by prescription, after a medical examination, for all women over age 18 and for younger women who had a child or an abortion. Subsequently, restrictions on the purchase of pills were lifted for unmarried women under age 18. Initially, laboratory tests could be performed only by gynecologists at a specialized outpatient clinic or hospital; later district industrial plant physicians (most of them general practitioners) were authorized to prescribe the pill and to decide whether laboratory tests were necessary (Kulcsar, 1984). Social insurance covered 85 percent of the cost. An IUD designed by Professor Szontagh and colleagues (1967) became available in 1971 for women who had already delivered a child or who were over age 18 and had been advised for medical reasons not to become pregnant (Hungary, 1977). Although voluntary sterilization was never prohibited, it was usually performed only for health reasons before legalization as a family planning method in 1987 (Hungary, 1987).

Knowledge-Attitudes-Practices (KAP) sample surveys, known as TCS in Hungary, of married women of reproductive age were initiated in the early 1950s and continued in diverse formats (David and McIntyre, 1981; Hungary/CSO, 1975a, 1975b, 1984). After effective contraceptive practice became a matter of social policy in the second half of the 1960s, survey findings reflected a rise in the percentage of married women of reproductive age practicing some form of contraception from 58 percent in 1958 to 84 percent in 1966 and about 90 percent in 1974 and 1977. About 98 percent of couples newly married in 1974 said they were contracepting (Hungary/CSO, 1975a, 1979; Jozan, 1977).

Data from the 1975 reinterview of the 1966 marriage cohort showed that, after nine years of marriage, 47 percent of the women under age 35 and practicing contraception were taking the pill, 10 percent had an IUD, 8 percent used condoms, 5 percent practiced rhythm, and 25 percent relied on withdrawal. Many more women had tried and discontinued pills, and others were hesitant to begin usage because of actual or presumed side effects. In the 1977 TCS Survey, 49 percent of the women took the pill, 13 percent had an IUD, 6 percent relied on condoms, 4 percent practiced rhythm, and 23 percent used withdrawal (Hungary/CSO, 1979). Surgical sterilization was not included among the contraceptive methods in the surveys.

The 1977 Fertility Survey indicated that among married women use of the pill or the IUD varied little by residence, education, socioeconomic status, or employment. Women with between nine and 12 years of education and those on child care leave had the highest rate of reliance on the pill and IUD while the lowest rates were found among women with only a primary education and those working in agriculture or not employed outside the home. Access to modern contraceptives, simplified after the 1973 announcement of forthcoming abortion restrictions, appears to have had an almost equal impact on women in all walks of life (Wulf, 1980; World Fertility Survey, 1980). According to the 1986 survey, 73 percent of married women practiced contraception; of these, 80 percent relied on the pill or IUD. Some 25 percent of these women had experienced at least one abortion.

The 1986 cross-sectional survey of 15–39-year-old married women also showed that among those who had completed at least eight years of education, 73–75 percent practiced contraception compared to 48 percent among those attaining a lower educational level. Smaller differences in contraceptive practice were noted between women working at a skilled level (74 percent) and those holding manual labor jobs (66 percent). Residents of Budapest had a lower level of contraceptive practice (66 percent) than women living in towns and villages (74 percent), which may reflect the lower fertility and higher abortion rates in the capital city. Whereas the use of IUDs was lower in the rural agricultural population, use of oral contraceptives was higher among urban married women holding jobs requiring more than manual skills (Hungary/CSO, 1987).

In the 1993 Family and Fertility Survey conducted in cooperation with the Economic Commission for Europe, 95 percent of sexually active fecund but nonpregnant married women practiced contraception. Over 38 percent of all the women interviewed relied on the pill, 17 percent on the IUD, 8 percent on condoms, 6 percent on withdrawal, 3 percent on periodic abstinence or other methods, and 5 percent used nothing. Some 7 percent of the respondents described themselves as infertile, 6 percent were pregnant, 2 percent were not sexually active, and 8 percent declined to respond (Kamaras and Kowalska, 1995).

PLANNED AND ACTUAL FERTILITY

To complement the retrospective cross-sectional surveys described in the section on Contraceptive Practice, the CSO conducted two prospective longitudinal surveys of fertility planning and attitudes, based on the married cohorts of 1966 and 1974.

In 1966, separate interviews with each spouse of a random national sample of 5.2 percent of all couples marrying that year (4,822) were conducted by marriage registrars on the day these couples registered their intention to marry. In 1969, in 1972, and in 1975, district nurses reinterviewed all the women from the original sample who could be traced and who were available and willing to respond. The follow-up rates were 96 percent of the 1966 respondents in 1969 and 92 percent in 1972. Because the aim of the survey was to describe changes in fertility behavior and attitudes occurring with increasing marital duration, questions on family planning were asked and/or processed only for women still living with their husbands, born in 1927 or later, and for whom questionnaires were fully completed on each occasion. In 1969 this constituted 82 percent of the original 1966 sample of women, declining to approximately 75 percent in 1972.

The reliability of surveys on planned and desired family size can be examined by comparing perceived and actual changes expressed over time. Findings from the 1966 marriage cohort showed that family size desires and plans changed with increasing marital duration, casting some doubt on the reliability of responses to retrospective questions (Hungary/CSO, 1975b, 1976). For example, when asked in the 1969 reinterview to recall how many children they had planned at marriage, only 70 percent of the respondents gave the same figure reported three years previously. In two-thirds of the instances of differences, women "remembered" wanting more and in one-third they "remembered" wanting fewer children.

In conventional one time retrospective surveys, where the time span between the interview and the period for which information is recalled is even longer, there may be a still greater likelihood of forgetfulness and "after the fact" rationalization. For example, in the 1972 reinterview, after six years of marriage, only 64 percent of respondents still desired to have the same number of children envisaged at marriage. However, those couples who had expressed plans for having two children at marriage were least likely to have changed their minds (regardless of recall accuracy) while the majority of those who had originally planned either three or just one child claimed in 1972 that they had planned on two children in 1966.

The two-child convergence is further confirmed by data from the 1966 TCS and the 1977 TCS on the family size that is considered "ideal" independently of the couple's personal plans or desires. In these eight years, the proportion of respondents considering two children as ideal rose from 60 percent to 68 percent, while the number preferring the three-child family dropped from 30 percent to 22 percent, and the number viewing one child as ideal doubled from 3 percent to 6 percent (Klinger and Szabady, 1978). In a 1974 public opinion survey it was noted that the proportion of those considering a number of children fewer than what they already have to be "ideal" increased while the proportion of those deeming a larger number of children "ideal" decreased (Pongracz and Molnar, 1976). Moreover, when answering the question on the ideal number of children, respondents tended to project what they believed to be "ideal" for others rather than for themselves (Molnar, 1976). Respondents expressed the view that the ideal number of children could be reached if the housing problems of newly married couples could be alleviated and if there were resources to place more children in nurseries and kindergartens (Pongracz and Molnar, 1975).

SOCIAL DIFFERENCES IN FAMILY PLANNING AND FERTILITY

The longitudinal marriage survey conducted between 1974 and 1990 afforded a special opportunity to consider the relationship between fertility and abortion and for the description of social differences on the basis of a complete fertility history. Results indicated that in the course of her life a woman had on the average 2.90 pregnancies of which 2.00 ended with live births, 0.55 with pregnancy interruption, and 0.45 with another fetal loss. These findings show a more favorable ratio between births and abortions than the reported data, which probably reflects some denial in the interviews. The average fetal loss seems high but probably includes induced abortions that were reported by the respondents as spontaneous. Nevertheless, the social and regional differences deserve attention.

The number of children was the lowest and the number of abortions and other fetal loss highest in Budapest. Among rural women the number of children was highest, but the number of abortions was only half as high and the incidence of other fetal loss was lower than in the capital or in provincial towns. The total number of pregnancies per female was 3.2 in Budapest compared to 2.86 in rural areas. Contraceptive practice was apparently more effective in the rural population than among the urban populace.

Religiosity

Questions concerning religion or religiosity were not included in the various CSO sample surveys on family, family planning, and fertility, . The longitudinal survey of 1990 marked the first opportunity to obtain information on this topic. Women who described themselves as religious and believers had on average more children compared to those who were not religious or believers.

In Budapest religiosity or its lack showed the smallest differences in the average number of children compared to rural families. The mother's educational level was one of the main variables explaining the differences in the number of children. With an increase in educational level the average number of children decreased. Religiosity did not eliminate these differences; it only moderated them. Among women of an indicated educational level, the religious ones had usually more children than the nonreligious women.

THE WOMAN QUESTION

Between 1905 and 1914 books and journals on women's issues began to be published in Hungary (Fodor, 1994). Hungarian women were active in the European suffrage movement, fought for better educational opportunities, and were part of the urban intellectual and political scene (Bollobas, 1993). After World War II the Democratic Association of Hungarian Women (DAHW), founded in 1945 as a multiparty umbrella organization, maintained links with smaller groups. When the Communist Party gained power in 1949, all civil associations were banned, including religious and secular women's organizations. The DAHW was coopted by the Communist Party in 1955 when it became part of the National Patriotic Front, an organization supposedly representing the public interest (Fodor, 1994).

After 1956 the National Council of Hungarian Women was established. The council developed a circle of experts who did extensive research on issues concerning women and of interest to the Communist Party. The head of the council was a member of the Party's Central Committee, but the organization itself had limited influence within the party. With the official women's movement and the women's sections of the trade unions in the hands of party officials, grassroots organizations were suppressed. Women's civil organizations that flourished before World War II and even afterwards disappeared and were not replaced until the beginning of the 1990s (Fodor, 1994).

The Communist Party Central Committee's key document on women, prepared in 1970, devoted 13 pages to issues concerning "wage-earner" women, emphasizing the need for full equality in the workplace. In three pages at the end of the document it touched lightly on women in the family (Hungary/*Documents*, 1980; Fodor, 1994). During the 40 years of communist rule, the party's emancipation effort was largely restricted to the workplace. It involved other areas of life only to the extent that it facilitated or hindered women's participation in the labor force. While the daily need to "reproduce the labor power" of wage earners and the rearing of the next generation could not be disregarded, the word "husband" was conspicuous by its absence in the 1970 document (Fodor, 1994).

By 1989 Hungarian women's employment had reached 80 percent of women aged 15 to 54 years (excluding students). Among women under age 40, employment was over 90 percent, including those on paid maternity leave. There was virtually full employment for mothers with young children; about 90 percent of 3–6-year-old children were in preschool programs (Adamik, 1991, 1993). However, women's income averaged 25 percent less than that of men in all employment categories (Toth, 1993).

The first multiparty election resulted in a significant decline in the representation of women in Parliament from about 25 to 30 percent during the communist regime to less than 8 percent of all MPs in 1990 (Adamik, 1997). Although unemployment has been increasing, fewer women are unemployed than men (Fodor, 1997). In 1992 the unemployment rate was 7.7 percent for women and 10.5 percent for men. By 1994, the proportion was 10.9 percent of women and 15.6 percent for men. Unskilled women were underrepresented among the unemployed population. Closings and privatization were more prevalent in heavy industries where women were less likely to be employed (Frey, 1992). There was a higher unemployment rate among young, less educated, and often Gypsy women (Adamik, 1993).

The Hungarian Women's Association (HWA) arose in June 1989 from the ruins of the Communist Party's women's organization. It represents Hungary in the United Nations Commission on the Status of Women and was the first CEE women's organization to be admitted to the International Council of Women. Its primary focus is on the collection and dissemination of information. A Government Commission against Sex Discrimination was established in July 1997 to guarantee equal opportunities for women. The HWA contends that ideological and conservative trends, the "fast and ill-advised elimination" of the family allowance system, and the criticisms of families with two earners have created an environment that renders "the situation of women today more uncertain than a decade ago" (Asbot-Thorma, 1998). Lack of grassroots organization and mass support is perhaps the greatest problem facing women's organizations. The hypocrisy of the official women's movement in the communist era still casts a long shadow over women's organizational activities.

SEX AND SOCIETY

While "sexual equality" was an everyday phrase in communist ideology, children were socialized into conservative sex roles (Toth, 1993). This was apparent in the images of men and women in school texts. Men were the breadwinners and women were perceived as only secondary earners. Boys received vocational training and girls were expected to help with household chores.

Sex and Marriage

Sex and marriage gradually received increasing public attention (e.g., Volgyes and Volgyes, 1977). One of the early books in the field, *In the Interest of Women* (Hirschler, 1958, cited in David and McIntyre, 1981), sold over one million copies in several editions in Hungary and in Czechoslovakia. In response to the need for more and better information, a Budapest telephone service was established to answer specific questions and provide advice on sexual problems (Dornberg, 1974).

As previously noted, the October 1973 Decision of the Council of Ministers and associated regulations and instructions directed that all Hungarian citizens under age 35 must participate in marriage counseling before a marriage license can be issued. Counseling is provided by local marriage consultation units and also at obstetrical and gynecological clinics. A physician advises on family planning, discusses contraceptive methods, and provides supplies. The registrar is instructed to refuse to issue a marriage license if the couple cannot produce written certification proving that they participated in the obligatory consultations.

The Hungarian Scientific Society for Family and Women's Welfare was organized in March 1976 within the framework of the Federation of Hungarian Medical Societies. It became a member society of the International Planned Parenthood Federation (IPPF) that same year (Jozan, 1977). The society facilitates research on family welfare, disseminates information, and seeks to develop public opinion conducive to a harmonious family life, the protection of women and children, and healthy reproduction.

Sexuality Education

Although the introduction of family life education into Hungarian schools was launched as part of the October 1973 Decision of the Council of Ministers, sexuality education was never formally included in mandatory national school curriculae. Provision of such classes was left to the discretion of headmasters, some of whom rigidly refused to adopt it (Forrai, 1996). In 1991 the Hungarian Sexual Education Foundation and the Semmelweis University of Medicine developed a sexuality education and AIDS prevention project. After preparing educational materials (until then largely nonexistent in Hungary), training courses were organized for physicians, nurses, and teachers. After research demonstrated the low level of knowledge about sexuality and contraception among adolescents, the Ministry of Culture and Education promulgated a new national curriculum for 6–18 year olds that includes methods and materials based on the Semmelweis program (Forrai, 1996).

Adolescents

A 1993 survey found that half of Hungarian women experienced their sexual debut at 18.7 years or earlier. The median age ranged from 18.5 years in the youngest cohort (20–24) to 19.2 in the oldest (41–44). The median level for men varied from 18.0 to 18.5 years (Kamaras and Kowalska, 1995). In a later survey, 40 percent of 14–19 year olds indicated they were sexually active (Forrai, 1996).

In recent years a change occurred in female teenagers' attitude to pregnancy. Between 1980 and 1990 the number of pregnancies decreased to a small extent. In 1980 there were 96.9 births and abortions per thousand 15–19-year-old females; in 1990 it was 71.6. A greater change could be observed, however, in the decision on whether or not to carry a pregnancy to term. Whereas in 1980 some 73 percent of the pregnancies ended with live births and 27 percent with induced abortions, in 1990 only 56 percent ended with childbirth and 44 percent with abortion. In other words, parallel with the decrease in children born to teenagers, the incidence of teen abortions increased by about 60 percent. The very slowly decreasing incidence of adolescent pregnancy reflects how much work still needs to be done. It may also be related to the 1987 increase in the age requirement for marriage from 14 to 18 years for girls and 16 to 18 for boys. A review of births plus abortions ratios among teenagers under age 18 in 29 developed countries showed that Hungarian teens were second only to black American women under age 18 (Westoff et al., 1983).

FUTURE PERSPECTIVES

The May 1998 parliamentary elections brought to power the Hungarian Civic Party, a mainstream right-wing organization spawned by the 1989 student dissident movement. It is governing in coalition with the Smallholders, a more hard-line right-wing party (Spolar, 1998). The new government plans to strengthen family and child allowances. A new Ministry of Social and Family Affairs was created, headed by the former lay secretary of the Conference of Cathodic Bishops. Should access to legal abortion be restricted, Hungarian society will likely find a way to cope (as it always has), while practicing ever more effective contraception.

chapter 9

Poland

Anna Titkow

EDITOR'S INTRODUCTION

Over the last century the experience of Poland has diverged significantly from that of other Central and Eastern European (CEE) countries. For more than 120 years, up to 1918, Poland did not appear as a state on the map of Europe. After World War I Poland regained its independence as a republic. In 1939 a nonaggression pact between Nazi Germany and the Soviet Union cleared the way for both countries to invade and occupy Poland. When Germany attacked Russia in 1941, all of Poland came under German occupation.

After the end of World War II a communist system assumed governmental control in 1947. The harshness of the Stalinist regime and total state control of industry and agriculture sparked a workers' revolt in 1956 that led to some reforms. The teaching of sociology as a separate discipline had been abandoned in 1950 but was reestablished in academic institutions and at the Polish Academy of Sciences during 1956–57. The "renaissance" of Polish demography began in 1956 when the government removed the "stamp of secrecy" from population statistics (Rosset, 1973; Kosinski, 1977a).

Gradually mounting labor unrest in the 1970s led to the strike of 1980, which established the independent Solidarity Union. It formed the nucleus of resistance, supported by the Roman Catholic Church in what many observers deemed a marriage of convenience (Michnik, 1994). After a period of martial law and continued loss of communist authority, the 1989 multiparty elections, the first free elections in CEE countries since 1939, overwhelmingly endorsed Solidarity Union candidates. By 1991, however, so many parties had emerged that the Solidarity Union won only 6 percent of seats in the lower house of Parliament (Sejm). In the September 1993 elections, Poland's ex-communists, now called Social Democrats, won the largest number of Parliament seats and, together with the Polish Peasants Party and several other parties, formed a "radical left" coalition. Most observers blamed the rejection of the previous government not on a yearning for past communist certainties but on discontent with economic "shock therapy," the widening gap between rich and poor, and on popular alarm at the growth of extreme nationalism and the imperious behavior of the Catholic Church (*The Economist*, 1994a).

Poland is unique in the way the abortion situation became dramatically politicized and changed during a period of unusual social transition (Standish, 1998). The Solidarity-led government had bought the support of the Catholic parties by overturning 35 years of access to abortion on request and pushing a restrictive abortion law through Parliament, making provisions for allowing Catholic instruction on school premises, and sharply restricting women's rights to control their own fertility (David and Titkow, 1994).

On 19 November 1995 Aleksander Kwasniewski, a former communist, was elected president of Poland, defeating the incumbent, Lech Walesa. Six years after the end of communism, the CEE country that had had the strongest anticommunist opposition and the weakest Communist Party, had a government led by former communists. In part at least, it was a rejection of the church's perceived interference in secular affairs (Ash, 1996a; Husarska, 1995). Liberalized abortion legislation was signed by President Kwasniewski on 20 November 1996, repealing the 1993 prohibitions that, along with Ireland's, had been the strictest in Europe.

In May 1997 the Constitutional Tribunal ruled the new abortion law unconstitutional but left it in effect for a six-month grace period. In the September 1997 parliamentary elections, Solidarity Electoral Action, a coalition of about 30 parties and conservative trade unions, emerged victorious and formed a new government. The old 1993 law was reimposed in December 1997 (Morris, 1998).

It is within this historical context that the chapter examines Polish demographic trends, abortion realities and politics, contraceptive availability and practice, the woman question, perceptions of sex and society, and future prospects. There is particular reference to the influence of the communist system and the Catholic Church on abortion legislation, contraceptive practice, and sexuality education, as well as the effects of the antiabortion legislation.

The chapter was translated by Anna Kinecka with funding from the MacArthur Foundation. Jerzy Holzer, Andrzej Kulczycki, Wanda Nowicka, and Jon O'Brien made valuable contributions. With apologies to our Polish colleagues, diacritics were omitted to avoid editorial and printing errors.—H.P.D.

DEMOGRAPHIC TRENDS

Population Policies

With the relatively slow spread of modern industry and its large rural population, Poland's fertility level remained among the highest in Europe for the first three decades of this century, followed by a substantial decline in the 1930s. Several authors (e.g., Latuch, 1973; Ziolkowski, 1974; Dziewonski, 1977) suggest that the period from 1945 to the early 1950s marks the first of several distinct population policy eras. It was pronatalist, stemming from a desire to compensate for war losses and the Marxist ideological predisposition to view population growth as a sign of social vitality. As shown in Table 9.1, the crude birth rate soared to 30.7 in 1950 and to 31.0 in 1951, levels never again reached in subsequent years.

Following the onset of difficult economic circumstances in the mid-1950s, rapid population growth became a "scapegoat to be blamed for all of the inadequacies of national development" (Ziolkowski, 1974). The second population policy period began with a campaign encouraging couples to have smaller families. Fertility regulation was advocated along with the model of the emancipated working woman freed from the burden of a large progeny. Advertisements for condoms, the first domestically produced contraceptive, appeared in newspapers. Folksongs praised contraception (Okolski, 1983). Birth rates declined from 29.1 per 1,000 population in 1955 to 16.3 in 1967, abetted in part by the availability of legal abortion on request. The total fertility rate dropped from 3.60 per woman in 1955 to 2.32 in 1967.

The third population policy era, beginning in the late 1960s, was marked by a return to a "moderately" pronatalist orientation (Ziolkowski, 1974). A series of pronatalist measures was initiated in the early and mid-1970s, beginning with an improvement of family allowances at the end of 1970 and further revision and extension of maternity leave in January 1972. Cash birth payments were increased in 1978. Child care subsidies for persons on family leave were extended to 24 months in 1982. Meanwhile the Catholic Church actively promoted a pronatalist ideology. As indicated in Table 9.1, the crude birth rate moved to higher levels during the 1970s, reaching 19.7 in 1983 before gradually declining again to 18.2 in 1985. Meanwhile, the total fertility rate continued its decline, falling to 2.28 per woman by 1980 before rising to 2.42 by 1983 and then decreasing again to 2.08 in 1989.

The fourth and most recent population policy era began in 1989-90 during the postcommunist period when Poland had a population of 38 million. The birth rate had fallen to 14.9 per 1,000 population in 1989 and reached an all time low of 10.7 in 1997. The total number of births dropped from a post-1959 high of 723,600 in 1983 to a post-1945 low of 412,600 births in 1997 when the total fertility rate fell to 1.51 per woman (David and McIntyre, 1981; Okolski, 1988; Poland/GPC, 1984, 1988, 1992, 1993; Skwiecinska, 1993; Poland/CSO, 1997; COE, 1998).

ABORTION—FACTS AND POLITICS

1818–1989

Poland's Penal Code of 1818 reflected a relatively humanitarian approach to abortion. According to Article 130, abortion was considered a crime, punishable

Table 9.1
Birth, Abortion, and Fertility Rates and Ratios: Poland, 1950–97

Year	Live Births	Legally Induced Abortions	Birth Rate	Legal Abortion Rate[e]	Legal Abortion Ratio[f]	Total Fertility Rate
1950	763,100	na	30.7	na	na	3.70
1955	793,800	1,400	29.1	0.2	2	3.60
1956[a]	779,800	18,900	28.1	3.0	24	3.50
1957	782,300	36,400	27.6	5.8	46	3.48
1958	755,500	44,200	26.3	7.1	58	3.36
1959[b]	722,900	79,000	24.7	12.7	109	3.22
1960	669,500	150,400	22.6	24.2	225	2.98
1965	547,400	168,600	17.4	24.3	308	2.52
1967	520,400	155,000	16.3	20.4	281	2.32
1970	547,800	148,200	16.8	19.9	271	2.20
1975	646,400	138,600	19.0	17.6	214	2.27
1980[c]	695,800	138,000	19.5	16.7	192	2.28
1981	681,700	132,900	19.0	16.2	195	2.24
1982	705,400	139,000	19.5	17.6	197	2.34

1983	723,600	131,000	19.7	16.9	181	2.42
1984	701,700	128,400	19.0	16.5	183	2.37
1985	680,100	135,600	18.2	16.7	199	2.33
1986	637,200	129,700	17.0	15.9	204	2.22
1987	607,800	123,500	16.1	14.9	203	2.15
1988	589,900	108,400	15.6	12.7	178	2.13
1989[d]	564,400	82,100	14.9	9.5	146	2.08
1990	547,700	59,400	14.3	7.0	108	2.04
1991	547,700	30,900	14.3	3.6	56	2.05
1992	515,200	11,600	13.5	1.3	22	1.93
1993	494,300	1,200	12.8	0.1	2	1.85
1994	481,300	800	12.5	0.09	2	1.80
1995	433,100	600	11.2	0.06	1	1.61
1996	428,200	500	11.1	0.06	1	1.58
1997	412,600	3,200	10.7	0.04	7	1.51

[a]Abortion law liberalized (April 1956). [b]More permissive abortion regulations issued (December 1959). [c]Administrative instuctions on abortion slightly restricted (December 1980). [d]Restrictive instructions on abortion implemented (March 1989). [e]Per 1,000 women age 15–44. [f]Per 1,000 live births.

Sources: David and McIntyre (1981), Henshaw and Morrow (1990), U.N. (1995a), Poland/CSO (1997), and COE (1998).

with up to 10 years' hard labor, if performed without the woman's consent. In other cases, including self-induced abortion (Article 323), it was classified as a misdemeanor, punishable with up to three years' house detention (Plakwicz and Zielinska, 1994). The legal prohibitions had strong social support until 1929 when the physician Tadeusz Boy-Zelenski launched a press campaign calling for the liberalization of abortion. His stand was endorsed by a circle of open-minded physicians, women's groups, and intellectuals.

In 1932 legal exceptions were introduced. Abortion was permitted if the pregnancy resulted from a criminal act or placed the woman's health or life at risk. The regulations required the presentation of a certificate on the need for abortion signed by two doctors, apart from the physician performing the termination. As this condition was difficult to fulfill, especially in rural areas, the law was frequently ignored. The penalty for illegal procedures was set at three years' imprisonment for a woman and five years for a practitioner, provided the abortion was performed with the woman's consent (Plakwicz and Zielinska, 1994).

Between 1945 and 1956 abortion remained officially banned, methods of family planning were not promoted, and sexuality education was forbidden. Following the lead of the Soviet Union, a law legalizing abortion on social as well as health grounds came into force on 27 April 1956 (Poland, 1958). The 1956 law was further liberalized by the Instruction of December 1959, granting the pregnant woman the right to abortion on social grounds alone (Poland, 1960). In the absence of a clear definition of social reasons, *de facto* induced abortion became available upon a woman's request. It could be performed in maternity hospitals, private physicians' offices, and in medical cooperatives by obstetricians, surgeons, and gynecologists. In state hospitals it was free of charge, covered by social insurance. Patients were entitled to sick leave. No gestational age was specified as a legal time limit. Whether the procedure was actually performed became the physician's responsibility (Holzer, 1991).

Abortion Registration

When speaking of abortion in Poland it is important to distinguish between the *de jure* situation and the *de facto* realities. The limited access to modern contraceptives and the lack of information and sexuality education were largely responsible for the dependence on abortion as the major method of fertility regulation. According to specialists, abortion data published by the Central Statistical Office (CSO)and the Government Population Commission are seriously deficient. Only abortions performed in hospitals and clinics had to be registered. Pregnancies terminated for a fee in doctors' private consulting rooms were not always reported. "Other abortions," including spontaneous and illegally provoked abortions, were supposed to be reported on admission to the hospital. It is likely that there was considerable underreporting, reflecting in part a continuous and significant shift from public hospitals to private facilities (Okolski, 1988). For instance, before 1956 it was estimated that about half-a-million clandestine terminations were performed annually. However, only 121,800 total abortions were officially registered in the first year after legalization, including 36,400 legal procedures (Table 9.1) and 85,400 "other abortions" (not shown).

Subsequent estimates by Polish demographers are discordant. Based on the 1977 CSO survey of women's fertility, Smolinski (1980) suggested that the annual total number of abortions was between 400,000 and 450,000, or two-thirds of the annual number of live births. Okolski estimated that 620,000 pregnancy terminations were performed annually, meaning that one in two pregnancies was interrupted and that annually 75 per 1,000 women of reproductive age terminated their pregnancy. Other demographers estimated the annual number of terminations at 400,000–600,000 (Lech, 1974) and 500,000 (Danek, 1988). It seems reasonable to assume that in 1985 there were between 400,000 and 620,000 abortions compared to the official figure of 136,600 abortions reported by the CSO (1993). Since there are no reliable data on total registered abortions (Holzer, 1996), Table 9.1 cites only officially reported legal abortions, rates, and ratios (with acknowledged limitations).

Fragmentary research in Warsaw hospitals conducted in the late 1960s (Tuchol-ska-Zaluska, 1975) and during 1978–80 (Sternadel, 1982), together with available demographic data, suggest that *de facto* abortion was the principal means of fertility regulation, regardless of society's expressed attitudes.

Attitudes

Attitudes to abortion evolved over time. For the last 10 years or so the following circumstances were most frequently cited in surveys as sufficient for obtaining an abortion: danger to the woman's life and health, serious and irreversible damage to the fetus, and pregnancy resulting from a criminal act. The importance of these circumstances tended to fluctuate. The level of acceptability of abortion on eugenic grounds was most stable, always at 85 percent. Justifiability of a termination for health reasons declined from 98 percent in 1979 to 85 percent in 1993. There was a somewhat smaller decrease in the popularity of the argument for terminating a pregnancy resulting from a criminal act from 85 percent in 1979 to 78 percent in 1993 (Titkow, 1984; Public Opinion Research Center, 1993).

Circumstances defined as "difficult living conditions of the family" (cited in the 1956 law) continued in the 1979–93 period to be considered as sufficient grounds for abortion by 50 percent of respondents. In 1993 this percentage rose to 65 percent. Justification based only on already having a large number of children gradually became less acceptable in the 1979–1993 period, declining from 46 percent to 25 percent.

The evolution of attitudes to abortion reflects two interesting developments. First, the proportion of individuals in favor of legalized abortion on request without restrictions rose from 26 percent in 1979 to 38 percent in 1993. Second, the popularity of the belief that pregnancy termination is unacceptable for any reason declined from 24 percent in 1979 to 4.1 percent in 1989. The latter view is linked directly to the restriction of women's reproductive rights. Despite the continuing bitter antiabortion campaign there has been a slight increase in the number of individuals agreeing that a student or working woman has a right to terminate a pregnancy if it complicates her life (from 27 percent in 1979 to 32 percent in 1993). In addition, in 1989, 38 percent of the interviewed 18–49-year-old women believed that a pregnancy could be terminated when the woman was very much opposed to having the baby (Titkow, 1984; Duch, 1994; *Studio Gazeta*, 1992; Demoskop, 1992; Cichomski and Sawinski, 1993). This was at a time when only 7.5 percent of 18–49-year-old women deemed abortion socially acceptable (Duch, 1991).

In 1960, the 1956 abortion law was supported by 60 percent of the adult Polish population (Malanowski, 1960). In 1990, at a time of exceptional intensification of the antiabortion debate, 59 percent of the population objected to the imposition of a legal ban on terminating pregnancy; 53 percent of churchgoers shared this opinion (Public Opinion Research Center, 1990).

The Abortion Context

Few issues are as politically delicate in Poland as abortion. The debate on abortion, revived in 1946, was silenced in 1950 at the height of Stalinism, and then continued again after the regulations on pregnancy termination were liberalized in the Soviet Union in 1955. Coverage of the controversy highlighted only arguments in favor of legalizing abortion (on health and social grounds) to end unsafe clandestine terminations dangerous to women's well-being. After the approval of the 1956 abortion law, immediately condemned by the Roman Catholic Church as sanctioning infanticide, discussions continued in 1958–59 on the subject of possible modifications. Arguments of the opposition were suppressed by the media although they drew attention, apart from questions of ethics, to such practical matters as the persistently high—even after legalization—incidence of postabortion complications. Supporters of abortion called for regulations to simplify formalities required to obtain legal procedures.

There were no changes in abortion policy from 1960 to 1989 despite fluctuations in attitudes and the public atmosphere surrounding abortion. In the 1960s families with many children were ridiculed while at the same time there was gradually mounting pressure on the government to restrict women's freedom in resorting to abortion. In 1970 the Episcopate sent a petition to the government calling for the annulment of the 1956 law. Physicians, demographers, and sociologists expressed concern that legalization of abortion had resulted in its being used irresponsibly as the only form of fertility regulation with mounting health risks (contraceptives were in short supply). When the government proposed more stringent criteria for abortion in 1973 the move was rejected by the Parliament Health Committee, which argued, contrary to facts, that the law was functioning properly. Choosing the "lesser evil" would help to avoid an increase in unsafe illegal terminations certain to occur if the abortion law were made more restrictive.

1973 brought a breakthrough in the controversy. The campaign of ridicule directed at large families was suspended and a new pronatalist policy was introduced. In 1974 the Government Population Commission was appointed, followed in 1976 by the Council for Family Matters. Opposition to the abortion law continued to mount and became increasingly public.

The government, political authorities, and specialized institutions such as the Family Planning Association (TPR/TRR) differed from the Roman Catholic Church in their involvement and actions. Misjudging the seriousness of the growing antiabortion campaign, the Ministry of Health in 1980 and 1981 proposed to toughen regulations on the criteria for pregnancy termination without restricting access to abortion. While the Council for Family Matters ignored the issue, it was discussed in 1983 in the annual report of the Government Population Commission. The Family Planning Association, unable to promote the use of modern contraceptives effectively, limited itself to issuing pleas for prudence and moderation (Okolski, 1988).

The proposed restriction of the abortion law was understood unequivocally as the prohibition of abortion on social grounds (in keeping with the demands of the Roman Catholic Church). It was overlooked that reliance on abortion could be lessened by implementing a realistic sexuality education program and by making modern contraceptives more readily available.

The strategy adopted by the Roman Catholic Church was pragmatic and effective. It launched a long-term, consistent campaign intended to inculcate the transcendent value of human life from the time of conception. The program was organized on four levels: religious classes for children and for teenagers, premarital training for young couples, and courses for newlyweds. The subject of abortion, sexuality, and family life was introduced in secondary school from the narrow perspective of the hierarchy. Particularly important for the church were the premarital courses on theology of marriage and pregnancy planning. Several thousand lecturers were trained for this purpose. Care was also taken to influence the medical community through an active ministry among medical students. In 1975–76, under the patronage of Cardinal Wojtyla (now Pope John Paul II), national scientific seminars of physicians and theologians were organized on the subject of pregnancy termination (Okolski, 1988).

The aim of the campaign was more than parish level promotion for the protection of the human fetus. Seminars were organized on natural methods of family planning. The Movement for the Protection of Conceived Life was launched. Material assistance and counseling were offered to pregnant women and young mothers. Catholic youth organized in the Oasis movement were mobilized. The Society for Responsible Parenthood was established (later suppressed under the martial law). In 1980–81 calls against the 1956 abortion law began to appear in the media (Okolski, 1988; Kozakiewicz, 1997).

1989–98

Imposition of martial law and its repercussions intensified the influence of the Roman Catholic Church on the public and on the opposition leaders. It was time for a "now or never" offensive. In answer to the plea for respecting conceived life a draft of a law recriminalizing and penalizing abortion was submitted to Parliament as early as May 1989 (Nurt, 1989).

After 1989 it became increasingly difficult to gain access to legal abortion, especially in state hospitals (Standish, 1998). During this time any woman who wanted an abortion had to present a request in writing, consult two physicians and a state-approved psychologist, who would decide to grant or deny the application. To avoid this process, more women sought private, unregulated, and expensive services (Kulczycki, 1995). The number of reported "other abortions" increased while the number of births declined. Meanwhile, the moral controversy over abortion developed into a political dispute on the nature of the state and the mutual relationship between the state and the Roman Catholic Church. Anyone opposing the recriminalization of abortion was suspected of harboring procommunist sentiments and branded a potential murderer.

Lacking an intelligible political program, the substitute theme of abortion helped to deflect public attention from such vital issues as salaries, employment, poverty, the constitution, and the social security system. Manipulation was semantic. Neutral terms were replaced by synonyms. The fetus became: "the conceived child," "the unborn child," "the baby," "the child in the mother's womb" (Matuchniak-Krasuska, 1995). As early as May 1990 (two-and-a-half years prior to the enactment of the antiabortion law), the Minister of Health decided that physicians were no longer obliged to perform abortions in state hospitals. A tacit instruction suggested that deaths of women resulting from abortions performed outside hospitals be registered under other headings. In May 1992 the Physicians' Guild enacted the Medical Ethics Code prohibiting termination of pregnancy except in cases of rape and danger to the life of the mother (*British Medical Journal*, 1992). Amendments to this code in December 1993 permitted the physician to proceed on matters of abortion in keeping with the principles of medical science and his/her own discretion but did not weaken the restrictive posture of a part of the profession. This is best illustrated by the case of a raped woman who was three times refused admission to a hospital even though she presented a prosecutor's certificate on the need for an abortion (Skipietrow, 1993b).

Two draft laws on abortion were submitted to Parliament in 1992. The first draft, proposed by the National Christian Union, was entitled "On the Legal Protection of the Conceived Child" and referred to the moral authority of the Roman Catholic Church. Anyone (apart from the physician saving the mother's life) implicated in the death of the conceived child was committing a crime and was to be liable to punishment by two years' imprisonment. The draft was signed by 171 parliament members, about 45 percent of the total membership.

The second draft, "On the Right to Parenthood," was submitted by the Parliamentary Group of Women. It provided for access to abortion if the pregnancy or the delivery endangered the life or health of the woman (before 20 weeks), in case of serious and irreversible damage to the fetus with an incurable disease endangering its life (before 24 weeks), if the pregnancy resulted from a criminal act (before

20 weeks), or if the living circumstances of the woman were very difficult (before 12 weeks). The draft also regulated questions of sexuality education, accessibility to contraceptives, and prenatal examination. It had the signatures of 91 members of parliament.

Nearly 200 members of parliament, almost 50 percent of all members, signed neither of the two drafts. After a debate Parliament adopted a more liberal (as compared to the National Christian proposal) version of the law, permitting abortion only if the life or health of the pregnant woman was at risk, if the pregnancy resulted from rape or incest, or if there was irreversible damage to the fetus. Termination of pregnancy for other reasons was to be punished by up to two years' imprisonment for the service provider but no punishment for the woman. Abortions could be performed only in state hospitals, not in private physicians' offices. In this manner, socioeconomic circumstances and personal reasons, motivating 97 percent of pregnancy terminations performed before the enactment of the antiabortion law, were "outlawed."

Implemented on 17 March 1993, the law required the national and provincial governments to ensure access to all contraceptives and the Ministry of Education to implement a course on sexual life, family values, and conceived life as well as on methods and means of responsible procreation. The ministry's failure to implement these instructions is discussed in the section on Sex and Society.

The *de facto* realities generated by the antiabortion law are perhaps best described as the "law looking the other way" or the "gray zone." The lack of reliable statistical data was matched by untold stories of personal tragedy and anguish. Women who had the right to legal abortion were often refused services even when they had all the necessary documents. Women in a difficult personal situation sought pregnancy termination at any cost, regardless of whether they broke the law, often risking their health and life (David and Titkow, 1994; Kulczycki, 1995; Nowicka, 1995, 1996a).

The enactment of the antiabortion law was not accompanied by well prepared enforcement regulations. For instance, it remained unclear whether single mothers were entitled to social welfare (Nowakowska, 1993). Also, the economic irresponsibility of the decision makers became quickly apparent. Under the new law all pregnant women with incomes below the lowest retirement pension were granted the right to an allowance starting from the fourth month of pregnancy until the baby was six months old, equivalent to 28 percent of average salary. Only an estimated 10,000 women were expected to apply, but 80,000 did before the end of 1993. For 47,000 women there was not enough money. The Economic Committee of the Council of Ministers then proposed that the allowance be reduced to 14 percent of the average salary, and be granted only from the eighth month of pregnancy until the baby was two months old (Skipietrow, 1994; Tomala 1994).

Reported Implementation

On 15 April 1994 the ministries of Health, National Education, Justice, and Labor and Social Policy presented to Parliament their initial reports on the country's new restrictive abortion law. Officially reported legal abortions performed in hospitals declined from about 30,900 in 1991 and 11,600 in 1992 to 777 in 1993. Of the 777 procedures, 735 were approved on grounds of threat to the women's life or health and nine were court sanctioned on grounds of rape. Reasons for the remaining 33 procedures were not made public (Rich, 1994; David and Titkow, 1994). (As shown in Table 9.1, the CSO reported 1,200 abortions for 1993). By 1996, officially reported legal abortions had declined to 500, rising to 3,200 in 1997 (COE, 1998).

The April 1994 ministerial reports noted that recorded miscarriages, which had decreased from more than 72,000 annually in the early 1980s to 51,800 in 1992, increased to 53,000 in 1993. The number of newborn babies abandoned by their mothers in Warsaw rose from 22 in 1992 to 49 in 1993 (Poland/MOH, 1993; Nowicka, 1996a). Infanticides recorded a new high of 59 in 1992, declining to 45 in 1995 (Federation for Women and Family Planning, 1996). Reported mortality from abortion increased from 3.3 percent of total maternal mortality in 1989 to 5.0 percent in 1991. For 1994, the World Health Organization reported 30.2 percent of total maternal mortality due to presumably illegal procedures (WHO, 1998).

Pregnant women seemed to evade health care statistics. Of 31 women filing rape reports with the State Prosecutor's Office, 30 were granted court permission for a hospital abortion but only 9 applied and obtained the procedure. It is not known how the remaining 21 women coped (Konarska, 1994). The State Prosecutor's Office investigated 53 reports of alleged clandestine abortions. Of these, 39 were dropped due to "lack of evidence that a crime had been committed." Six women were not pregnant (Skwiecinska, 1994a).

One case attracted international media attention and resulted in Poland's first abortion trial since the enactment of the 1993 antiabortion law. It involved a pregnant 37-year-old divorced mother with one child who was unable to afford raising another. The woman was reported to the authorities by her former lover "to teach her a lesson" for having sought an illegal abortion after a state clinic refused to perform the procedure. Her physician and former lover faced criminal charges that were later dismissed (Perlez, 1995a). Subsequently, other physicians were investigated (Nowicka, 1996b). A gynecologist committed suicide rather than face prison following a police raid (Standish, 1998).

The Ministry of Justice announced that no report was filed on the death of a woman allegedly resulting from a case of illegal abortion described in the press (Swiatkowska, 1994). However, 56 cases of infanticide were registered plus 6 female suicides, including that of a physician who had performed abortions (Folaron, 1994; Labuda, 1994).

According to the Ministry of Health, the low number of court procedures against infraction of the antiabortion law reflected the size of the clandestine abortion problem. In the absence of any reliable estimates, the extent of clandestine procedures was best described in colloquial parlance as "thousands in the East and hundreds in the West." This colloquialism referred to "abortion tourism," which in combination with the domestic abortion underground, represented the basic form of "silent abortion." In the opinion of Lew-Starowicz (1994), chairman of the Family Development Association, the population was coping with the new situation by reducing frequency of intercourse and improving the level of effective contraceptive practice. The birth rate declined from 13.5 per 1,000 population in 1992 to 10.7 in 1997 (Poland/CSO, 1997; COE, 1998). Trends in births and legal abortions over the years 1950 to 1997 are shown in Figure 9.1.

During 1992–94 the price of abortion increased 10 times. In 1992 a hospital termination cost 350,000 zloty as compared to 1½ million to 3 million zloty at a private clinic. After the implementation of the restrictive 1992 Medical Ethics Code, prices soared to 8 million zloty, equal to nearly double the average monthly income (Poland/CSO, 1997). In 1996 the cost of a clandestine abortion in Warsaw ranged from 13 million zloty up to 25 million zloty (or U.S.$580 to U.S.$1,116 at the rate of 22,400 zloty to one U.S. dollar). The average monthly income was 4.7 million zloty (Poland/CSO, 1997). For those having the money, access was quite easy, thanks to euphemistic newspaper advertisements, offering "the full range of gynecological services" or "Gynecologist—everything (the cheapest)." Abortion tourism flourished. There was some question whether it was legal for travel agencies to advertise such services as "Gynecology—inexpensive treatment." Some newspapers refused to accept these ads for ideological reasons. The cost of a tourist abortion ranged between 6.5 and 21 million zloty, depending on how far advanced the pregnancy was and the type of anesthesia used. Terminations were least expensive in Ukraine and Belarus. They also carried the highest risk of complications (as noted in reports from hospitals near the border). Austria, Germany, and the Netherlands were the most expensive and safest (Nowicka, 1996a).

The Parliamentary Group of Women persisted in its struggle to liberalize the law on abortion. At its initiative a new draft for an amendment of the law, signed by 202 members (mainly from the Democratic Left Alliance), was submitted to Parliament. The draft provided for access to abortion for economic and personal reasons and for the right to terminate a pregnancy in private physicians' offices. The draft was adopted in June 1994 by the lower house and in July by the upper chamber of Parliament but vetoed by President Walesa. Some MPs who had voted in favor or abstained in June complained of having been admonished by bishops' missives that if they supported the new draft the "blood of the innocents would fall on their and their children's heads" (Cichocka, 1994b; Nowakowska et al., 1994).

The political elites practiced moral double speak. On the same day they rejected the liberalization of the abortion law, parliamentarians also refused to accept reports by the ministers of Health, Education, and Labor and Social Policy, criticizing their failure to provide sufficient assistance to pregnant women, implement an appropriate program of sexuality education in the schools, and ensure satisfactory access to contraceptives (as required by the 1993 abortion legislation). The report by the Ministry of Justice was accepted because "the Ministry had desisted from persecuting at all costs those responsible for performing illegal abortions" (Nowakowska et al., 1994). As Ewa Letowska (lawyer and former ombudsman) noted in a TV commentary, the hypocrisy of the law undermined the relationship between citizens

Figure 9.1
Births and Legally Induced Abortions: Poland, 1950–97

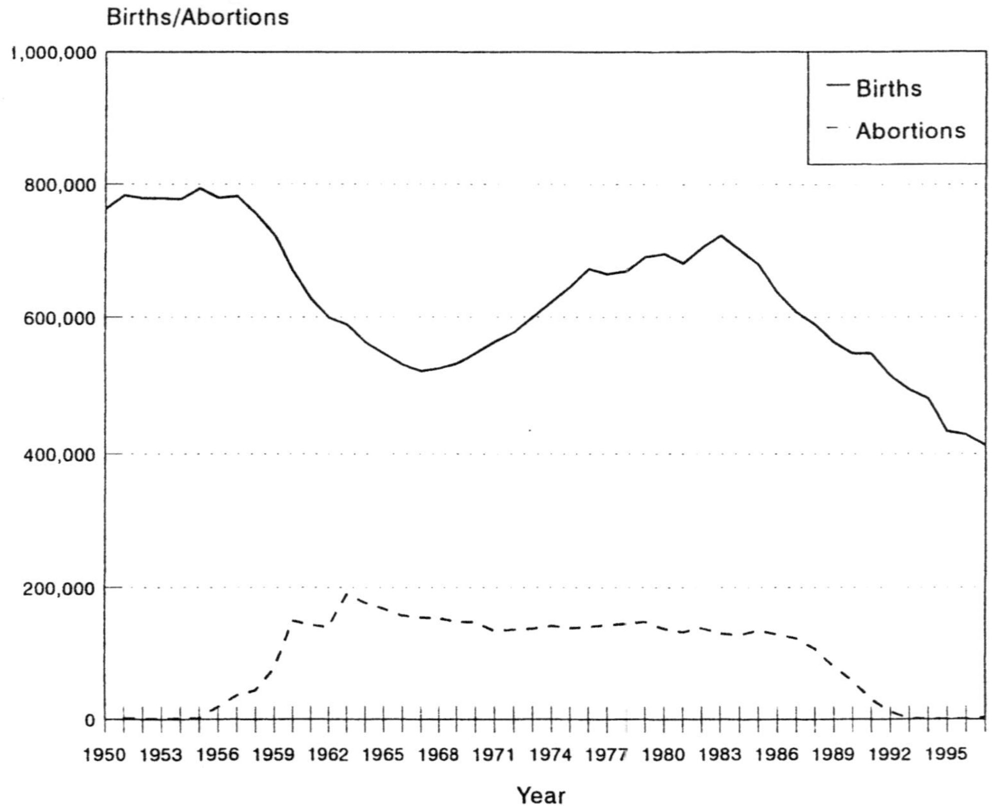

Abortion law liberalized in April 1956. Restrictive instructions on abortion implemented in March 1989.

Sources: See data from Table 9.1.

and the state. Subsequently, Barbara Labuda, head of the Parliamentary Group of Women, was threatened with expulsion by her party, the Union of Freedom, for, among other breaches of party rule, neglect of parliamentary business to concentrate exclusively on religious and ethical issues (*The Lancet*, 1994). For more than two years after the passage of the restrictive 1993 law, Parliament failed to implement policies in accord with its belief that "abortion is best reduced through improving health education and contraceptive practice rather than by legal-administrative means" (Kulczycki, 1995).

With the November 1995 defeat of President Walesa, and with the concerted efforts of many nongovernmental organizations, particularly the Federation for Women and Family Planning, the atmosphere for reversing the restrictive abortion law changed. On 30 August 1996 the lower house of Parliament, the Sejm, over strong Catholic Church opposition (Kozakiewicz, 1997), voted to liberalize the 1993 legislation. On 3 October Parliament's upper house, the Senat, sided with the opposition to keep the existing statute. On 24 October the Sejm overrode the Senat by 228-to-195 votes. The bill was signed into law by President Kwasniewski on 20 November 1996.

Scheduled to go into effect in January 1997, the new legislation stipulated that women could have abortions until the twelfth week of pregnancy if they were financially or socially unprepared for childbirth. The law did not specify how such status was to be determined. Abortions could again be performed in licensed private clinics. However, women seeking abortions were required to obtain counseling from an authorized physician, who would not be performing the procedure, and adhere to a three-day waiting period afterward. In addition, the law provided for the development of a sexuality education curriculum for the schools and for subsidizing the cost of oral contraceptives. Anyone found to have performed an abortion illegally could be imprisoned for up to two years (Gold, 1986; Kozakiewicz, 1997). Implementation was resisted by medical associations (Murphy, 1997). It was further stymied when the Constitutional Tribunal ruled the law unconstitutional on 28 May 1997 on the basis that it failed to protect human life from conception (Perlez, 1997). The restrictive 1993 law was reimposed by Parliament on 17 December 1997, effective 23 December.

CONTRACEPTION

Availability

At the beginning of the 1950s and in the 1960s the condom and vaginal jelly were widely available. Subsequently, local production of condoms declined, decreasing from 31,404,000 in 1969 to 19,858,000 in 1983. Local production of vaginal jelly was completely discontinued in 1983. By 1978 the limited domestic production of hormonal products, supplemented by imports, satisfied the demand of only 2.3 percent of women in the 15–49-year age group. IUDs, imported on a small scale (1 million), could only be purchased in foreign currency stores (Pewex).

If the supply of condoms produced at that time had been distributed equally among all 20–59-year-old men, there would have been only 2–3 per person per year (Okolski, 1988). That contraceptives were in short supply was noted in 1984 by the Government Population Commission. However, nothing was done to remedy the situation.

In the period preceding the enactment of the 1993 antiabortion law and the enforcement of the Medical Ethics Code of 1992 (allowing abortion only in case of a criminal act or for health reasons), contraceptives were gradually withdrawn from pharmacies (Snitow, 1993). Despite a 1991 law requiring pharmacies to sell contraceptives, pharmacists, participating in "Pharmacists for Life," refused to sell contraceptives and destroyed their limited inventories.

The pharmaceutical market is in turmoil, partly due to new regulations reducing subsidies and excluding contraception from national health insurance coverage (Markowska, 1998). Technically, only gynecologists are allowed to prescribe oral contraceptives. Imported oral hormonal contraceptives may be purchased on prescription at full price. Nonprescription contraceptives include the "morning after" pill, intravaginal jelly, and spermicides. In 1994 the average monthly cost of contraception was equivalent to six pairs of panty hose (*Kobieta I Zycie*, 1994). Thus far efforts by the Parliamentary Group of Women and the Federation for Women and Family Planning to make insurance cover 70 percent of the price of prescribed contraceptives have not succeeded.

Contraceptive Practice

Results of surveys of reproductive behavior and attitudes are difficult to compare because different methodologies were employed at different times and because CSO mail questionnaires used during 1975–87 may have been of doubtful validity (Smolinski and Kuciarska-Ciesielska, 1989). Nevertheless, certain findings are worth noting.

A survey of a representative national sample of the adult population in the late 1950s found that 55 percent of the respondents used some form of contraception (Malanowski, 1960). In the 1963–68 period, 53 percent of town and city women said they used withdrawal. In the 1975–87 period contraceptive practice improved but the popularity of "natural methods" remained very high, even in comparison to other Roman Catholic countries (Okolski, 1988).

Data on contraceptive practices of newlyweds from 1975 showed that 66 percent did not use any method of fertility regulation. By 1985 nonuse dropped to 40 percent but four times as many brides as in 1975 were pregnant at the time of marriage (Smolinski and Kuciarska-Ciesielska, 1989). A follow-up of the women married in 1975 showed that 12 years later the ratio of individuals not practicing some form of contraception had declined to 11 percent. Simultaneously, the number of married couples who mentioned using modern methods grew from 10 percent to 49 percent, reflecting greater use of condoms (Smolinski and Kuciarska-Ciesielska, 1989; Kuciarska-Ciesielska, et al., 1988).

In general, during 1977–87, 70 percent of married couples used withdrawal or the rhythm method or a combination of the two. In 1987 one of these techniques was used by 51 percent of married couples. Oral contraceptives were used by 6.9 percent in 1977 and 15 percent in 1987. By 1987 the IUD was used by 8 percent of town and city women and 4 percent of rural women (Smolinski and Kuciarska-Ciesielska, 1989).

Results from a 1989 representative national sample of 18–49-year-old women showed that fertility regulation was practiced by 77.5 percent (Duch, 1991). Only natural methods (thermic, Billings, the rhythm method) were used by 7.4 percent, as compared to 36 percent exclusively practicing modern contraception. The remainder used a combination of different methods or withdrawal (65.6 percent). The condom continued to be widely used (30 percent) in contrast to the pill (9.6 percent) and the IUD (14.1 percent). Slightly different results were obtained in 1991. They showed declines between 1989 and 1991 in use of the rhythm method (50 percent to 35 percent), withdrawal (62 percent to 46 percent), the pill (9.6 percent to 6.3 percent), and the IUD (14.1 percent to 7.1 percent). In both years the same percentage of individuals (31 percent) used condoms (Kowalska, 1991).

Such divergence in results is not easy to interpret. A two-year interval between inquiries is too brief for any radical change in the popularity of basic contraceptive methods to occur. The decline in the popularity of the pill and the IUD may have resulted from budgetary constraints and supply shortages at the time. Unclear and confusing terminology together with mounting ideological pressure directed at all forms of contraception (except natural methods) may also have been responsible for evasive and untruthful answers. This suggestion is confirmed by results of sample surveys conducted in 1993, two weeks after the adoption of the antiabortion law. No less than 81 percent of respondents maintained that they used no contraception. However, some 30 percent of the interviewed 25–39 year olds admitted to trying to avoid an unwanted pregnancy (Konarska, 1993b).

Factors Influencing Practice

Survey findings suggest that the main factor discouraging effective contraceptive practice was not cost but the limited range of available products and suppliers, especially in rural areas (Demoskop, 1991). A 1993 inquiry ranked the reasons for the low level of contraceptive practice as follows: the absence of an appropriate mind set (30 percent), lack of information (21 percent), embarrassment (20 percent), high cost (8 percent), and shortage of suppliers selling contraceptives (2 percent) (Konarska, 1993a).

All the research results indicate that preference for a specific method of fertility regulation tended to be closely related to the intensity of the religious convictions of the respondents (while level of education was inversely proportional). Nevertheless, it does not follow that strong catholic convictions necessarily go hand in hand with acceptance of natural fertility. On the contrary, in 1979, 80 percent of churchgoers considered that preventing unwanted pregnancies was desirable (Titkow, 1984). Ten years later, when evaluating a given method of fertility regulation, 7 percent and 10 percent of respondents were guided by moral and religious considerations, respectively. The principal criteria mentioned by 70 percent of respondents included side effects, method effectiveness, and problems associated with its application. Women tended to prefer natural family planning methods and were more likely to terminate a pregnancy in case of method failure (Duch, 1994). There is an intriguing contradiction between the declared importance of the effectiveness of the selected contraceptive method and the widespread use of the least effective techniques.

In 1993, after the implementation of the antiabortion law, 56 percent of a nationally representative sample of adults disagreed with the position of the Roman Catholic Church banning all methods of pregnancy prevention apart from natural ones. Only 16 percent agreed entirely while 19 percent were ready to accept this prohibition with reservations (Konarska, 1993a).

It appears that, regardless of the political situation, Polish society realized its reproductive goals over several decades with minimal resort to modern contraceptive techniques. For the majority, ideology had little to do with it. Rather, the virtual lack of effective contraceptive practice can be ascribed to the sphere of political institutions skillfully manipulating the cultural taboo that encompasses sexuality in Poland.

With the restrictions on access to abortion, sexuality became a source of fear for many women. Anxiety about an unwanted pregnancy and dependence on natural methods of family planning promoted by the church but believed to be ineffective resulted in an avoidance of sexual life. Women who, against the edicts of the church, use a pill or an IUD are often troubled by a guilty conscience. The price of receiving priestly absolution is removal of the IUD and/or discontinuation of pill use (Nowicka, 1996a).

THE WOMAN QUESTION

Communist Era

There have been women's groups since the founding of the Women's League in 1913. In 1918 Polish women were among the first in Europe to be granted voting rights (Standish, 1998). After World War II, the communist government restructured the Women's League as an instrument of government policy. Under communism public discourse was always officially "pro-woman." The state's language and ideology stressed equality between women and men. However, Marx's concept of false consciousness, which led to the Western feminist practice of "consciousness raising" was not thought applicable in an already communist state (Hauser et al., 1993).

Between 1945 and 1954, when every pair of hands was needed to rebuild a war-ravaged country, streets were full of posters showing beaming women tractor and tram drivers. Then, in 1954–58, women were encouraged to return to their traditional role of mothers and housewives. From 1958 until the late 1960s the productive and reproductive views of the woman question were combined. In the 1970s a lively pronatalist policy was encouraged, favoring a shift in child rearing responsibility from medical and educational establishments back to the parents, meaning the woman's shoulders (Sokolowska, 1977, 1980). Longer maternity leaves and increased child benefits were introduced (Holzer and Wasilawska-Trenker, 1985). The 1976 Constitution included the following provision: "The Polish People's Republic supports women's position in society, especially as mothers and working women." The 1980s, particularly their onset, saw the introduction of a three years' unpaid maternity leave and a demographic boom of several years' duration.

Polish women are used to status inconsistency, to the discrepancy between their level of education and their position in the social structure, public life, and in the centers of power. They have also become accustomed to rewards coming from such a situation, including their high prestige in a society shaped by the combination of tradition, history, and "managerial matriarchy."

Transitions

After the June 1989 elections Polish women found themselves in a special situation created by a combination of diverse elements: historical and cultural traditions, the character of socioeconomic changes occurring after World War II, and the new reality in which a supposedly democratic system oppresses half the nation.

The complete break with the postwar past left an ideological vacuum ready to be filled with a new ideology, that of the Roman Catholic Church, binding for all. This bode ill for the status of women, especially as the leaders of the national Christian organizations (who played a considerable role prior to the 1993 parliamentary elections) created the impression that their principal political, strategic, and personal goals were to corral Polish women into the world of "Church, Kitchen, and Children."

The essence of the process, inspired and controlled by the Roman Catholic Church's intent to reduce women's functions to sex-social roles, materialized in the Law on Family Planning, Human Fetus Protection, and Conditions of Abortion, adopted in January 1993 and substantially reimposed in December 1997. Two elements of the transition period, the evolution of women's status and the political appeal of abortion, were closely interrelated. However, beliefs and actual behavior regarding fertility regulation changed little despite an increase in political debate and the imposition of severe restrictions on access to legal abortion. No recent debate in Poland has been comparable in emotional intensity, pervasiveness of discussion, or in the depth of divisions provoked (Fuszara, 1993; Matuchniak-Krasuska, 1995).

In all, the post-1989 changes in the political system were not positive for women (Kuratowska, 1992; Titkow, 1994a). Some survey results show that Polish women believe they have little money, their economic situation is unstable, and their living standard has declined (Public Opinion Research Center, 1993; Regulska, 1994).

Criminalization of abortion (in effect not as drastic as initially intended) is one of the efforts to morally and legally sanction the philosophy depriving women of the right to decide on their destiny. It is an attempt to prevent a situation in which the role of a professional and publicly active woman would be equally socially acceptable as that of mother, wife, and homemaker. Under such a philosophy it is easier to burden women with the costs involved in the systemic change: for instance, to solve the problem of unemployment.

While Polish women have been exploited in the past, it has not been in a manner unequivocally threatening their self-esteem and dignity. There is considerable evidence that, in the period of transition, a dramatic shift occurred in this subtle, special balance between a diversity of factors defining the status of Polish women in society. Postcommunist Poland has seen the rise of an unequivocal pattern of pressure directed against women (Titkow, 1992, 1993a, 1993b, 1994a, 1994b, 1995). One confirmation is Poland's failure to comply with international conventions protecting reproductive and human rights (Nowicka, 1993).

Women entered the period of transition with education similar to that of men—11 years on the average (Domanski, 1992). Women constituted 47 percent of the group with completed higher education and 59.5 percent with secondary education. A much smaller proportion of women were in the group of individuals who completed basic vocational training in 1992. Women accounted for 46.3 percent of all employed persons and 51.3 percent of the total population (Poland/CSO, 1997). Despite their relatively high level of education, women are discriminated against in terms of earnings, professional career, and participation in public life. For example, in the period from 1982 to 1991 women's earnings amounted on the average to 66 percent of those received by men (Domanski, 1992).

Polish women usually occupy medium managerial posts: kindergarten and school principal, head of a health center (54 percent of physicians are women). According to 1992 CSO data, in 1983 women filled 11 percent of all leadership positions; in 1991 this percentage declined to 9 percent. After the 1990 elections women held 11 percent of senior positions in local authorities, and 6 percent in provincial self-governments (Tess, 1994). The number of women in Parliament is about half of what it was in the communist era but, unlike in the period before 1989, women members of Parliament are genuinely active in political life.

Among the unemployed, 42 percent were women in 1990. In 1993 it was 53 percent (Sucholinska, 1993) and in August 1997 it reached 63 percent (Poland/CSO, 1997). In 1991 the number of jobs available to women was seven times lower than that for men (Olczyk, 1991). The seriousness of this situation is aggravated by the fact that 60 percent of all unemployed women are those with the lowest qualifications whereas offers of employment require young, pretty, and computer-literate women fluent in languages. Women's unemployment shows a tendency to increase and become chronic (Janowska et al., 1992; Titkow, 1992; Knothe, 1994; Velkoff, 1995).

Under Polish employment regulations the disabled, the young, and women are supposed to be under special protection. Women used to be protected in connection with their procreative role. Nonetheless, under the 1989 law, a pregnant woman or one on maternity leave can be made redundant if the enterprise employing her goes bankrupt or closes. In such a case, she is entitled to a benefit payment equal to maternity allowance for a period corresponding to the duration of maternity leave. Pregnant women made redundant are usually unaware that they are entitled to this benefit (probably because they are not properly informed), which is more favorable than the one for the unemployed (Fuszara and Tarnowska, 1995).

Unemployment breeds a tendency to engage in illegal behavior. It happens that when a pregnant woman is illegally dismissed from her job, there is a pretense that she is leaving of her own accord. Unemployment data also show that laws made under the former system have ceased to function as a social protection net. A number have been made nonoperative by the market economy with the attendant sharp competition in the labor market. A case in point is a law making it illegal to hire women for certain jobs: restricting the choice by curtailing the opportunity to choose (Fuszara and Tarnowska, 1995).

The "blessing" of early retirement at age 60 is turning against women. They are now being retired despite their good physical condition and capabilities and even against the interests of employing companies. Still in force is the regulation enabling women to take days off to care for a sick child or a family member. But this greatly undermines the "general worth" of women in the labor market.

Polish women are in a novel predicament. Conditioned to using a well-developed range of social welfare opportunities for available jobs they are now being hit by unemployment and a reduction of social welfare and other social benefits that once formed part of the previous sociopolitical system.

As yet there is no mass-scale women's rebellion nor appreciable strengthening of the existing women's organizations. This is not surprising, one of the reasons being the imprint left by several decades of "real socialism," that is, *state-owned society*. However, in comparison to the period before 1989, there is much more cooperation between diverse women's organizations. The best example is the Federation for Women and Family Planning, composed of six "ideologically" disparate organizations. Their joining together in 1992 was motivated by the need to integrate and upgrade the impact of action on family planning. The federation recognizes that the right of the woman whether, when, and how many children she wants to have is a basic human right. Lacking that she will not be able to participate fully in the life of the nation or develop her own individuality according to her own will. With the support of international organizations, the federation seeks to restore legal and safe abortion, disseminate knowledge on effective contraception, and support initiatives to improve the quality of women's health care (Federation for Women and Family Planning, 1994; Simpson, 1995; Nowicka, 1996a).

SEX AND SOCIETY

Perceptions of Sexuality

Polish society has a traditional attitude toward sex, deeming it acceptable only within marriage. The explicit ban by the Roman Catholic Church on premarital sexual relationships, even between fiancees, is accepted among devout churchgoers by 41 percent (for young women) and 28.5 percent (for young men). However, a 1989 nationally representative sample of 18–49-year-old women indicates that cohabitation, for instance, is generally accepted (Duch, 1994). Such figures suggest that divergences between the spheres of sex, marriage, and procreation will continue to grow.

The survey data confirm that sex is an important element of married life and is considered by almost three-fourths of respondents as a marital duty. The same percentage views sexual compatibility as one of the main preconditions for a happy marriage. Relating these findings to Polish couples' current position at the bottom of the European ranking list of frequency of lovemaking (Mlynarczyk, 1994), one could say, unkindly, that Polish couples seldom excel either in their "marital duties" or in their perception of the marriage partner. Or, was this perhaps the effect of living in fear of an unwanted pregnancy, budgetary constraints on access to modern contraceptives, or awareness that abortion was not available (except on medical, eugenic, or criminal grounds)? Or, in the rural areas, a lack of sufficient knowledge of effective methods of contraception or not being in the habit of using them?

Family Planning Association

Following his establishment of Poland's first family planning clinic in 1928, Boy-Zelinski in 1929 organized the League for the Reform of Morals, later transformed into the Society for Popularizing Conscious Motherhood and the Reform of Morals. Opposed by the government and boycotted by the clergy, the society closed its doors at the outbreak of World War II (Jakubowska, 1976). Reconstituted in 1957, it was admitted to associate membership in the International Planned Parenthood Federation and elected to full membership in 1965.

In 1970 the society was renamed the Family Planning Association (TPR). In 1981 its name was changed to Family Development Association (TRR). The name changes reflected an evolving shift from specialization in reproductive health and sexuality to general family health. In the late1970s, with only 64 centers at its disposal in the entire country and minimal financial support, the TPR/TRR limited itself to promoting the condom and vaginal jelly, providing condom vending machines, distributing leaflets to teenagers, and maintaining a telephone "hotline" (Okolski, 1988).

The TPR/TRR's reduced importance was the result of lack of finances, poor management, and diverse private attitudes toward fertility-regulating practices, characteristic of the prevailing political situation. Plans for upgrading the level of sexuality education and expanding the network of family planning counseling centers could never be implemented. There is little expectation of improving family planning resources. The official argument for reducing the remaing number of operating counseling centers (from 20 to only 5) and outpatient clinics (from 9 to 8) are budget cuts by the Ministry of Health. In reality, it may be a penalty for promoting modern methods of contraception (Cichocka, 1993).

Sexuality Education

In the 1970s Polish specialists distinguished between sexuality education, understood as teaching moral and ethical precepts, and sexuality information, intended to facilitate a rational sex life (Kozakiewicz, 1972, 1976, 1978). In this chapter sexuality education is used in the latter sense, that is, as sexual enlightenment.

The Polish experience shows that economic development, processes of urbanization and modernization, as well as achievement of legal and educational equality by women, are not necessarily synonymous with or a guarantee of an enlightened sex life. It requires both some basic knowledge of sex and the habit of using suitable forms of contraception, dependent on their availability. This disparity was generated over several decades of almost nonexistent activity by the state institutions responsible for family planning and public education, influenced by the pressure of the Roman Catholic Church to transform its sex ethic into nationwide law.

Around 1965, 70 percent of the population agreed that it was proper to educate teenagers about sexuality. This goal was not easy to implement as parents usually were (and still are) reluctant to raise this subject with their children and have nowhere to turn for assistance (Malewska, 1966, 1968). Usually, they were hoping that the school would do the job for them. However, these expectations were not fulfilled in a satisfactory manner. By 1978 only 19 percent of urban and 9 percent of rural primary schools held classes on sexuality education during "hours at the free disposal of the class teacher." At the secondary school level, courses on family life education were implemented in only 7 percent of grammar schools, 1.5 percent of technical schools, and less than 0.5 percent of vocational schools (Kozakiewicz, 1978).

Sexuality education in the schools, stipulated by Article 4 of the 1993 Law but strongly opposed by the hurch, was never implemented (Nagorski, 1993). The vice minister of education officially declared that his ministry did not intend to introduce an independent program of sexuality education in the school curriculum. He explained that "pro-family education" was provided in other subject areas such as literature or biology (Kowalska, 1993a). This view was seconded by the vice minister for health, who announced that the purpose of sexuality education was to delay sexual debut, promote marital fidelity, and provide the undisciplined with protective means and help for avoiding sexually transmitted diseases (Skipietrow, 1993a).

At its June 1994 session, the Parliamentary Health Commission rejected the annual report of the Ministry of Education, citing the ministry's failure to implement the sexuality education program. In actuality there was no program (Perlez, 1995b). The teacher could remain silent on the subject of contraceptive methods or AIDS and still be in compliance. The subject of modern contraception was not included in the primary school curriculum, while in secondary schools only natural methods were mentioned. The main ancillary textbook approved by the Ministry of Education was described by specialists as "containing evident nonsense and making education and the school system a laughing stock" (Janikowska, 1994; Skwiecinska, 1994b, 1994c; Konarska, 1994). One example of information: "Fertility, this generous gift of nature, is sometimes destroyed by contraceptives" (Newman, 1994). The textbook was approved for use in schools in August 1992 when the highest positions in the Ministry of Education were held by active members of the National Christian Union.

In 1994 specialists on sexology working with teenagers maintained that young people's knowledge of sex consists of a mixture of elements of pornography, snatches of popular knowledge, and fragments of the so-called values system (Cichocka, 1994a). Teacher training programs and materials were being developed by the Polish Sexological Society. A book entitled *Modern Sex Education* by Dr. Z. Lew-Starowicz, President of the Sexological Society, was roundly denounced by the Catholic media (Husarska, 1995). Meanwhile, members of the Polish Society of Natural Family Planning Educators instituted teacher training courses throughout the country (Poland/GPC, 1993). In December 1997 parliament voted to ban sexuality education as a separate subject in state schools. On 16 December 1998 parliament eliminated the never implemented requirement that high schools offer a mandatory course. Henceforth the minister of education will decide whether and what kind of sexuality education courses should be offered at any given school.

Access to and quality of information on sex and contraception for adults is no better. The subject is usually avoided by the media. When it is mentioned, it is often "bunk" (Skwiecinska, 1994a); for example, the pill and the IUD are frequently said to be dangerous to the woman's health. Following pressure exerted by the Roman Catholic Church, an entire chapter on other than natural methods of fertility regulation was expunged from the new Medical Home Manual (Nowicka, 1993). Poles do not hide their biological urges—every newsstand and videoshop caters to them—but with few exceptions, sexuality education went underground (Newman, 1994).

Adolescents

Teenagers are more daring. For example, in the period 1964–89, the percentage of adolescents ready to accept premarital sex rose from 8.4 percent to 29.4 percent (girls) and from 19 percent to 40 percent (boys). Those who considered premarital sex "out of the question" declined from 60 percent in 1964 to 30 percent in 1989 for girls and from 34 percent to 23 percent for boys (Malewska, 1966; Titkow, 1984; Duch, 1994).

In a later survey of 15-year olds, 36 percent of boys and 18 percent of girls admitted being sexually active. One-third had used contraception at last intercourse. Boys were more likely to do so (39 percent) than girls (29 percent). Almost all boys (94 percent) and the majority of girls (73 percent) had seen or handled a condom (Poland/MOH, 1990). Still, for far too many teenagers, the solution to an unwanted pregnancy remains unwanted wedlock.

FUTURE PERSPECTIVES

It is fair to say that the Polish population has little support for attaining a wanted model of reproduction. For example, there is little chance to learn about and make use of modern techniques of fertility regulation. Medical doctors are not much help because their views and actions on abortion and acceptable forms of modern contraception have not been responsive to most women's and couples' needs.

In the struggle for their rights, the people also cannot count very much on forms of direct democracy offered by a referendum. This is best illustrated by Parliament's rejection in January 1993 of the draft for a referendum on recriminalization of abortion despite the support given to the draft by hundreds of thousands of signatures collected by the Civic Committee for the Referendum and the earlier rejected draft proposed by the Parliamentary Group of Women (Milewicz and Nowakowska, 1993).

More alarming are such apparently minor cases as the expulsion of leaders of the medical school students' "Sami sobie" (For our own benefit) movement by the Ministry of National Education from the program of health promotion. These leaders tried to encourage young people to adopt reasonable and safe sexual practices (Kowalska, 1993b).

The negative role of the Roman Catholic hierarchy in the process of promoting modern family planning and fertility regulation techniques is extremely significant and consistent. One can hardly dispute the opinion of Poland's ombudsman, Professor Tadeusz Zielinski (1994), that probably not since 1918 has the Roman Catholic Church in Poland had a stronger influence than now.

The church's impact on Polish attitudes, though significant, should not be overrated (Kissling, 1991). The church is not flourishing. Public trust in the church fell from 90 percent in the late 1980s—when, with a Polish pope at its head, the church had the prestige of an embattled government-in-exile—to about 50 percent in 1994. During the battle over Poland's restrictive abortion law in 1992, the church came dangerously close to replacing the communists as the oppressor in the minds of many citizens. In public opinion polls the army replaced the church as the country's most trusted institution (David, 1991, 1992a; Fuszara, 1993). "Wounds made by the Communist Sexual Hypocrisy" (Kot, 1994), works shown by artists at the December 1993 "Ars Erotica" exhibition, first ever in Poland, may be augmented by the "wounds" caused by the sexual hypocrisy of the transition period.

Commenting on the demographic situation, Holzer (1984) stated that the government was interested primarily in developments occurring during its term in office. For instance, if fewer babies were born, less emphasis was placed on housing construction, family welfare, and subsidies for the educational system. This comment applies to the entire post-1945 period, including recent years. Obviously, the link between reproduction and the general strategy of the government also relates to women's share of the labor market. The strategy has been and continues to be highly fluid, treating women in an instrumental manner depending on day-to-day political and economic targets.

Certain developments reinforce hopes for a better future. Antiabortion restrictions stimulated greater use of more effective contraceptives (Public Opinion Research Center, 1993). Mass media reports are encouraging actions to upgrade maternity clinics standards of childbirth. Delegates to the 1994 United Nations Population Conference in Cairo were able to meet and listen to a representative of a nongovernmental organization, the President of the Federation for Women and Family Planning. Several years ago, under the former system, the forum would have heard only the official position of the representative of the Polish government. Subsequently, the Polish Committee of Nongovernmental Organizations was invited to contribute to the official government report prepared for the 1995 Beijing Conference on Women.

The 1996 reversal of the 1993 restrictive abortion legislation was itself reversed in 1997. The post-1989 shifts and changes in political life in Poland suggest continuing instability regarding longstanding controversies related to women's status and reproductive rights. Meanwhile, more effective contraceptive practices are taking root.

chapter 10

Romania

Adriana Baban

EDITOR'S INTRODUCTION

Romania has long been a "special situation" in studies of population policies and the responses they elicit in reproductive behavior. It has been of considerable interest to me for over 25 years (David, 1970). During most of that time, Romania pursued a rigidly enforced pronatalist policy, banning the importation of contraceptives, strictly prohibiting most abortions, and imposing a tax on childless couples. In few other countries have the effects of legislation restricting access to legal abortion been so dramatic, demonstrating the need to ensure women's rights and reproductive health.

Nicolae Ceausescu's drastic pronatalist population policy was already well established by the time of the 1974 World Population Conference convened by the United Nations in Bucharest. During the days of the conference, the United Nations Center for Demography (CEDOR) was officially dedicated, focusing initially on the relationship between population and development, the main topic of the conference deliberations. In the eyes of many observers, of which I was privileged to be one, the United Nations's (UN) decision to locate CEDOR in Bucharest implicitly legitimated Romania's population policy, glossing over the severe restrictions on access to legal abortion. (CEDOR was closed 20 years later in 1984 following espionage allegations.) It was not until 1997 that a new Demographic Research Center was established at the Academia Romana in Bucharest under the direction of Vladimir Trebici.

Only in Romania did uprisings against communism descend into violence (Tagliabue, 1997). After 1989 the once-prosperous Romania advanced only moderately from its descent into one of the poorest, most repressed, and deeply isolated nations in Central and Eastern Europe (CEE). The faces of power remained the same, led by President Ion Iliescu, a former senior communist official (Perlez, 1994b). Indecisive policies resulted in economic stagnation, worsening of women's rights, and very limited governmental commitment to family planning to reduce the increasing resort to abortion.

The 1996 local elections foreshadowed changes in the power of the ruling party. The Minister of Health, who had been Ceausescu's personal physician, resigned in August 1996. In November 1996 Romanians moved closer to democratic reforms with the election of Emil Constantinescu as their new president (*The Economist*, 1996b; Perlez, 1996; Spolar, 1996d). By the end of 1997 the inflation rate had surpassed 150 percent and the value of the currency had dropped 84 percent (Romania/NCS, 1998). A year later, political instability and procrastination about reforms do not bode well.

This chapter will present the Romanian experience in historical perspective, noting demographic trends and the problems with official statistics. The impact of severe restrictions and population policies on fertility-regulating behavior, including fertility planning, abortion, and contraceptive practice, will be discussed. There follow comments on the woman question in a patriarchal society, sexuality education, adolescents' and young adults' perceptions, and society's responses to new-found freedoms and reproductive rights. The chapter benefited greatly from the constructive criticism of Gail Kligman, Daniela Draghici, Roseanne Murphy, Florina Serbanescu, and Patricia Stephenson.—H.P.D.

HISTORICAL TRENDS

Romania, a country situated to the west of the Black Sea, shares borders with Moldova, the Ukraine, Hungary, the Federal Republic of Yugoslavia, and Bulgaria. In 1918 Romania became a unitary state with sizeable Hungarian and German minorities. It is an island of Latin languages where Hungarian, German, Jewish, and Byzantine cultures blend together with Romanian ways of living. The strongest religious influence is that of the Orthodox Church, followed by the Catholic Church. In 1998, Romania had a population of 22.5million people (COE, 1998). About 55 percent of the population live in cities (Muresan, 1996).

The destiny of Romania's identity was strongly influenced by its geographic position at the crossroads between Europe and the Orient. The domination of the Ottoman Empire in the southeastern part of the country until 1877, of the Austro-Hungarian Empire in the northwestern part until 1918, a monarchy of Western European tradition between the two world wars, a communist dictatorship for 45 years, and a pluralistic democracy since 1990, are all historical factors impinging on Romanian attitudes and mentality. Demographic policy was usually overdetermined by the ideology and goals of those in power.

Demographic Trends

At the beginning of the twentieth century, Romania was one of the least developed and most agrarian countries in Europe. About 80 percent of the population lived in rural areas. Prior to World War I, the birth rate was close to 40 per 1,000 population. The gradual decline in births in the immediate post–World War II period was associated with educational reforms providing more opportunities for women, forced industrialization, and agricultural policies that removed incentives to produce children for farm labor (David and McIntyre, 1981; World Bank, 1992; Romania/National Committee, 1994).

As shown in Table 10.1, while Romania still had one of the highest CEE total fertility rates (TFR) per woman (3.17) in Central and Eastern Europe in 1950, the birth rate fell rapidly from 26.2 in 1950 to 14.3 in 1966, at that time the lowest in Europe. TFR declined to 1.90 by 1966. First- and second-order births constituted 73 percent of all births compared to 43 percent in 1938 (Prisacaru et al., 1992). The sharp decline in births was attributed primarily to the passage in 1957 of what was then the most liberal abortion law in Europe (Mehlan, 1965; Ghetau, 1978, 1979). When abortion availability was suddenly curtailed and divorce made more difficult in 1966, the birth rate rose abruptly to 27.4 in 1967, but gradually declined again to 14.3 in 1983. With further severe restrictions on access to abortion and modern contraception, the birth rate rose only slightly to 16.0 in 1989. Following legalization of abortion and improved availability of contraceptives since January 1990, the birth rate declined to 10.2 in 1996 before rising to 10.5 in 1997 (COE, 1998). The relationship between public policies and the private behavior of Romanian women and men will be discussed in subsequent sections.

Living Conditions

Living conditions had been deteriorating steadily during the mid-1970s. In the early 1980s Ceausescu decided to repay all foreign debt. Living standards, already near the bottom in Europe, plummeted further as fuel, food, and raw materials were exported. Health care was slashed and education budgets gutted (Spolar, 1996a). Most food staples were available only with ration cards. Each family was allowed a specified amount of sugar, oil, and flour per month. The black market flourished. Hot water, gas, and electric services were interrupted several times each day. Because the delivery of energy to industry was a priority, women had to cook either late in the evening or before four o'clock in the morning. Considering the power shortages, refrigerators and household appliances were often useless. By the end of the 1980s the per person consumption of kilowatts had been precisely determined. Only one 25 kilowatts bulb was allowed per room (Hausleitner, 1993). Higher wattage bulbs were not available, not even on the black market. It was not a propitious time to have children.

After the events of December 1989 over 600,000 people migrated from villages to towns, seeking jobs. By the early 1990s the rural population declined from over

Table 10.1
Birth, Abortion, and Fertility Rates and Ratios: Romania, 1950–97

Year	Live Births	Total Registered Abortions	Birth Rate	Abortion Rate[f]	Abortion Ratio[g]	Total Fertility Rate
1950	426,800	na	26.2	na	na	3.17
1955	442,900	na	25.6	na	na	3.09
1956	425,700	na	24.2	na	na	2.89
1957[a]	407,800	na	22.9	na	na	2.73
1958	390,500	129,500	21.6	31.0	332	2.59
1959	368,000	578,000	20.2	139.1	1,571	2.43
1960	352,200	769,800	19.1	186.1	2,186	2.33
1965	278,400	1,112,700	14.6	252.2	3,997	1.91
1966[b]	273,700	973,400	14.3	218.2	3,556	1.90
1967	527,800	205,800	27.4	46.0	390	3.66
1968	526,100	220,200	26.7	48.9	419	3.65
1969	465,800	257,500	23.3	56.8	553	3.20
1970	427,000	292,400	21.1	64.0	685	2.89
1973	378,700	375,800	18.2	80.8	992	2.43
1974	427,700	334,600	20.3	71.8	782	2.70
1975	418,200	359,400	19.7	77.1	859	2.62
1976	417,400	383,200	19.5	82.4	918	2.55
1979	410,600	403,800	18.6	87.9	983	2.49

1980	398,900	413,100	18.0	90.2	1,036	2.45
1981	381,100	427,100	17.0	94.0	1,121	2.37
1982	344,400	468,000	15.3	103.4	1,359	2.17
1983	321,500	421,400	14.3	91.4	1,311	2.00
1984	350,700	303,100	15.5	64.6	864	2.19
1985[c]	358,800	302,800	15.8	63.9	844	2.26
1986	376,900	184,000	16.5	38.5	488	2.39
1987	383,200	182,400	16.7	37.8	476	2.42
1988	380,000	185,400	16.5	38.0	488	2.31
1989	369,500	193,100	16.0	39.2	523	2.19
1990[d]	314,700	992,300[e]	13.6	199.3	3,153	1.83
1991	275,300	866,800	11.9	172.3	3,149	1.56
1992	260,400	691,900	11.4	139.1	2,657	1.51
1993	250,000	585,800	11.0	166.2	2,343	1.44
1994	246,700	530,200	10.9	105.1	2,149	1.41
1995	236,600	502,800	10.4	99.5	2,125	1.34
1996	231,300	456,200	10.2	90.2	1,972	1.30
1997	236,900	347,100	10.5	68.6	1,465	1.32

[a]Abortion on request legalized (September 1957). [b]Abortion restricted (September 1966). [c]Abortion further restricted (December 1985). [d]Abortion legalized (December 1989). [e]Includes legal abortions and the number of spontaneous or illegally induced abortions requiring aftercare or treatment at hospitals. [f]Per 1,000 women age 15–44. [g]Per 1,000 live births.

Sources: Ghetau (1978), David and McIntyre (1981), WHO (1990), UNFPA (1990), Henshaw and Morrow (1990), World Bank (1992), Romania/MOH (1993), U.N. (1995a), Muresan (1996), COE (1998), and Romania/NCS (1997).

80 percent to 46 percent of the population (Romania/CBS, 1993). It was aging and increasingly female (Trebici, 1991). However, in 1997, for the first time in Romanian history, the rural-urban migration flow reversed (Ghetau, 1998).

POPULATION AND ABORTION POLICIES: PUBLIC POLICY AND PRIVATE BEHAVIOR

1945–65

Romania's population and abortion policies have ranged from liberal legislation to rigid pronatalism and severe restrictions on abortion (David and McIntyre, 1981; Hord et al., 1991; Kligman, 1992, 1998). In 1948 Article 482 of the Penal Code was revised, outlawing abortion. In 1955 Decree No. 456 modified the 1948 text, permitting abortion if the pregnancy represented a danger to the woman's health or if one of the parents suffered from a grave hereditary disease. In 1957 Government Decree No. 463 followed the lead of the Soviet Union by legalizing abortion on request (Romania/MOH, 1957). The objective was to give women the right to decide whether and when to have children, and to discourage resort to unqualified abortion practitioners (Mehlan, 1965).

Abortion centers were organized in large- and medium-sized hospitals. Outpatient facilities were attached to industrial plants having a sizeable female work force. Women could request and receive termination of an unwanted pregnancy within the first three months of gestation without seeking prior approval of an abortion commission. Bureaucratic formalities were kept to a minimum. While name, age, occupation, and number of previous births and abortions were recorded in a register, no effort was made to verify this information. Abortion quickly became socially as well as legally acceptable. The typical fee was 30 lei (less than U.S.$2.00 in 1957), of which the physician received about half. Doctors worked in shifts and were permitted to perform up to 10 abortions per day. Pregnancies of more than three months' duration could be terminated only in hospitals on medical certification. Performing an abortion outside an approved facility was a punishable offense (Cristea et al., 1961).

In 1958, the first full year of abortion availability on request, 129,500 abortions were registered, rising to 578,000 in 1959 (Table 10.1). The abortion ratio rose from 332 abortions per 1,000 live births in 1958 to 1,571 in 1959. During the first six months of 1962, there were 14 abortions for every live birth in Filantropia Hospital in Bucharest (Mehlan, 1965). Official statistics, shown in Table 10.1, indicate that the total number of registered abortions (including women receiving legally induced abortions, plus women admitted to hospital for aftercare or treatment of complications of spontaneous and illegal abortions) nearly doubled from 578,000 in 1959 to 1,112,700 in 1965 (Ceterchi et al., 1975; Romania/NCS, 1997). The total abortion ratio increased to 3,997 abortions per 1,000 live births in 1965, the highest ratio ever officially reported from any country up to that time. Put another way, about 80 percent of all conceptions were terminated by abortion in 1965 (Berelson,

1979). The abortion rate rose from 31 per 1,000 women 15–44 years of age in 1958 to 252.2 in 1965, the highest rate ever recorded in Romania.

During the decade following the 1956 legislation, abortion was widely available at low cost and in a manner preserving the woman's privacy, with the result that abortion became a socially accepted major method of fertility regulation (Muresan and Copil 1974; Trebici, 1976). Although posters with instructions about contraception were widely displayed in abortion centers, and physicians were technically required to offer instructions in the use of contraceptive methods, little instruction actually occurred.

1966–73

In 1966 the Ministry of Health informed the Plenum of the Romanian Communist Party Central Committee that in 1965 there were four abortions for every live birth (Romania/MOH, 1968). In a speech to the June 1966 National Conference of Romanian Women, then Premier Nicolae Ceausescu attributed the low birth rate (14.6 in 1965) at least in part to the easy divorce laws and appealed to the women to strengthen the family. Although Ceausescu made no reference to abortion, the Conference Report identified liberalized abortion as a problem and suggested that revisions in the law could be expected. Official comment on the adverse fertility situation peaked in September 1966, coincident with the release of the March 1966 Census results. Health officials cited the "excessive" rise in the number of abortions, commented that the overwhelming majority of reasons for abortion requests were "superficial" and "frivolous," and enumerated potential risks of repeated abortions (Wright, 1975).

On 1 October 1966, one decade after abortion had been liberalized, the Council of State abruptly reversed abortion policy by Decree No. 770, effective 30 days later on 1 November 1966 (Romania, 1966). The availability of abortion on request was suddenly limited to (a) women over 45 years of age, (b) women already supporting four or more children, (c) women whose life, in the judgment of a special medical commission, was endangered by the pregnancy, or who were faced with the risk of congenital malformation, or whose pregnancy resulted from rape or incest, or who were "physically, psychologically, or emotionally incapacitated." More than 100 narrowly defined medical indications were included in Instruction No. 819 of 19 October 1966, with leeway remaining only in the psychiatric area (Romania, 1966, 1967). A legally authorized abortion had to be performed by a gynecologist in an authorized hospital within the first trimester of pregnancy. Thereafter abortion was permissible up to the sixth month only after it had been established that a grave pathological condition threatened the woman's life.

The Penal Code was revised and penalties stiffened for providers and seekers of illegal abortion. By Decree No. 771, Article 428 was amended and subsequently became Article 188 (Law No. 15 of 21 April 1968), stipulating a range of penalties for abortions performed under other than legally sanctioned conditions. Physicians risked loss of license and prison terms of up to 12 years. Article 186, as further modified by Law No. 6 of 1973, directed that self-induced abortion was punishable by imprisonment from six months to two years or with payment of a fine. The Penal Code did not differentiate between married and unmarried women. Article 187 provided penalties for unauthorized possession of abortifacients or instruments for inducing abortion. Physicians failing to notify competent authorities promptly after having performed an abortion under emergency conditions were liable to imprisonment of one to three months. The 1966 Decree also mandated that every woman coming to a hospital after "spontaneous abortion" had to be reported to the district attorney. The physician was required to present a written report on whether there were indications of induced abortion (Romania, 1966, 1967, 1976; Ceterchi et al., 1974).

The Preamble to the October 1966 Decree refers to the "great prejudice to the birth rate and rate of natural increase" resulting from the practice of abortion as well as the "severe consequences to the health of women." As subsequently reported by Muresan and Copil (1974), "the government adopted new complex socioeconomic measures designed to alleviate the overdependence on abortion." The objective was "to stimulate the natality and reach 18–19 per 1,000, a rate corresponding to the economic progress of the population," and likely to assure Romania "a population of about 24–25 million inhabitants by 1990." Thus, the explicit purpose of the 1966 Decree was to implement a new demographic policy and to "eliminate abortion as a form of birth control" (Muresan and Copil, 1974: 367–368).

At the same time, obtaining a divorce was made much more difficult. The early 1960s had seen a steep rise in divorces, especially among younger couples. By 1965 the divorce rate per 1,000 population was 22 percent of the marriage rate for that year (Muresan and Copil, 1974; Trebici, 1976, 1978). By Decree No. 779 of 8 October 1966, obtaining a divorce required both a large filing fee (ranging from 3,000 to 5,000 lei [U.S.$170 to U.S.$280] depending on income) and a compulsory six-month (one-year when there were children under age 16) period of trial reconciliation (Ceterchi et al., 1975). The number of divorces granted fell from 37,000 in 1965 and 26,000 in 1966 to 48 in 1967 with a modest resurgence in the next few years (David, 1970; Sora et al., 1983). Proof of the mother's insanity was one of the few grounds for fathers to obtain custody of their children (Kligman, 1998).

The antiabortion law was accompanied by limited pronatalist incentives, including paid medical leaves for pregnancy; social care for mothers with more than three children under age 18; and increases in financial allowances from 100 to 131 lei for each child up to age 16. Maternity grants of 1,000 lei (then roughly U.S.$85), previously paid only for tenth or higher order births, were now paid for all third and higher order births. However, the maternity leave of 16 weeks was the shortest of any CEE country. In 1972 and 1979 family allowances were raised slightly. While paid maternity leave was provided for child illness up to three years of age, a woman applying for such leave would usually find herself first on the personnel reduction list. One-child families were punished by being made ineligible to purchase three-

or four-room apartments. Women who delivered and reared 10 or more children were awarded medals proclaiming them "Heroine Mother." In his final speech on 21 December 1989, Ceausescu promised to increase family allowances with a further 100 lei (then equivalent of U.S.$1.00).

1973–89

In February 1973 the age for eligibility for legal abortion on request was reduced by Decree No. 53 from 45 to 40 years. Analysis of data collected between 1966 and 1972 suggested that the proportion of all live births contributed by women in the over age 40 category was insignificant (Kligman, 1998). It was a politically important gesture at a time when the U.N. was negotiating arrangements for convening the 1974 World Population Conference in Bucharest. 1974 was also the year in which Ceausescu ascended to the presidency of Romania.

In 1983 insertion of IUDs for medical reasons was prohibited and surgical sterilization was forbidden except for very strictly defined medical reasons approved by a medical committee. As noted by Kligman (1998), 1984 was the year that ushered in the Orwellian policies that became the basis for Margaret Atwood's novel, *The Handmaid's Tale.* Ceausescu accused medical personnel of failing to supervise pregnant women adequately. The Romanian State Security Police (Securitate) and representatives of the Prosecutor's Office began investigations of alleged illegal abortion at every obstetrical-gynecological clinic.

In December 1985 Ceausescu again restricted access to legal abortion to women over 45 years of age. Having four children was no longer sufficient grounds for abortion on request. To qualify, a woman had to have five living children of her own, all under the age of 18 years (Ceausescu, 1985). He proclaimed that "the fetus is the socialist property of the whole society. Giving birth is a patriotic duty. Those who refuse to have children are deserters, escaping the law of natural continuity" (Ceausescu, 1986: 217).

The extent to which the state attempted to control each woman's reproductive life, the political and legal machinery created to enforce its demographic policies, and the gradual distortions of statistical reporting systems have been well documented by Kligman (1995, 1998). For example, women between the ages of 16 and 45 and working in or attending state facilities were required to undergo annual medical examinations to verify that their reproductive health was satisfactory. In some institutions more frequent examinations were demanded. Those who refused to appear were told that they could lose their right to dental and medical care, pensions, and social security. It was an implicit threat, not an explicit one, meaning that only "obedient" women perceived it as a compulsory rule. To meet the norm, doctors had to perform 50 to 60 gynecological examinations a day to cover a factory population in a specified period of time. Factory physicians received their full monthly salaries only if plant employees had achieved a state-stipulated monthly birth quota. Special taxes on monthly salaries of childless individuals over the age of 25 years, regardless of sex or marital status, originally imposed in 1977, were increased. The sums subtracted depended on employment in selected state sectors (Kligman, 1995). Securitate agents frequently appeared in maternity wards and obstetrical-gynecological clinics. State organs coordinated efforts to discover and destroy networks through which clandestine abortions were arranged. Show trials were held of physicians, midwives, and nonmedical abortion service providers. Unlawful abortion was considered such a grave transgression that the 1986 general amnesty, which reduced sentences for most crimes, listed illegal abortion among the most serious offenses along with murder (Romania, 1986). Women known to have had abortions, whether induced or spontaneous, were officially registered at medical dispensaries and examined at three-month intervals to discourage repeated pregnancy terminations. The repressive regulations forced many women suffering medical complications from clandestine abortions to stay away from hospitals, thus risking permanent injury to their health or, often, death.

Despite very difficult circumstances, there were courageous gynecologists who became adept at manipulating both the official rhetoric and statistics. Some would tell a pregnant woman, who was adamant against having a child, that they thought she was pregnant but were not absolutely certain. They suggested that she return in one month time, and, if her pregnancy were confirmed, it would then be officially registered. This "warning" gave the woman one month to resolve her situation (Kligman, 1998). Other physicians managed, for sizeable fees, to diagnose a medical condition that qualified the woman for a legal abortion. For most physicians, however, the institutional framework of surveillance, the political climate, and the culture of fear and suspicion within which they practiced, served as a deterrent to their humanitarian concerns.

Many women persisted in their quest for terminations of unwanted pregnancies, regardless of potential consequences for health or partner relations. Interviews conducted after 1990 revealed a sizeable proportion of women who had had 20 to 30 abortions (Chaze, 1992; Johnson et al., 1993; Baban and David, 1994; David and Baban, 1996; Baban and David, 1997; Kligman, 1998). Estimates suggest that less well educated women experienced about five life time abortions (Serbanescu, 1994).

1990–97

Following the overthrow of Ceausescu in December 1989, the new government confronted the epidemic of maternal mortality and morbidity without delay by reversing the restrictive legislation on 26 December 1989, its first day in power. A new law was passed, effective 1 January 1990, authorizing the importation, production, and sale of modern contraceptives, and permitting abortion on request through the first trimester when performed by an obstetrician-gynecologist (Romania/MOH, 1989). At first, the procedure cost 30 lei (the price of a loaf of bread); a few months later in 1990, the price was increased to 500 lei (approximately three-fifths of an average weekly income). To make abortion more expensive, the government raised the price of state-provided procedures in July 1993 to 3,200 lei which, at that time, was the equivalent of 5 percent of the average monthly salary. Terminations without charge were provided for women having four children, for medical reasons, and for women having no income. Privately practicing physicians charged prices ranging from 10,000 to 20,000 lei, assuring privacy and personal attention. In 1990 a doctor performed about 30 abortions per day; in 1994, the average number was about 10 per day. Both vacuum aspiration and curettage were used. As shown in Table 10.1, 992,300 total registered abortions were reported in 1990, almost 200 per 1,000 women aged 15 to 44 years (Romania/MOH, 1993). Abortions subsequently declined to 866,800 in 1991 and 347,100 in 1997, when the abortion ratio stood at 1,465 per 1,000 live births (Romania/MOH, 1994, 1995; COE, 1998). The annual fluctuations in births and abortions are shown in Figure 10.1.

With the establishment of private practice in mid-1991, the figures reported to the Ministry of Health were no longer complete. It is quite likely that the actual abortion ratio was closer to 3:1 in 1994 than the reported 2:1. In Bucharest, the ratio was 6:1 (Serbanescu, 1994). The population's access to contraceptives (when and where available) is still more or less controlled by state-employed, low-paid gynecologists, many of whom have more of an economic interest in performing abortions than in providing contraceptive counseling. The Ministry of Health is still reluctant to allow general practitioners to provide family planning services and does not permit them to insert IUDs. Health care providers seldom initiate discussions about sexuality and contraception (Johnson et al., 1996).

DEMOGRAPHIC AND HEALTH EFFECTS

Data Collection

To monitor implementation of pronatalist policies and their effects on economic and social development, the Plenum of the Central Committee of the Communist Party proposed the establishment of a National Commission of Demography (NCD) within the Council of State. The National Assembly enacted a law on 18 March 1971, establishing the Commission (Romania/NCD, 1971). Members were appointed by Decree No. 197 of the Council of State on 29 May 1971, representing a broad range of specialists and relevant state organizations. To strengthen the work of the NCD, 40 district commissions were organized in 1973. The presidents of these district commissions were appointed NCD members (Muresan and Copil, 1974). Results of studies were to be published in the *Official Bulletin*. However, when there was lack of agreement between the findings and communist ideology, documents were marked "top secret" and forbidden to be published. The commission functioned for twelve years after which it was consigned to oblivion and publicly eclipsed by the Higher Council on Health.

As more fully described by Kligman (1998), two institutions managed the collection and analysis of vital statistics, the Central Board of Statistics (CBS) and the Computing and Health Statistical Centre (CCSS) within the Ministry of Health.

Figure 10.1
Births and Total Registered Abortions: Romania, 1958–97

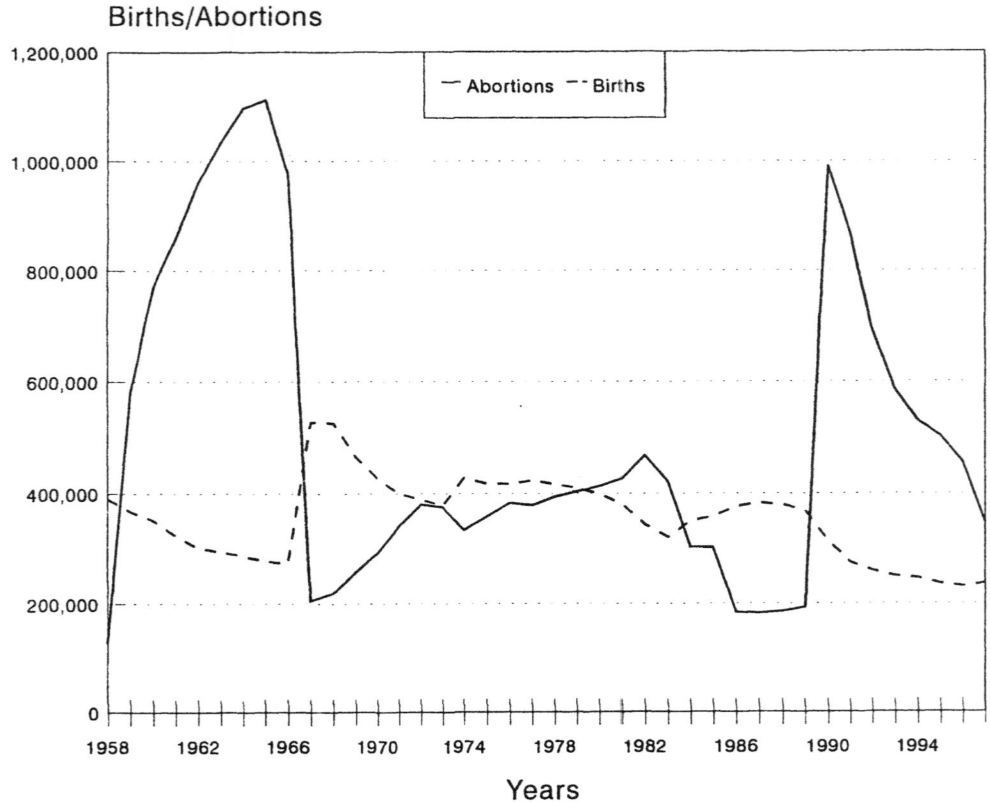

Abortion restricted in October 1966. Abortion legalized in December 1989.

Sources: See data from Table 10.1.

The Centre analyzed demographic data, especially with respect to natality, mortality, abortion, and infant and maternal mortality. It sent its reports to the Central Board of Statistics. Beginning in 1982, the number of abortions cited in the Ministry of Health reports was consistently lower than the data received independently by the Central Board of Statistics (CBS). No one contested the data presented by the Ministry of Health. By the mid-1980s, the *Statistical Yearbook* produced annually by the CBS had become a fairly slim volume. In 1986, the section on Population and the Labor Force consisted of only four pages of data (pages 16–20).

Another constraint on registering abortions was that the salary of medical staff was related to the number of abortions they reported. If it exceeded the plan, the salary was reduced. Since various ailments, such as measles, recurrent fevers, syphilis, malignant tumors, and so on, qualified for legal abortion, physicians took advantage of these provisions to accommodate a woman's choice (Kligman, 1998). Women arriving at hospitals after incomplete illegal abortions were often classified as having experienced "spontaneous abortions." In sum, Romanian statistics, especially those from the 1980s, were often unreliable and must be analyzed with caution.

Natality

The Romanian experience was unique in the sense that abortion availability was restricted suddenly and severely, with little advance warning. At the same time, access to modern contraceptives was reduced and divorce made more difficult to obtain. As shown in Table 10.2 and Figure 10.2, the birth rate rose month by month from its record low of 12.8 per 1,000 population in December 1966 to a peak of 39.9 in September 1967 before declining to 21.5 in December 1968. The dependence on abortion was illustrated by the proportionately sharper birth increase in the urban areas (from 11.4 in 1966 to 26.9 in 1967) compared to rural regions (from 16.1 to 27.6). The apparel industry had to manufacture emergency supplies of maternal clothes and infant wear, while government housing offices were besieged by couples demanding larger apartments to accommodate their expanding families. Meanwhile, as shown in Table 10.1, total registered abortions declined from 1,112,700 in 1965 to 973,400 in 1966 and 205,800 in 1967.

As couples began to adjust to the new realities, the birth rate declined from 27.4 per 1,000 population for 1967 to 18.2 in 1973. Improvements in family allowances and birth grants by Decree No. 414 of July 1973 may have helped boost the birth rate to 20.3 in 1974. As shown in Table 10.1, the birth rate subsequently decreased to 18.6 in 1979 and 14.3 in 1983, equal to the 14.3 reported in 1966 (Romania/MOH, 1979; Romania/CBS, 1993). Meanwhile, the total number of abortions increased from 220,200 in 1968 to 403,800 in 1979 and 468,000 in 1982 when the abortion ratio stood at 1,359 per 1,000 live births. From 1980 to 1983, abortions exceeded live births (Romania/MOH, 1990).

The short-term effects of abortion restrictions on births in 1958–96 is shown in Figure 10.1. As noted in Table 10.1, tightening of abortion restrictions in 1985

Table 10.2
Monthly Birth Rates: Romania, 1966–68

	1966	1967	1968
January	12.7	15.4	29.2
February	14.8	15.7	29.2
March	15.1	16.6	29.4
April	15.4	17.8	27.8
May	15.2	20.7	26.5
June	14.8	29.9	25.9
July	14.3	38.7	26.0
August	14.2	38.5	26.0
September	14.1	39.9	27.7
October	14.9[a]	36.1	26.3
November	12.4	21.1	24.2
December	12.8	27.7	21.5

[a]Abortion policy restricted.

Sources: Romania/CBS (1974) and David and McIntyre (1981).

resulted in only a slight rise in the birth rate from 15.8 per 1,000 population in 1985 to 16.5 in 1986 and 16.7 in 1987 before declining again to 16.0 in 1989. Romanian couples had found means of coping, mostly through clandestine abortions. Physicians performed abortions under other diagnoses that did not require official notification. And, among registered abortions, alleged spontaneous abortions occupied first place.

The estimated total fertility rate (TFR) gradually declined from a mean of 6.6 per woman in 1899 and 3.17 in 1950 to 2.73 in 1957. As shown in Table 10.1, after the liberalization of abortion in 1957, TFR dropped further to 1.90 in 1966. Following rerestriction of abortion in October 1965, the TFR rose to 3.66 in 1967 before declining again to 3.20 in 1969 and 2.55 in 1976 (Ghetau, 1978).

The demographic effects of the restrictive abortion legislation have been extensively discussed (e.g., David and Wright, 1971, 1974; Teitelbaum, 1972, 1974; Berelson, 1979). In the opinion of Ghetau (1979), the abortion restrictions seemed to be achieving the targets set in 1966—an increase in the birth rate to around 18-19 per 1,000 population and a population of 24–25 million by 1990. In actuality, these goals were never attained. The subsequent tightening of abortion restrictions in 1985 increased the TFR only slightly to a peak of 2.42 per woman in 1987.

With access to legal abortion restored in 1990 and confronted by the economic difficulties of the transition, Romanian couples decided on smaller families. The birth rate decreased from 16.0 per 1,000 population in 1989 to 11.9 in 1991 and 10.5 in 1997. The abortion rate rose from 39.2 per 1,000 women aged 15–44 years

Figure 10.2
Monthly Birth Rates: Romania, 1966–68

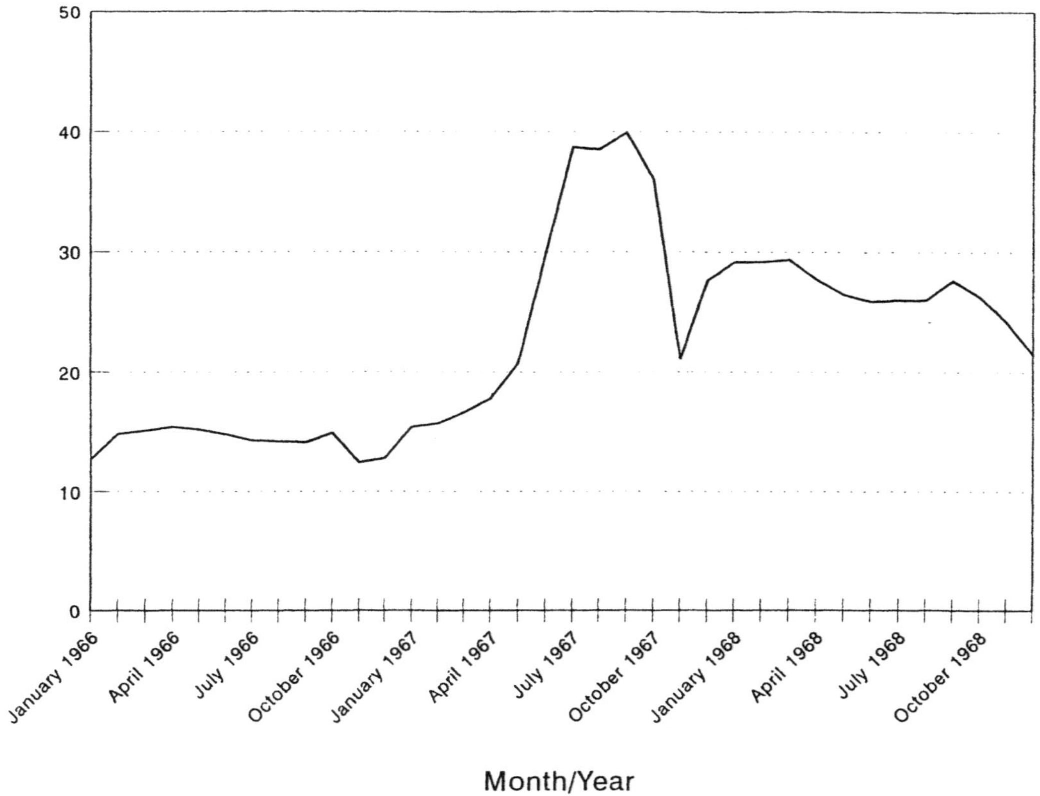

Abortion restricted in October 1966.

Sources: See data from Table 10.2.

in 1989 to 199.3 in 1990 before declining to 68.6 in 1997. The TFR dropped from 2.19 per woman in 1989 to 1.83 in 1990 and 1.32 in 1997. Marriages decreased from 9.2 per 1,000 population in 1977 to 8.3 in 1990 and 6.5 in 1997 while divorces rose from 1.2 per 1,000 population in 1977 to 1.4 in 1991 and 1.5 in 1997 (Prisacaru et al., 1992; COE, 1998).

Maternal Mortality

All maternal deaths were registered by the police and were deemed accurate compared to other statistics, such as infant mortality (Johnson et al., 1996). Maternal mortality rates (MMR), maternal deaths due to abortion, and nonabortion MMR are shown in Table 10.3 and Figure 10.3. During 1960–65, there was an average of 76 maternal deaths per year that resulted from "septic abortion" (Legge, 1985). The number of such deaths rose from 64 in 1966 to 173 in 1967, the year after the new abortion restrictions were implemented, rising further to 192 in 1968. In contrast, Mehlan (1965) reported only one to two deaths per year from legal procedures during the initial years of the availability of abortion on request. By 1981 the number of maternal deaths attributed to abortion increased to 456, reaching a peak of 545 in 1989 before declining to 181 in 1990 (after relegalization) and 50 in 1997 (Rochat, 1991; Badea et al., 1993; Mutler, 1994; Romania/MOH/CCSS, 1995; Romania/NCS, 1997; Ghetau, 1998).

Serbanescu (1994) estimated that about 45 percent of all pregnancies were unwanted in the period before 1989. Many clandestine abortion providers were medically untrained, often practicing in unhygienic conditions. During the 20 years from 1969 to 1989, there were an average of 341 maternal deaths per year associated with clandestine procedures (Muresan, 1991; Stephenson et al., 1992). During the four years of 1969–72, there were only 11 deaths that resulted from legal procedures—less than 1 percent of all abortion-related maternal deaths.

As also shown in Table 10.3, the maternal mortality rate rose from 85.1 deaths per 100,000 live births in 1965 to 148.8 in 1984, when 86 percent of maternal deaths were due to abortion. By 1989 there were 169.6 maternal deaths per 100,000 live births, 10 times higher than the highest figure previously recorded in Europe. Of that number, 545 deaths or 87 percent were abortion-related (Pierotti, 1991; Stephenson et al., 1992; WHO/EURO, 1991, 1994). In 1990 the MMR declined to 83.6 per 100,000 live births with 69 percent still due to abortion. By 1997 the MMR had dropped further to 41.4 per 100,000 live births, of which 51 percent were due to abortion (Ghetau, 1998). While progress in reducing maternal mortality has been considerable, Romania continues to have the highest rate in Europe, in part because the rate for nonabortion-related deaths has remained fairly stable. It is widely believed that clandestine abortion continues to be the major cause of abortion-related maternal mortality (Serbanescu, 1998). Some rural women and poor urban women still tend to prefer untrained providers whose charges are less than those of physicians.

Table 10.3
Maternal Mortality Rates and Ratios: Romania, 1960–97

Year	Total Maternal Deaths	Maternal Mortality Rate[a]	Maternal Deaths Due to Abortion	Abortion-Related Maternal Mortality Rate[a]	% Maternal Deaths Due to Abortion
1960	298	84.6	96	27.2	32
1965	237	85.1	47	16.9	20
1966	235	85.9	64	23.4	27
1967	511	96.8	173	32.8	34
1968	509	96.7	192	36.5	38
1970	497	116.4	315	73.8	63
1975	516	123.4	385	92.1	75
1980	527	132.1	441	110.6	84
1981	534	140.1	456	119.6	85
1982	602	174.8	521	151.3	86
1983	547	170.1	471	146.5	86
1984	522	148.8	449	128.0	86
1985	493	137.4	425	118.4	86
1986	571	151.5	488	129.5	86
1987	575	150.0	491	128.1	85
1988	591	155.5	524	137.9	89
1989	627	169.6	545	147.4	87
1990	263	83.6	181	57.5	69
1991	183	66.5	114	41.4	62
1992	157	60.3	100	38.4	64
1993	133	53.2	86	34.4	65
1994	149	60.4	94	38.1	63
1995	113	47.8	59	24.9	52
1996	95	41.1	51	22.1	54
1997	98	41.4	50	21.1	51

[a]Per 100,000 live births.

Sources: Ghetau (1978, 1998), David and McIntyre (1981), UNFPA (1990), WHO (1990), World Bank (1992), Romania/MOH (1993), U.N. (1995a), and Romania/NCS (1997).

Figure 10.3
Maternal Mortality Rates: Romania, 1965–97

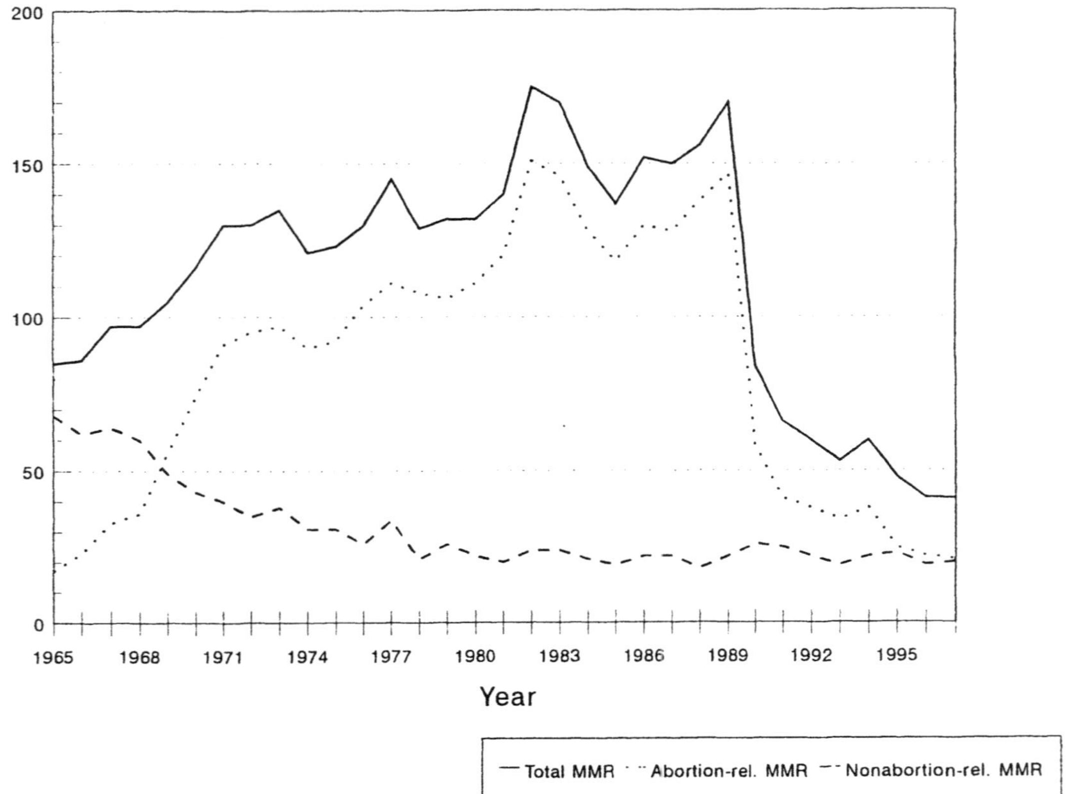

Abortion restricted in October 1966. Abortion legalized in December 1989.

Sources: See data from Table 10.3.

Precise figures on postabortion morbidity do not exist. Local women and health officials report that unsafe abortion, induced by the woman herself or by an untrained person, involved potentially very dangerous techniques, such as scraping the uterus with a rubber tube, uterine lavage with a caustic fluid, introduction of foreign bodies into the uterus, or external trauma (Stephenson et al., 1992; Baban and David, 1994). Such methods often result in damage to the uterine cervix, chronic infection, and severe anemia that, in turn, increase the risk of postpartum hemorrhage and infection, infertility, preterm birth, and low birth weight (Royston and Armstrong, 1989; WHO,1990; UNFPA, 1990).

Infant Mortality Rates

Infant mortality rates (IMR) were also affected by the abortion restrictions. After declining from 81.5 per 1,000 live births in 1956 to 46.6 in 1966, the rate rose to 59.5 in 1968, among the highest in Europe (Coroi and Gorgos, 1980). In 1968 maternity hospitals were overwhelmed by admissions and a large proportion of the staff were themselves on maternity leave. As the number of births declined, the infant mortality rate fell to 54.9 in 1969 and to 23.9 by 1983 (Wright, 1973, 1975; Romania/MOH, 1993).

With the further restrictions on abortion, IMRs were no longer published. In 1986 a 15–30-day delay was imposed on registering premature and low birth weight babies, a ruse to avoid acknowledging deaths in the first month of life. Inquiries suggest that infant mortality reached 28.9 per 1,000 live births in 1987 (Kligman, 1992). Research studies indicate that of the 327,000 deaths under age one, officially registered between 1967 and 1989, about 45 percent could have been saved with better medical care (Trebici, 1991). By 1997 IMR had declined to 22.0 per 1,000 live births (COE, 1998).

Of those children who survived unintended conceptions and unwanted births, thousands were abandoned to state care in public orphanages. The plight of these children has been well reported (e.g., Battiata, 1990; Nachtwey, 1990). Epidemiological studies of psychological and neurological disorders among such children showed higher levels of mental retardation than in the general population (Grigoroiu-Serbanescu, 1990). With the economic crises in the late 1990s child abandonment persists (Perlez, 1994a; *RFE/RL Newsline*, 1997).

FERTILITY PLANNING

Studies of family demographic trends were published in the 1960s (Caramelea, 1968; Trebici, 1968, 1971, 1979). Several reflect a well-established preference for small families (Mehlan, 1965; Potts, 1967; Trebici, 1976). A 1964 study showed that in Bucharest the total fertility rate for professionally active women was 1.2 children, compared to 1.7 for economically inactive women (Ilea et al., 1967). Another study of working women indicated that 61.5 percent of high income women were childless, compared to 13.4 percent of low income women (Pescaru, 1966). Matioc and Costa (1963) reported 28 percent of marriages to be childless.

Based on 1966 census data, a representative sample survey was conducted in 1967 of 13,000 women aged 15 to 49 years and married for the first time. For the country as a whole, the total fertility rate per woman was 1.9; for women with only primary education the rate was 2.6; for women with secondary education 1.2; and for women with university education 0.7. An inverse relationship was noted between the size of localities and fertility. The number of live-born children per typical woman age 15 years or older in 95 towns under 10,000 population averaged 1.98, compared to 1.15 per woman in the Municipality of Bucharest (Romania/CBS, 1969; Romania/MOH, 1970; Ilea et al., 1974; Romania/NCD, 1976).

A follow-up 1975 MOH Survey showed that rural women considered 3.3 children optimal, compared to 2.3 children for urban women. In general, economically inactive housewives and farm wives preferred larger families than women with higher education and greater technical skills (Trebici, 1978). After 10 years of marriage, 11 percent of urban families had no children, compared to 5 percent of rural families (Mesaros, 1975).

An informal nonrepresentative 1970 survey in Bucharest reflected a preference for the one-child family among educated persons (David and Wright, 1971). The preference for small families, especially among more educated urban women, was confirmed by subsequent reports that 84 percent of women with university degrees had only one child, as did 79 percent of secondary school graduates, compared to 45 percent of elementary school graduates (Mesaros, 1972; Moskoff, 1980).

The restrictions on fertility regulation contrasted with women's desire to limit family size. An informal study conducted in 1989 by the Institute of Hygiene and Public Health in Bucharest demonstrated a prevailing desire for small family size. Of the 5,000 respondents, 55 percent wanted only one child, 28 percent wanted two children, 12 percent wanted three children, and only 5 percent wanted four children (UNFPA, 1990). Average desired family size was 1.67 children.

CONTRACEPTIVE PRACTICE

Historical Trends

Coitus interruptus, backed by legal or illegal abortion, has long been the traditional method of fertility regulation in Romania. Condoms and contraceptive creams were manufactured locally. Oral contraceptives and IUDs were imported mostly from Hungary and the former German Democratic Republic. Demand was limited primarily to "educated women." Little information was disseminated about contraceptive practice. While the use of contraceptives was not legally prohibited in October 1966, official importation largely ceased. At the time of the 1974 World Population Conference in Bucharest, modern contraceptives were available "for medical reasons," but were very hard to find and could no longer be privately purchased even when prescribed by a gynecologist. Only a few hospitals and research centers were authorized to continue importation of oral contraceptives. Even these exceptions were eliminated in 1985 when the sale of contraceptives in pharmacies was prohibited. Gynecologists who performed the few legally permitted abortions in hospitals were not allowed to insert IUDs afterward. The entire procedure was delineated by decrees and closely watched by the authorities to ensure compliance (Kligman, 1998). Modern contraceptives were simply not available, except for very high ranking party and government officials. Rhythm and *coitus interruptus* continued to be the country's major methods of contraception (Moskoff, 1978, 1980).

Although residents living along the northwestern border with Hungary had access to a greater variety of smuggled contraceptives, the lack of quality control, the high price, and the illegal medical procedure of IUD insertion created considerable barriers. While it was possible to find illegally produced domestic IUDs, pills, and spermicides made from cocoa butter and quinine on the black market, their cost was prohibitive for most women. Even a condom, when located, was said to be very expensive for Romanian circumstances. Permission for sterilization could be granted only by a special medical commission in cases where the woman met even more stringent criteria than those for legal abortion (David, 1990). Some physicians report that they refused to perform essential caesarean sections for fear of being accused of sterilizing the woman during the operation (Kligman, 1998).

Post-1990 Surveys

A national household probability survey of reproductive health issues was conducted in 1993 with a sample of 4.861 women aged 15–44 and regardless of marital status (Serbanescu and Morris, 1995; Serbanescu et al., 1995). Only 34 percent of respondents said that their most recent pregnancy was planned, 12 percent considered it mistimed, and 51 percent deemed it unwanted. Thus, almost two-thirds of women reported unintended pregnancies at time of last conception. Differences in educational level did not show a strong relationship with pregnancy planning. Responses reflected a high level of misinformation about modern contraceptives, contributing to low contraceptive prevalence. As of late 1992, only 57 percent of women living in union reported that they or their partner were using contraception; 43 percent relied on traditional methods (withdrawal and rhythm) and 14 percent mentioned modern methods. This was an improvement, however, from less than 1 percent reliance on modern methods in 1990 and about 5 percent in 1992 (Hopulele-Petri and de Graaf, 1992).

The 1993 survey noted that, after abortion was relegalized, contraceptive prevalence increased 20 percent but augmentation of traditional methods, rather than a change to modern methods, accounted for 70 percent of the increase. Major reasons for the continued high rates of unintended pregnancy were limited sexuality education, mistrust and misinformation about modern methods, lack of trained providers, and a shortage or uneven distribution of contraceptive supplies (Serbanescu et al., 1995).

Similar observations on contraception and abortion emerged from a structured interview study of Romanian women conducted between November 1991 and April 1992 (Johnson et al., 1996). While the real costs and accessibility of abortion and contraception were much closer than in the past, confusion, myths, and misunderstandings about modern methods continued. The prevailing public opinion is that modern contraceptives are harmful. It is still widely believed that IUDs cause cancer; that pills cause liver and heart disease and growth of body hair; and that tubal ligation causes fat, loss of sexual pleasure, and, sometimes, madness (Johnson et al., 1993). Clandestine and self-induced abortions persist, at times with tragic consequences.

Misconceptions are often shared by health care professionals who are reluctant to promote modern methods. Romanian doctors continue publishing articles about the dangerous side-effects of pills and IUDs (van Dam, 1993). This is mainly due to their lack of family planning training and limited access to current medical literature, but also because abortion services are far more remunerative for gynecologists than is the provision of contraceptive supplies and counseling (David, 1992a). In the time it takes to counsel one woman, four or five privately paid abortions can be performed. The system of "under-the-table-payment" for special services, a continuation from the practices of the communist regime, will likely be difficult to reverse. Women who repeatedly resort to abortion need postabortion counseling services that are not always available in places where abortions are provided (Bolton et al., 1997).

Despite women's strong desire to prevent unwanted pregnancy, surveys suggest that four out of five traditional method users do not want to change their current practice. As a major factor in their decision, besides the fear of side-effects of modern contraceptives, women mention partner resistance or preference for traditional methods, the difficulty of obtaining modern methods, and cost. One-fourth of women users of traditional methods indicated a physician's recommendation as an important reason for their continuance of these methods (Serbanescu and Morris, 1995). The influence of religion upon women appeared insignificant, both in contraceptive practice and in abortion decisions (Chaze, 1992).

In 1990 international donations of contraceptive commodities made it possible o distribute them at low or no cost. This situation changed as foreign exchange for purchases abroad was restricted and manufacture of local products was delayed. In 1995 the price of a monthly supply of pills was equivalent to the price of an abortion. There is concern that raising the price of abortion would encourage poor women to return to self-induced abortion or to seek cheaper procedures from unqualified persons. Within the context of the economic deterioration of everyday life and of the increase in unemployment, many women, particularly those with many children, cannot afford to pay the monthly cost of pills, which is not covered by health insurance. Misperceptions about voluntary sterilization persist (Bolton et al., 1997). Previous legislation, which allowed voluntary sterilization only for women with five or more children, for women over age 45, or for limited medical reasons, has yet to be modified (Serbanescu et al., 1995). Male sterilization, often confused with castration, is seldom discussed.

Family Planning Association

In February 1990 a group of 20 volunteers and physicians organized a nongovernmental organization, the Society for Education in Contraception and Sexuality (SECS). It has been affiliated with the International Planned Parenthood Federation (IPPF) since 1992 and has received considerable international donor support (Köö, 1992; Draghici, 1997). As of 1997 SECS had seven family planning clinics in Bucharest and in five other cities. They became model programs based on women-to-women services using trained general practitioners. In addition, SECS organized training programs for medical personnel, psychologists, social workers, and teachers oriented to the prevention of sexually transmitted diseases. Of particular importance was the opportunity to work with physicians who lacked prior training in sexuality education.

The small number of family planning clinics in the country, constraints in economic resources and management logistics, and shortages or uneven distribution of contraceptive supplies could be additional reasons for the limited impact of the family planning program in reducing resort to abortion. Governmental family planning clinics are located only in large urban areas (Köö, 1998). Rural areas are deprived of easy access. In the absence of an efficient advertising system, lack of media interest, and limited governmental commitment, family planning clinics are not well known. The deteriorating economic situation is seriously threatening their sustainability.

THE WOMAN QUESTION

Patriarchal Values

In the period before World War II Romania was a largely rural country with a strong peasant culture. Romanian society was characterized by deep social divisions between urban and rural populations, between ethnic groups, and between those with education or property and those without. The overwhelming majority (80 percent) of Romanians were peasants (Kligman, 1985, 1988, 1991; Fischer and Harsanyi, 1994). This meant that traditional patriarchal values and strictly defined gender roles dominated much of society, enduring well into the post–World War II socialist era.

The patriarchal rural family fostered the ideal of a subordinate and submissive woman. The man's superiority was intrinsic to this tradition as was the suppression of women's sexuality. This view was not universally shared. In 1895 Eugenia Ianculescu-Reuss edited the first feminist magazine, *The Woman's Rights*, and served as President of the League for the Rights and Responsibilities of Women. *Romanian Women*, which also began publication in 1895 under the editorship of

Andela Xenopol, challenged the stereotype of maternity as being the only societally accepted field of women's expression. Ideologically committed to the demise of the patriarchal society, the Romanian Communist Party actually built on its values while espousing an egalitarian spirit in theory. Some women, such as Ana Pauker and Elena Ceausescu, attained and wielded considerable political power (Fischer, 1985). For them, equality meant treating women like men (Kligman, 1994). Unlike men, however, women were expected to bear a triple burden: productive work, reproductive labor, and housework. Help from husbands was more evident in towns, particularly among better educated, intellectually sophisticated couples.

Within the context of patriarchal traditions, gender communication was codified and sexuality was viewed mostly in terms of procreation. Women's virtue was often linked to their sexual passivity. They were not supposed to enjoy sex. Intimacy was made more difficult by the lack of privacy in crowded apartments and houses. Men were allowed more freedom. Alcoholism and adultery were tolerated as long as the man fulfilled his obligation of providing for the family. The peasant psychology continued to flourish in provincial towns and working class cultures of the cities as the collectivization of agriculture and forced industrialization led to the ruralization of the town rather than the urbanization of the village (Gilberg, 1979).

In the interaction between the legal rights gained by women and the difficulties encountered in translating those rights into everyday realities, the major barriers were the traditional patriarchal system and women's biological role as the bearer of children. There was little recognition of the complexities of the parent-child relationship and its consequences for a working mother, the family, or the labor force. Nor was there an awareness that aspirations for a better quality of life would motivate limiting family size. Rather, it was believed that socialism would create ideal conditions for the family and that couples would wish to have more children.

Patriarchy and State Policy

By the mid-1960s the state and party had adopted the attitudes of the peasant culture, the state's rhetoric about the family resonated with familiar cultural patterns, and sexuality was eliminated from public discourse (Kligman, 1995). The decrees became tougher year by year as the state sought increasing control over reproduction. Women felt vulnerable and at risk since it was they, not their sexual partners, who were punished if the police discovered a clandestine abortion (Harsanyi, 1993; Kligman, 1998). Unmarried couples were forbidden to share a hotel room. Police checked hotels for illicit sexual activity at odd hours, often at 6:00 in the morning. A woman so discovered was arrested for prostitution while the man was allowed to leave unscathed (Harsanyi, 1993).

As the state grafted its policy onto the traditional patriarchal system, the pseudo-puritanical atmosphere was reinforced by a socialist reinterpretation of the peasant tradition of female modesty. Signs of feminine beauty or desirability gradually disappeared, being proclaimed harmful and indecent. Movies were censored so that not even a kiss or bare arm could be shown. Girls were educated to fear men, avoid pleasure, and see sex as a "necessary evil" accepted only for the sake of establishing a family (Harsanyi, 1993). Partner relations suffered as growing numbers of women refused to have sex for fear of an unwanted pregnancy (Puia and Hirtopeanu, 1990; Baban and David, 1994, 1996). Some women suggested that the only positive benefit was that men had fewer extramarital relationships, fearing an unwanted pregnancy (Johnson et al., 1996).

Communist education built on patriarchal traditions and instilled in young women the idea that sexuality outside the family and without purpose of procreation was immoral and led to insanity. The axiological system legitimated as supreme values the woman's personal sacrifice for the sake of the community and the effacement of her individuality. Conventional behavior was obedience to father, husband, and state, coupled with suspicion of the motives of others. A divorced woman, a childless married woman, or an attractive unmarried woman were incompatible with the image of the ideal communist woman: an asexual mother, a work and life comrade of man. Social conditions imposed and legitimated a woman's dependence on men despite claims of economic egalitarianism. The woman as a frontline worker in socialist competitions, heroine mother, outstanding housewife, and active member in community affairs, represented diverse sides of the "new socialist individual—multilaterally developed."

In 1966 nearly 80 percent of women aged 20 to 57 were economically active, representing 45 percent of the total labor force (David and McIntyre, 1981). This situation caused the regime to consider women not only a basic production factor, but also as a reproductive source of labor. The necessity of workforce renewal legitimated the strong pronatalist policy pursued for 23 years. The right to work was changed into an obligation. The person who did not join was called a social parasite and prosecuted (Harsanyi, 1993).

While the proportion of women in political positions and in Parliament increased substantially in the 1970s and 1980s, the price paid for such promotion was considerable. The women had to be politically quiescent and socially acceptable: married with children and asexual. Dressing elegantly or applying makeup were frowned upon. In actuality, women promoted under Ceausescu had less power than their male colleagues, regardless of formal titles. They had to demonstrate that they subordinated their particular interests to the needs of society as a whole—that is, the Communist Party. The National Council of Women neither represented nor served women. It functioned according to directives from above, supporting policies detrimental to women. As noted by Fischer and Harsanyi (1994), communist ideology reinforced for women the value of work outside the home but denied them the promised public facilities, convenience goods, and appliances to help with household tasks. Their right to be different from men or to express their sexuality was destroyed in the name of equality and the party's economic goals.

The great gap between woman's theoretical emancipation and the circumstances of everyday life undermined the slogan "woman's equality to man." Woman's professional choice was not always made according to personal development, but according to the new socioeconomic socialist context in which the husband's income could no longer assure family support and extra income was vital. To diminish the apparent conflict between woman's three roles (worker, wife, and mother), the socialist state created social and legislative facilities. But, like many other measures that had been taken, they were only theoretically or ideologically expressed. The reduced number and the low quality of nurseries and kindergartens, the lack of modern time-aving facilities for domestic work, and the everyday craving for scarce food made a joke of the "social protection" assurance of the socialist state. While the law permitted women to raise a child to the age of six without loss of job seniority, in actuality, the family could not dispense with the woman's income for such a period of time. In addition, workers were often summoned on Sundays for "patriotic work" to fulfill the "great five-years' plan," further extending Romania's official six-day work week and reducing family time.

The Transition Process

In the elections of May 1990, the first electoral campaign in more than 50 years, most major parties ignored women's issues. Women were elected to only 1 of 117 senate seats and 13 of 383 places in the Chamber of Deputies. In 1992 there were 16 women deputies and four women senators (Romania/MOL, 1993). As the president of the Romanian Association of University Graduate Women declared, the only agreement among so many conflicts in Romania was to ignore women (Chiochirca, 1994). In 1998, there were no women ministers in the Romanian government and less than 5 percent of parliament members were women (Köö, 1998).

The low involvement of women in politics is a natural response to the coerced and manipulated involvement they experienced earlier. When questioned, women express their revulsion against Elena Ceausescu and Ana Pauker and the hypocrisy of the former National Council of Women. Activists agree that most women wish to make different choices, to focus any extra energy on their families, or on newer economic opportunities that have appeared during the transition. Two incomes are necessary to cope with inflation and to hedge against rising unemployment. "The Elena Ceausescu Syndrome" and the antifeminist activities of the National Council of Women discourage many women from participating in political life (Campeanu, 1994).

The number of unemployed women doubled from 1992 to 1993. The female unemployment rate (61.9 percent) is considerably higher than that of men (38.1 percent). It is twice as high in the age group aged 25–29 years (Carlson, 1994). The differential is the result of higher layoffs and lower entrance rates (Earle, 1994). In June 1993, only 53 percent of the unemployed received unemployment benefits, 39 percent received social assistance, and 8 percent received nothing (Romania/MOL, 1993). With very limited social protection, there is a "feminization of poverty" (Kligman, 1994). Domestic violence continues to be widespread in all socioeconomic groups (Baban and David, 1997). There are no crisis centers or shelters, and media coverage is virtually nonexistent. The woman question has not become part of public discourse (Raduly, 1997). While traditional role models remain strong, there is also a growing recognition of the importance of the political process in shaping women's lives, developing new options, and securing women's rights and reproductive health.

SEX AND SOCIETY

Perceptions of Sexuality

Communist ideology led to pseudo-puritanism and hypocritical morality, eliminating sexuality from public debate in 1966 and submitting it to rigid regulations. To speak about sexuality as a human problem was deemed "vulgar libertinism, taken over from the West" whose "society was completely rotten" (Baban and David, 1997). Lack of contraceptives, fear of unwanted pregnancies, and the anxieties associated with an illegal abortion placed an awesome burden on couples' sexual lives. The silence on sexuality led to guilt, embarrassment, and distorted perceptions in partner relations.

Different women in different socioeconomic and marital circumstances and at different levels of sophistication were said to cope differently with the burdens imposed on them by an increasingly autocratic patriarchal regime, often depending on the support they had from husbands, family, and friends. Since it was impossible at the time to do research on how women actually coped with their diverse life situations, interviews were conducted in 1993 with a group of 50 women of diverse age, socioeconomic status, and cultural background. Each woman was asked at length about her knowledge of sexuality, attitudes expressed in the parental home, and experiences in married and unmarried partner relations (Baban and David, 1994, 1996; David and Baban, 1996).

In analyzing the interviews, five major themes were identified, running like a thread through nearly all the narratives:

1. Sex was seldom discussed at home. Parental prudery, ignorance, and prohibitions fostered a perception of sexuality as a taboo topic, resulting in misconceptions about sexual facts, overvaluation of virginity, a distorted relationship between boys and girls, and an interpretation of premarital sex as sinful and dirty. Young women were ill prepared to cope with their sexuality. While some of these conditions can be found in many cultures, nowhere in recent history was the combined authority, patriarchal tradition, and an autocratic state as heavy a burden for women as in Romania.
2. Sexuality was perceived as a source of personal stress with different meanings at different stages of women's sexual lives. Initial embarrassment, inadequacies, and shyness were followed by fears of unintended conceptions and unwanted pregnancies. Lack of access to modern contraceptives and fear of using contraception heightened stress, denied women the capability of controlling their own fertility, and impaired partner relations. Although sexuality posed a threat, marriage was perceived as a desirable social condition, providing an identity and a basis for attaining socially accepted motherhood.

3. Repeated unintended and unwanted pregnancies were traumatic and often catastrophic life events, further engendering in many women a sense of powerlessness and helpless dependence on their partners and on chance, a fate that robbed them of their dignity and undermined their sense of femininity.
4. While the reality of the unwanted pregnancy was viewed as a traumatic experience during the Ceausescu era, the decision to abort was usually based on practical life considerations and seldom evoked ethical concerns. Implementing the decision and finding a clandestine provider generated the greatest personal stress and anxiety about risks to physical and psychological health. Abortion was perceived as the ultimate and desperate means of a woman's asserting control over her own fertility and preserving the well-being of her small family. Few women reported short-term postabortion guilt or regret; most experienced relief following the resolution of the preabortion stress. The greatest psychological cost was the impairment of partner relations associated with postabortion reluctance to engage in sexual activities.
5. The economic struggle of daily life contributed to the erosion of partner relations. The narratives reflect poor interpersonal communications in emotional, professional, and sexual areas, while retaining a sense of cooperation in jointly confronting and resolving family problems. Fissures at the inner core were evident in the shared solidarity against the world while remaining solitary within the relationship. Discussions with Romanian colleagues suggest that more sophisticated intellectual couples coped more successfully.

At the political and social level, conscious distortions of existing reality were a part of daily life. They also appeared at the individual level, sometimes in an almost unconscious way. In the area around Cluj (where the study was conducted) the 50 women's attitudes and perceptions of sexuality were influenced by an amalgam of strongly pronatalist policies, difficult socioeconomic living conditions, and the patriarchal rural mentality flourishing under Ceausescu's regime. Although adapting to urban behavior after they had been transplanted, people preserved their old village values. Thus, despite the liberalization of abortion and greater availability of modern contraceptives, beginning in 1990, self-induced abortions and resort to untrained abortion providers continued.

A 1993 survey related to sexual myths, conducted among 186 women between 30 and 50 years of age, showed that 93 percent believed that men have a greater interest and readiness for sex than women; 81 percent held that men should take the initiative in making love; and 69 percent thought that satisfaction of men's sexual pleasure had priority over their own. One-half of subjects suggested that it was better for a woman to begin sexual life after marriage, while one-quarter accepted initiation of a sexual relationship with a future husband before marriage. Only 6 percent considered that a sexual relationship was necessary before marriage. However, almost all the women deemed it important that the man's sexual experience be more extensive and longer than their own. The survey group proved to have "traditional morals," giving greater "sexual rights" to men than to women (Baban, 1994).

Sexuality Education

In the 70 years between 1874 and 1944 seven books specifically oriented to sexuality education were published in Romania (Cressin, 1977). After the ascent of the communist regime, sexuality education was never mentioned. School courses were called "health education" and reduced to superficial aspects of hygiene, physiology, and human plumbing. Teen-oriented publications did not include themes on sexuality. Discussions about sexuality were not encouraged. Premarital counseling centers in public clinics were restricted to meeting the medical requirements of the marriage law.

With the fall of Ceausescu, the "sexual revolution" came to Romania. Film censorship was abandoned, pornography flourished, and prostitution and domestic violence increased (WHO/EURO, 1994). When sexuality education was not included in the post-1989 curriculum of the Ministry of Education, local nongovernmental organizations initiated programs with pupils, teachers, doctors, and nurses (Vinereanu, 1991; Chaze, 1992; SECS, 1993). Teaching materials were developed with international assistance and distributed in schools, in enterprises with a large proportion of women workers, on military bases, and in other public places. While constrained by the resistance of teachers, local school boards, and lack of adequate teacher training, the Ministry of Education in the mid-1990s adopted a sexuality education program developed by the Youth for Youth Foundation with international assistance (Draghici, 1997).

Young Adults

There is growing public awareness about the sexual and reproductive health problems of adolescents and young adults. Little communication exists between parents and children. Information about sexuality is obtained mostly from friends, the media, and pornographic movies, often leading to lack of confidence and low self-esteem (Liiceanu, 1994). In a 1993 survey of 1,595 women 15–24 years old, more than half (58 percent) reported that they had never had sexual intercourse. This included 83 percent of those aged between 15 and 19 years and 30 percent of those aged between 20 and 24 years. Of the women reporting sexual experience, 37 percent had their first sexual intercourse before age 18, 34 percent between 18 and 19 years, and 29 percent between 20 and 24 years. Earlier age at sexual debut was inversely related to education level and occurred more frequently in rural areas (Serbanescu and Morris, 1995).

A nationally representative survey of 2,025 women and 2,047 men 15–24 years old was conducted in 1996 (Serbanescu and Morris, 1997). About one of every two young women and about seven of every 10 young men were sexually experienced before age 20. Only 26 percent of women and 14 percent of men had ever discussed methods of fertility regulation with their parents. About 39 percent of unmarried women and 35 percent of unmarried men reported using any method of contraception at sexual debut. Of these, the majority relied on traditional methods such as withdrawal. Contraceptive practice at most recent intercourse was more efficient with 35 percent of unmarried young women and 43 percent of unmarried young men using a modern method. The condom was mentioned most frequently (Serbanescu and Morris, 1997).

Subsequent research using the WHO Narrative Research Methodology (WHO, 1993) showed that young adolescents try to hide their emerging relationships from their parents (Meekers et al., 1997). Because parents tend to have more traditional conservative attitudes, communication about reproductive health issues is stifled. And, because many parents still try to control their childrens' sexual behavior, they seldom volunteer information just as their parents never did.

FUTURE PERSPECTIVES

The authoritarian pronatalist regime of Nicolae Ceausescu, superimposed upon traditional rural Romanian attitudes and values, created an atmosphere that deprived women of becoming equal partners with men in their expression of sexuality and in exercising their human rights to reproductive health. Lack of access to modern contraceptives and legal abortion resulted in personal tragedies for many. All too often, their partners, like their fathers, were domineering, denigrating, and demanding. Mothers were frequently nonsupportive and noncommunicative about sexual matters. It was a very difficult time to be a young woman or wife.

Along with the political and economic changes that began in December 1989, a social revolution occurred in the realm of women's reproductive rights. Abolition by the postcommunist government of the draconian law forbidding women access to both safe abortion and modern means of contraception was followed by notable improvements in women's health.

Liberalization of abortion legislation was to guarantee ready and equal access to safe procedures for women of all social strata. Some family planning clinics and women's reproductive health centers have been established but the concept of family planning remains poorly developed. To combat lack of information and suspicion of modern contraceptives and to diminish the tradition of using abortion to regulate fertility, programs of mass education and professional training were initiated. Sexuality education was introduced in schools although implementation is still sporadic. Surveys indicate that legislative changes did not significantly change traditional contraceptive practices and reliance on abortion. A wide gap remains between program aims and implementation. Despite improved health services, Romania continues to have the highest reported maternal mortality in Europe. Placing their trust in nonmedical abortion providers who may be more affordable or more familiar, the practice of clandestine abortion persists, especially among women who wish to terminate unwanted pregnancies beyond the legal gestational limit of 12 weeks. Postabortion counseling is very limited and prenatal services seldom address postpartum contraceptive needs although, officially, any gynecologist may prescribe contraceptives and insert IUDs (Serbanescu et al., 1995).

The Romanian experience of the Ceausescu era demonstrated that women's private behavior prevailed over severely restrictive public policies, regardless of personal risks or costs to health. While the severe prohibitions on access to safe abortion and modern contraceptives were lifted immediately after the fall of Ceausescu, much work remains to be done if access to modern contraceptives is to be demedicalized, if reliance on effective contraception is to replace dependence on abortion, and if gender shared responsible sexuality is to become a national norm.

chapter 11

Russian Federation and USSR Successor States

Andrej A. Popov and Henry P. David

EDITOR'S INTRODUCTION

The Russian Federation, the largest republic in the Former Soviet Union, became independent on 8 June 1990 when the Congress of the People's Deputies of the Russian Federation approved a declaration that Russian sovereignty should supersede that of the Soviet Union (*The New York Times*, 1990). Historically, Russia had been a sovereign state with Moscow as its capital since 1480 when Tatar domination ended. The Russian Empire, with St. Petersburg as its capital, emerged in 1721 under the rule of Czar Peter the Great. After the emancipation of the serfs in 1861, peasants flocked from rural areas to the cities. It was from the urban proletariat that Lenin drew support for the 1917 Bolshevik Revolution that toppled the Romanov Dynasty. In 1918 Moscow was selected as the capital again. Established immediately after the October Revolution, the Russian Soviet Federated Socialist Republic (RSFSR) became a Union Republic on 22 December 1922 when the Union of Soviet Socialist Republics (USSR) was created as a federal state. It spanned a multitude of diverse nationalities, traditions, and cultures with wide variations in reproductive behavior. For the next 70 years the Communist Party was the font of all power. The governmental organizations were the instruments for implementing party decisions.

Following some relaxation of Soviet controls during *glasnost* (1985) and *perestroika* (1987), the Soviet leadership in 1988 created new legislatures for Russia and the other union republics, permitting multicandidate elections. In 1991 Boris Yeltsin was elected President of the Russian Federation, the first executive in Soviet history to be elected by direct popular vote. On 25 December 1991 Mikhail Gorbachev resigned as President of the Soviet Union. The Communist Party collapsed. The implosion of the USSR and the peaceful creation of an independent

Russian Federation and 14 successor states was one of the most dramatic events of the twentieth century.

The Russian Federation is the world's largest country, stretching across 11 time zones from the Baltic Sea to the Bering Strait. Three-quarters of the country's 147 million citizens live west of the Ural mountains. The federation is comprised of regional jurisdictions with varying degrees of autonomy, including republics, districts, and metropolitan areas.

While the Soviet Union was usually treated as a single entity in international publications, the 15 former Soviet republics were generally classified into four groups: Slavic regions and Moldova, consisting of Belarus, Russia, and the Ukraine plus the non-Slavic country of Moldova (once a principality of Romania); the Baltic countries of Estonia, Latvia, and Lithuania, all annexed in 1940; the Caucasian republics of Armenia, Azerbaijan, and Georgia; and the Central Asian lands of Kazakstan, Kyrgyz Republic, Tajikistan, Turkmenistan, and Uzbekistan. As of 1997 United Nations publications list the Slavic republics (including the Russian Federation) and Moldova under Eastern Europe, the Baltic republics under Northern Europe, the Caucasus under Western Asia, and the Central Asian republics under South-Central Asia.

In the years following the demise of the Soviet Union, the search continues for a single term to describe the group of countries that emerged from the rubble (Goble, 1997). The terms proposed seem less appropriate as time passes. For example, reference to the "Former Soviet Union" diminishes the status of successor states and overstates similarities among several countries whose only commonality was shared Soviet occupation. The term "Newly Independent States" is without historical precedent. In the past, countries arising from the ruins of an empire were quickly viewed as nations like all others. For the purposes of this chapter, it was decided to focus on historical Soviet trends and on the Russian Federation with additional notes on the 14 other USSR successor states.

This chapter will (1) present evolving trends in population policy and demographics, (2) examine the effects of pronatalist measures and the health system; (3) review the impact of legalization, recriminalization, and reliberalization of abortion and the many problems associated with interpreting officially reported abortion data; (4) note the wavering policies on the availability of contraceptives and the contraception/abortion interrelationship; (5) discuss the woman question and the influence of the multiple burdens women had to bear; and (6) consider Russian traditions and changing attitudes to sex and society as reflected in sexuality education, adolescent sexuality, and the work of the Russian Family Planning Association. The reader is cautioned that, for reasons discussed in the chapter, cited official statistics leave much to be desired in terms of accuracy and confidence (Avdeev, 1998; Zakharov, 1998a, 1998b). Discrepancies exist between diverse sources. In this chapter particular reliance has been placed on the time series produced by official sources identified in the tables.

Parts of this chapter were drafted by Andrej A. Popov before his tragic accidental death in April 1995 at age 38. Dedicated to his memory, the chapter was completed by the editor. The suggestions of Libby Antarsh, Elena Ballaeva, Valentina Bodrova, Valery Chervyakov, Carl Haub, Timothy Heleniak, Stanley Henshaw, Igor Kon, Gulnara Kuzubaeva, Marina Malysheva, Olga Voronina, Michele Rivkin-Fish, and Sergei Zakharov are gratefully acknowledged. Sergei Zakharov was especially helpful in providing officially reported demographic data. Responsibility for any errors or misinterpretations in the final version rests with the editor.—H.P.D.

POPULATION POLICY

Few of the early Marxists produced papers that could be clearly identified as demographic. Marx and Engels believed that the socialist society would have at its disposal a constantly expanding productive capacity and that demographic processes would be brought under the rational control of a cooperative community (Engels, 1983 [1881]). In *The Origin of the Family, Private Property and the State*, Engels, 1972 [1884]) pointed to the family and its subjugation of women as a symptom of the exploitative social order and as a means of preserving private property, passed from male to male. In a 1913 article, Lenin advocated abolition of Czarist legislation prohibiting abortion and unrestricted dissemination of medical information on contraception. At the same time, Lenin opposed neo-Malthusianism, holding that a growing proletariat was an essential prerequisite for a victorious socialist revolution. Lenin was concerned more with addressing the social and economic causes of abortion than with preventing abortion by repression that compromised the health of women (Vostrikova, 1964; Podyashchikh, 1968).

The Soviet government never promulgated a population policy. When Premier Kosygin refused to sign the 1966 United Nations Statement on Population, he expressed the view that reproductive behavior was a private family affair that should not be subject to state planning. It was generally held that the solution to the world population problem depended on implementing basic social and economic reforms, but that demographic policies and family planning could make a contribution toward achieving a better balance between population and economic growth (Guzevaty, 1966, 1974; Ryabushkin, 1978).

Most Soviet demographers experienced great tensions between the ethical need for personal/private decision making on whether or when to have a child and the need to represent state interests in fostering economic growth (Perevedentsev, 1972; Valentei and Kwasha, 1974; Lapidus, 1978). Thus, the retirement age for women was reduced to 55 years (compared to 60 years for men), partially to encourage grandmothers ("babushkas") to stay home to care for their grandchildren, who often shared the same flat (Ballaeva, 1997). The term "population control" was rarely used, perhaps because in Russian it suggested bureaucrats stationed at the bedroom door (Besemeres, 1975).

DEMOGRAPHIC TRENDS, POLICIES, AND PUBLIC HEALTH

Historical Context

The total fertility rate in Russia was about 7.5 children per woman as estimated from the 1897 census, followed by considerable fluctuations and a sharply declining trend from 1910 onward (Coale et al., 1979; David and McIntyre, 1981; Ross and Frejka, 1998). The October Revolution dramatically changed the family life of Imperial Russia (Ransel, 1979). In 1918 a Family Code was promulgated, equalizing women's status with men's, removing marriage from church control, and making divorce easier to obtain. In 1926 a new Family Code was enacted for the RSFSR, which accorded *de facto* relationships equal rights with legally registered marriages and authorized divorce when either partner registered a desire for it (Schlesinger, 1949). Children received full legal rights whether born in- or out-of-wedlock. Other measures and official statements promulgated within a decade of the revolution reflected a determination to alter radically the institution of marriage, sexual norms, and family life (Lorimer, 1946, 1958; Peterson, 1964; Mace and Mace, 1963; Geiger, 1968; Bronfenbrenner, 1968; Juviler, 1978; Heitlinger, 1979; Besemeres, 1980; Yvert-Jolu, 1981).

At the beginning of the twentieth century early marriage was encouraged for social and economic reasons (Vishnevsky, 1996). During the late 1920s more conservative views regained favor, followed by a complete reorientation of official policy in the 1930s. Abortion on request, promulgated in 1920, was prohibited in 1936, divorce was made more difficult, and penalties increased for nonpayment of alimony. In 1941 a tax was introduced on the earnings of single persons and married couples without children. In 1944 legislation withdrew the legal recognition of quasi-conjugal rights formerly accorded to *de facto* wives and single mothers (text in Schlesinger, 1949). Though women were not encouraged to abandon their jobs for the home, traditional sex roles were reasserted, previously expressed radical views on the family were recanted, and sexual license was deemed a remnant of the capitalist past (Schlesinger, 1949). Wedding rings, once considered the yoke of women under capitalism, reappeared in the shops (Murstein, 1974).

Demographic research institutes were closed at the USSR Academy of Sciences in Leningrad in 1934 and at the Ukrainian SSR Academy of Sciences in Kiev in 1938. This was in part because specialists in demography and statistics were "reluctant" to accept the official dogma of "an absence of population problems" in the USSR. The Great Famine of 1933–34 created a demographic catastrophe, a fact hidden from the public. The 1937 Census reflected the terrible results of agricultural collectivization. Some Census organizers were purged and others were executed. Stalin declared the 1937 Census "wrong and harmful," ordering the findings burned (Zakharov, 1998a). Research on the family's role in society ended when a resolution of 4 July 1936 condemned the application of empirical and quantitative methods to social policy questions (Liegle, 1975). The terms "sociology" or "behavioral science" were deemed synonyms for a bourgeois science of society (Shalin, 1978).

After Stalin's death in 1953, the authoritarian structure of family legislation gradually crumbled. Exemptions to the childlessness tax were introduced in 1954, abortion on request was relegalized in 1955, and a revised Family Code reforming divorce proceedings was promulgated in 1968. While official policy on sexual matters remained conservative, private sexual behavior differed significantly from public Victorian prudishness (David, 1974; Smith, 1977; Heitlinger, 1979; Besemeres, 1980).

Following the Twentieth Party Congress in 1956, social science research gradually revived. In 1958 the Soviet Sociological Association was organized (Shalin, 1978). Research on the family accelerated during the 1960s with the founding of the Scientific Research Institute of the Central Statistical Office (1963), the publication of Kharchev's monograph on *Marriage and the Family in the USSR* (1964), and the establishment of the Center for the Study of Population Problems at Moscow State University in 1968.

Marriage and Divorce

Early and universal marriage has always been typical of Russia (Vishnevsky, 1996). Childbearing typically followed within a year of marriage (Willekens and Scherbov, 1995). Since the 1960s, the proportion of premaritally conceived pregnancies increased. For example, in his analysis of the archives of the Leningrad (St. Petersburg) Baby Registry, Golod found that in 1963 some 24 percent of couples had conceived their first child three months before marriage. The percentage rose to 28 percent in 1973, 38 percent in 1978, and to 49 percent in 1984 (cited in Kon, 1995: 169). During the 1980s the rate of nonmarital childbearing began to climb, reaching 18 percent of all births by 1990 (Hollander, 1997). By 1996, 23 percent of all births were out-of-wedlock (Zakharov, 1997). Within marriage, the interval between the date of marriage and first birth declined to less than six months in 1993 (Ivanova, 1996). Cohabitation gradually became more accepted but the proportion remains relatively small (Frejka, 1998). A 1989 poll showed that approval was greatest among younger persons with higher education. Public opinion was more tolerant of unregistered unions than of voluntary childlessness (Matskovksi and Bodrova, 1990).

The divorce rate per 1,000 population rose from 0.5 in 1950 to 4.6 in 1994 before declining to 3.8 in 1997 (Yvert-Jolu, 1981; Russia/Goskomstat, 1997; Vishnevsky, 1996; COE, 1998). In a 1997 representative survey in Rybinsk it was noted that the unmarried mother model is increasingly preferred by single women. They fear that the child's father will become an added economic burden. Many women in Rybinsk are the sole income earners for their families (Ballaeva, 1997).

Births and Total Fertility Rates

In the nineteenth century lack of widespread fertility regulation in combination with the overall high nuptiality level was responsible for the average Russian woman giving birth to seven or more children during her reproductive life. In 1890, the birth rate was 50.3 per 1,000 population (Mitchell, 1975). At that time total fertility rates (TFR) in Russian European regions varied between 5.9 and 8.7 per woman (Zakharov, 1994). Beginning in the twentieth century, the birth rate dropped to 30.9 per 1,000 population in 1920 and to 23.8 in 1946. It has been suggested that the mechanism of fertility decline operating at the family level was influenced by the continuing stress associated with the population losses incurred during the civil war and epidemics (1917–23), the famines, purges, and forced agricultural collectivization of the 1930s, and the 25–27 million deaths during the Great Patriotic War (Feshbach, 1982a, 1982b; Haub, 1994). One result was repeated postponement of childbearing (Zakharov and Ivanova, 1996a).

As noted in Table 11.1, birth rates in the USSR declined from 25.4 per 1,000 population in 1957 to 15.9 in 1991, the last year of the union. Total fertility rates decreased from 2.82 per woman in 1957 to 2.27 in 1990. These figures masked considerable differences between the constituent republics (Clem, 1977, 1980).

As shown in Table 11.2, the birth rate in the Russian Federation decreased from 23.2 per 1,000 population in 1960 to 14.6 in 1970, rose to 15.9 in 1980, peaked at 17.5 in 1983, and then fell steadily to 13.4 in 1990 and 8.6 in 1997. Meanwhile, the TFR declined from 2.54 per woman in 1960 to 2.01 in 1970, to 1.89 in 1980 and 1990, and to 1.23 in 1997. Trends in births and abortions are presented in Figure 11.1.

Pronatalist Policies

The first explicitly pronatalist incentives were introduced in 1936 at the same time that the liberal abortion system was revoked (see section on abortion). Initial family allowance payments and nonmonetary awards focused on higher order births (three and above) and were paid annually for five years. In 1944 payments were increased, extended to third and higher order births, and paid on a monthly basis until the fifteenth birthday (Urlanis, 1974). A progressive cash birth bonus payment for second and higher order births was also introduced in 1944. Awards were established for "heroine mothers" (Lorimer, 1958). A tax was levied on the income of persons with fewer than three children, including childless single or widowed women. In 1948 all family allowances were cut in half. None of these measures had a significant effect on the RSFSR birth rate, which, after rising briefly to 26.7 in 1950, continued its decline to 15.8 in 1979 when the TFR reached a low of 1.89. While abortion liberalization in 1920 may have contributed to the fertility decline and rerestriction in 1936 facilitated growth, reliberalization in 1956 had no marked effect on overall fertility levels in the Soviet Union (Pressat, 1979).

On 2 September 1981 the Presidium of the Supreme Council of the USSR enacted a law "on measures of improving state financial assistance to women with children" (Popov, 1993). Involving a regional-differential demographic policy, it introduced a broad range of financial and social measures to stimulate the birth of second and third children in the European part of the Soviet Union. A partially paid

Figure 11.1
Births and Legally Induced Abortions: Russian Federation, 1960–96

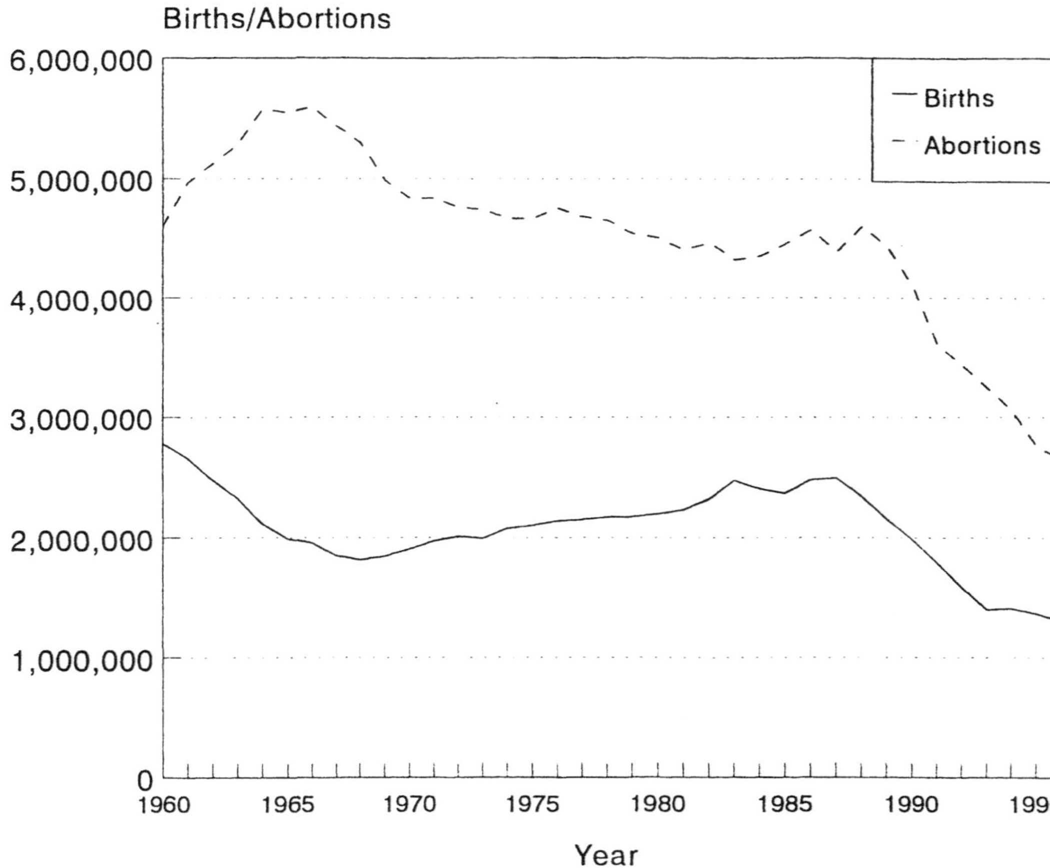

Sources: See data from Table 11.2.

Table 11.1
Birth, Abortion, and Fertility Rates and Ratios: USSR, 1957–91

Year	Live Births	Legally Induced Abortions	Birth Rate	Legal Abortion Rate[a]	Legal Abortion Ratio[b]	Total Fertility Rate
1957	5,163,700	5,109,000	25.4	84.7	989	2.82
1958	5,239,600	5,927,000	25.3	98.6	1,131	2.81
1959	5,264,500	6,014,000	25.0	101.1	1,142	2.84
1960	5,341,000	6,623,000	24.9	112.0	1,240	2.80
1961	5,191,900	7,109,000	23.8	119.6	1,369	2.67
1962	4,959,100	7,361,000	22.4	123.1	1,484	2.61
1963	4,757,700	7,617,000	21.1	126.7	1,601	2.52
1964	4,456,800	7,251,000	19.5	120.1	1,627	2.45
1965	4,253,100	7,260,000	18.4	119.6	1,707	2.45
1966	4,241,600	7,904,000	18.2	129.4	1,863	2.43
1967	4,093,100	7,878,000	17.3	127.7	1,925	2.40
1968	4,087,900	7,611,000	17.2	122.3	1,862	2.39
1969	4,087,000	7,256,000	17.0	120.9	1,775	2.42
1970	4,225,600	6,970,000	17.4	110.7	1,650	2.46
1971	4,314,800	7,262,000	17.8	112.5	1,683	2.46
1972	4,404,000	7,160,000	17.8	109.4	1,626	2.42
1973	4,386,200	7,160,000	17.6	108.2	1,632	2.41

1974	4,546,100	7,114,000	18.0	106.6	1,565	2.41
1975	4,611,500	7,135,000	18.1	105.7	1,547	2.40
1976	4,719,700	7,293,000	18.4	107.4	1,545	2.35
1977	4,693,400	7,238,000	18.1	106.1	1,542	2.30
1978	4,763,400	7,160,000	18.2	104.7	1,503	2.27
1979	4,807,100	7,009,000	18.3	102.4	1,458	2.24
1980	4,851,400	7,003,000	18.3	102.3	1,444	2.24
1981	4,961,400	6,834,000	18.5	99.6	1,377	2.27
1982	5,100,300	6,912,000	18.9	100.3	1,355	2.35
1983	5,391,900	6,765,000	19.8	97.7	1,255	2.40
1984	5,386,900	6,780,000	19.6	97.2	1,259	2.40
1985	5,374,400	7,034,000	19.4	100.3	1,309	2.46
1986	5,610,800	7,116,000	20.0	101.2	1,268	2.52
1987	5,599,200	6,818,000	19.8	97.0	1,218	2.53
1988	5,381,100	7,228,000	18.9	103.2	1,343	2.45
1989	5,062,200	6,974,000	17.6	99.8	1,378	2.34
1990	4,853,000	6,194,000	16.8	89.8	1,276	2.27
1991	4,642,800	na	15.9	na	na	na

[a]Per 1,000 women age 15–44. [b]Per 1,000 live births.

Sources: USSR/MOH (1975–91), Goskomstat (1988), and Zakharov (1998b).

Table 11.2
Birth, Abortion, and Fertility Rates and Ratios: Russian Federation, 1960–97

Year	Live Births	Legally Induced Abortions	Birth Rate	Legal Abortion Rate[a]	Legal Abortion Ratio[b]	Total Fertility Rate
1960	2,782,400	4,600,000	23.2	161.2	1,653	2.54
1961	2,662,100	4,964,100	21.9	174.5	1,865	2.42
1962	2,482,500	5,121,700	20.2	178.5	2,063	2.31
1963	2,331,500	5,279,000	18.7	182.8	2,264	2.23
1964	2,122,000	5,572,300	16.9	191.4	2,626	2.14
1965	1,990,500	5,543,500	15.7	188.0	2,785	2.13
1966	1,957,800	5,591,000	15.3	187.3	2,856	2.07
1967	1,851,000	5,436,000	14.4	180.0	2,937	2.00
1968	1,816,500	5,298,700	14.1	173.9	2,917	1.98
1969	1,847,600	4,987,200	14.2	162.4	2,699	1.97
1970	1,903,700	4,837,700	14.6	156.3	2,541	2.01
1971	1,974,600	4,838,500	15.1	155.7	2,450	2.05
1972	2,014,600	4,765,900	15.3	153.3	2,366	2.02
1973	1,994,600	4,742,900	15.1	152.5	2,378	2.00
1974	2,079,800	4,673,900	15.6	150.3	2,247	1.99
1975	2,106,100	4,670,700	15.7	149.8	2,218	1.97
1976	2,146,700	4,756,800	15.9	151.6	2,216	1.97
1977	2,156,700	4,685,900	15.8	148.3	2,173	1.94

1978	2,179,000	4,655,800	15.9	146.3	2,137	1.90
1979	2,178,500	4,540,200	15.8	141.6	2,084	1.89
1980	2,202,800	4,506,000	15.9	140.2	2,046	1.89
1981	2,236,600	4,400,400	16.0	137.3	1,967	1.91
1982	2,328,000	4,462,800	16.6	140.4	1,917	2.04
1983	2,478,300	4,317,100	17.5	137.3	1,742	2.11
1984	2,409,600	4,347,000	16.9	139.7	1,804	2.06
1985	2,375,100	4,454,400	16.6	144.0	1,875	2.05
1986	2,485,900	4,579,400	17.2	148.1	1,842	2.15
1987	2,500,000	4,385,600	17.2	148.8	1,754	2.22
1988	2,348,500	4,609,000	16.0	145.5	1,962	2.12
1989	2,160,600	4,427,700	14.6	137.2	2,049	2.01
1990	1,988,900	4,103,400	13.4	125.6	2,063	1.89
1991	1,794,600	3,608,400	12.1	109.5	2,011	1.73
1992	1,587,600	3,436,700	10.7	103.7	2,165	1.55
1993	1,398,300	3,244,000	9.4	97.7	2,320	1.39
1994	1,408,200	3,060,200	9.6	92.1	2,173	1.40
1995	1,363,800	2,766,400	9.3	83.3	2,028	1.34
1996	1,304,600	2,652,000	8.9	79.8	2,033	1.28
1997	1,259,900	2,498,100	8.6	75.7	2,016	1.23

[a]Per 1,000 women age 15–44. [b]Per 1,000 live births.

Sources: Avdeev and Monner (1996), Goskomstat of Russia (1998), COE (1998), Zakharov (1997, 1998b, 1999), and USBOC (1998).

maternity leave was introduced, initially for one year and subsequently extended to three years. Having a "large family," officially defined as having three children, determined eligibility for benefits in public services and housing (Zakharov and Ivanova, 1996a). As shown in Table 11.2, there was an immediate response in the Russian Federation birth rate. It climbed to 17.5 per 1,000 population in 1983 and remained fairly stable at around 17.0 until 1988 when it decreased to 16.0. It then continued its decline to 14.6 in 1989 and 8.6 in 1997. Similarly, the TFR rose to 2.11 per woman in 1983 and reached 2.22 in 1987 before falling to 2.01 in 1989, 1.73 in 1991, and 1.23 in 1997.

Zakharov and Ivanova (1996a) hold that the pronatalist incentives introduced during the 1980s resulted in women having their first births at progressively younger ages and spacing them more closely, thus achieving their reproductive goals earlier. At the beginning of the 1990s there was a reversal in the trend toward reduced birth intervals; second and third births were postponed. However, as previously noted, births arrived ever earlier after marriage. "The shift to longer intervals between births seems to be an inevitable negative compensation for the pronatalist policy pursued during the 1980s" (Zakharov and Ivanova, 1996a: 53). At the same time economic conditions and changes in life styles, especially in the timing of marriage and family formation, also had some effects (Zakharov and Ivanova, 1996b).

Family Size Intentions

Research on family size intentions and actualities was initiated in the mid-1960s by the Laboratory of Demography of the Research Institute of the USSR Central Statistical Office (Belova and Darsky, 1968). When asked their husband's opinion on desired family size in the 1969 Soviet National Sample Survey, women usually responded that they thought their husbands wanted more children (except in the high fertility republics, where opinions were more likely to converge). Many women, however, dismissed their husbands' views as irrelevant and having little influence on their childbearing decisions (Belova and Darsky, 1975-76). With the economic collapse of the early 1990s both the desired number of children and the birth rate declined. The results of the 1994 VCIOM survey showed a lack of confidence in the future and continuing low fertility perspectives among Russian couples (Bodrova, 1994). In late 1997 many young urban couples were reluctant to have even one child (Hiatt, 1997).

Health Care

In the 1920s Soviet medicine was dominated by the perception that "illness" was a product of a capitalist society and that socialism would rid society of alcoholism, prostitution, drug abuse, and so on (Solomon, 1989). This orientation deemphasized the scientific and clinical approach and produced physicians poorly trained in modern methods. It was not until Stalin's decision to collectivize agriculture and industrialize the Soviet Union that the health care system adopted a more scientific approach to its mission of maintaining a healthy labor force (Sigrist, 1947). Health care was defined as a right of citizenship and became a public service provided by the state that employed all health personnel. Services were free, provided in state-owned facilities, and financed by the state budget and payments from industrial enterprises. Health care was oriented to improving public health. For example, obligatory examinations of women in their work places helped to cure many gynecological disorders from the 1960s to the 1980s. However, a physician was more than a medical doctor. He/she was a state representative with primary responsibility to the state, not to individual patients.

By the time of *perestroika* and *glasnost*, the health system had deteriorated severely (Chazov, 1992; Randolph, 1996). The low staff salaries, shortages of modern equipment and drugs, and the often very poor quality of care encouraged bribery to obtain better services (Paton, 1989; Field, 1990; Ryan, 1990; Feshbach and Friendly, 1992; Barr and Field, 1996). Conditions were exacerbated by the economic and political problems that followed. In many facilities patients were obliged to purchase medical supplies at nearby kiosks prior to obtaining in- or outpatient services (Erlanger, 1992; Haub, 1994; Cassileth et al., 1995; *The Economist*, 1994b; Rozenfeld, 1996). In 1993 a compulsory national health insurance plan was introduced to augment funding (Tulchinsky and Varavikova, 1996; Burger et al., 1998). Subsequently the state health care system became more interested in maintaining the physical well-being of women of fertile age while giving less attention to the reproductive health of older women (Ballaeva, 1997). In a 1998 survey, Russia ranked lowest among 27 European countries on an index of 14 different health indicators (*The Economist*, 1998d).

Infant Mortality and Life Expectancy

A review of Soviet publications suggests that infant mortality declined from 81 per 1,000 live births in 1950 to 21.2 in 1971, rising subsequently to 22.8 in 1974, the year publication of statistics ceased (Davis and Feshbach, 1980). Among the factors believed to contribute to the increase were alcoholism (mostly in men causing "drunk conceptions"), the breakup of extended families, and the declining budgetary support for the health system (Eberstadt, 1981, 1990, 1993, 1994a). After the Ministry of Health resumed publication of statistics, the infant mortality rate remained fairly stable around 18.0 between 1989 and 1992, a rate considerably higher than in other European countries. Moreover, the official definition of infant mortality differed from that recommended by the World Health Organization (WHO), resulting in an understatement in comparison to other countries (Anderson and Silver, 1986). In the USSR, babies born alive after only 28 weeks' gestation, less than 1,000 grams in weight and 35 centimeters in length, or who died within seven days of birth, were excluded from counts of infant deaths. The successor states are adopting the WHO standard.

During the 1970s life expectancy at birth started to decline. It stabilized in the 1980s but dropped dramatically between 1989 and 1994 from 64.2 to 57.7 years for men and from 74.4 to 71.2 years for women (Notzon, 1998). Likely explanations were widely explored (e.g., Chen et al., 1996; Shkolnikov and Mesle, 1996; Shkolnikov et al., 1998). In 1995 and 1996 life expectancy at birth increased 2.2 years for Russian men and 1.4 years for women, still the lowest in 40 years (Leon and Shkolnikov, 1998).

ABORTION

1649–1919

Traditionally, abortion in Russia was prohibited by law. Capital punishment for inducing abortion was introduced in 1649 (Sadvokasova, 1969; Popov, 1993, 1994b). During the reign of Peter the Great, capital punishment was abolished and punishment for abortion reduced. The 1845 Penal Code defined abortion as a premeditated act. Article 1462 stipulated that anyone convicted of performing an abortion without the woman's consent was liable to a loss of civil rights and four to six years of hard labor (usually in Siberia). Self-induced abortion or a procedure performed with the woman's consent also entailed loss of civil rights with resettlement in Siberia for an unspecified period (Engelstein, 1992). In practice, the law was rarely implemented as most juries rendered a verdict of "not guilty" (Kon, 1997).

At the close of the nineteeth century, every fifth pregnancy was believed to end in abortion (Sadvokasova, 1969). In the period before World War I, abortion became a matter of public health concern. Repeal of the restrictive law was urged a the 1913 Congress of the Pirogov Society of Russian Physicians as a means of reducing the large number of women coming to hospitals with complications from illegal procedures (Heitlinger, 1979; Engelstein, 1992). One of the leading Russian philosophers of the time, V. V. Rozanov, responded by declaring that a fetus is state property and that damaging national property requires strong punishment (cited by Vasilevsky and Vasilevsky, 1924; Popov, 1994b). This sentiment was echoed more than 70 years later by Ceausescu in Romania (see Chapter 10).

1920–34

On 18 November 1920, three years after the revolution, the Soviet Union became the first country to legalize abortion on request of the woman. A Joint Decree was issued by the Commmisariats of Health and Justice together with an explanatory commentary (USSR, 1920; Field, 1956). It stipulated that abortions could be provided only in state hospitals by licensed physicians without charge to the woman. Physicians charging a fee could be prosecuted. Those found guilty of inducing an illegal procedure could lose their licenses and were liable to criminal trial.

The 1920 decree, liberalizing abortion, was intended to be temporary. It was thought that, as social conditions improved and the state assumed more of the burdens of child rearing, the problem of unwanted pregnancy would disappear and reduce reliance on abortion (Callahan, 1970). Childbearing was deemed a woman's obligation and fertility regulation was considered antisocialist by some leading Soviet feminists (Kollontai, quoted in Holt, 1977 and Stloukal, 1993). However, there were also those who viewed abortion as a means of eugenic cleansing of the population and advocated compulsory surgical sterilization for persons suffering mental disease (Gens, 1927; Tomilin, 1927; Popov, 1994b).

Abortions were performed, often without anesthesia, by licensed physicians in the "abortoria" of local hospitals. On 3 November 1924 a fee was introduced for abortions performed for nonmedical reasons (Mace and Mace, 1963). The fee was waived for indigent women but special commissions controlled the number of free procedures that could be provided in each facility (Avdeev, 1989). Also in 1924, private abortion clinics were established in Moscow and Leningrad. An abortion industry emerged as abortion became the cheapest and most readily available method of fertility regulation. The public health system was adapted to abortion technology (Popov, 1993, 1994b).

As women were exhorted to enter the labor force and the organization of child care facilities lagged, the demand for abortions increased. In Moscow, the birth rate declined from 30.6 per 1,000 population in 1921 to 28.9 in 1926 and 14.7 in 1934 (Lorimer, 1946). Meanwhile, abortions per 1,000 live births rose from 190 in 1921 to 550 in 1926 and 2,710 in 1934 (Lorimer, 1946). The increase reflected the greater availability of abortions and the growing motivation to reduce fertility in the face of severe housing shortages, upheavals of collectivization, and difficult living conditions.

1935–54

In the early 1930s access to abortion was gradually restricted (Stloukal, 1993). Communist Party members were *de facto* denied access to legal abortion (Avdeev, 1989). Beginning in 1935, a new regulation was issued, prohibiting abortions for first pregnancies, restricting the time limit to the first three months after conception with a minimum of six months between procedures, and prescribing curettage without anesthetic as the only method of pregnancy termination (Mace and Mace, 1963). A 27 June 1936 decree prohibited abortion except when performed for compelling medical or eugenic reasons (USSR, 1936; Sigrist, 1947). Any physician performing an illegal procedure was liable to imprisonment of one to two years. Medically unqualified practitioners faced a longer sentence. A pregnant woman attempting to self-induce was to be reprimanded for the first offense and fined up to 3,000 rubles (approximately equal to a year's average income) for a second offense. Contraceptives were not prohibited but their acceptance was not promoted. A decree to increase the production of rubber products for prophylactic purposes was never fully implemented (USSR/Central Executive Committee and Council of People's Commissars, 1936).

The commentary accompanying the 1936 decree justified the liberal 1920 abortion law as a "regrettable necessity," related to the chaos of the civil war (1917–20). The view was expressed that the social climate had changed, that the health of Soviet women had to be protected from injuries resulting from abortion, and that conditions for mothers and children had improved. As previously noted, family allowances were initiated as were taxes for couples without children. The concept of the glory of motherhood was (re)introduced and divorces became more difficult to obtain. Efforts to stimulate births were part of a rise in nationalism amidst the looming threat of Nazi Germany (Heer and Bryden, 1967; Geiger, 1968; Madison, 1968; Dreijmanis, 1972).

The immediate effect of the prohibition of abortion was an increase in the USSR birth rate from 33.6 per 1,000 population in 1936 to 39.6 in 1937 (Lorimer, 1946). In the cities of European Russia, the impact was more dramatic, with increases of 92 percent in Moscow and 69 percent in Leningrad (Lorimer, 1946; Heer, 1968). Subsequently the birth rate declined as Soviet couples again resorted to illegal abortion. The personal agonies Soviet women experienced to avoid unwanted births have been well described (Engel and Posadskaya, 1998).

The nearly 20-year struggle against clandestine abortion was unsuccessful. Directives issued by the Commissariat of Health had limited impact (Chalidze, 1977); instructions to produce more condoms were no more successful (USSR/PCH, 1936; USSR/CEC, 1936). In the absence of reliable, convenient, and safe contraceptives, clandestine abortion flourished. Voluntary sterilization was prohibited except on strictly specified medical grounds (David, 1974).

1955–79

The rising incidence of illegal abortions produced growing concern (Shlindman and Zvidrinsh, 1973, cited in Stloukal, 1993). The severity of the 1936 restriction was softened by the Decree of 5 August 1954, absolving the pregnant woman from legal liability in terminating her pregnancy (Field, 1956). On 23 November 1955, two years after Stalin's death, the 1936 restrictions were repealed by a decree of the Supreme Soviet (USSR, 1955a). The reasons cited were to reduce "the harm caused to the health of women by abortions performed outside of hospitals," and "to give women the possibility of deciding for themselves the question of motherhood" (Field, 1956; Popov, 1993). The 1955 decree resembled that promulgated in 1920, stipulating that abortions could be performed only by qualified personnel in medical facilities. The decree was not accompanied by a commentary and was given little publicity. It was thought at the time that the relaxation of restrictions on abortion reflected recognition of the failure to enforce compulsory reproduction, and a preference for restoring women's reproductive rights in the postwar environment of the unequal sex composition of the Soviet population (Stloukal, 1993; Popov, 1994b).

On 20 December 1955 the Ministry of Health (USSR, 1955b; WHO, 1971) issued more specific implementation instructions, stating that abortions may not be performed if the health of the woman was threatened or when a previous pregnancy was terminated within the immediately preceding six months. When gestation exceeded 12 weeks, strong medical justification was required. Reinstatement of abortion on request did not result in a sudden decline of the birth rate, suggesting that legal terminations largely replaced illegal ones. The cost of abortions was set at five rubles, equivalent to about U.S.$7 at the time (Savage, 1988). On 18 March 1956, by Order No. 55-m of the Ministry of Health, the collection of abortion data was reinstituted but statistics were not officially released until 1988. The number of abortions reached a peak of 7.9 million in 1966 (Avdeev, 1994; Kulakov, 1995).

1980–87

After a woman had formally requested an abortion, a gynecologist (most often a woman) discussed her reasons for wanting the termination and advised about possible adverse consequences. If the pregnant woman persisted, and if she met the specified conditions for abortion, her application had to be approved. Termination was performed in a hospital, usually in the gynecological unit, often involving a stay of one to three days. The Soviet Labor Code specified that voluntary abortion qualified the woman to take unpaid sick leave (up to five days), thus protecting her against dismissal or disciplinary action for an unauthorized absence (Heitlinger, 1979). Vacuum aspiration was widely used, at least in the major cities, with light anesthesia, if any (Mandel, 1975). There was no charge for therapeutic abortion or when requested on social grounds by a single or working woman earning less than 60 rubles per month (Lydolph et al., 1978). After the twelfth week abortion was allowed only on medical grounds. There was no age limit but women under age 18 were expected to have parental consent. Husbands had no voice in decision making and were not always informed. As in other state facilities, conditions were often uncomfortable and demeaning.

In the 1970s and 1980s much emphasis in "the fight against abortion" was placed on potential health risks. In brochures, medical bulletins, and related educational materials legal abortion was officially described as only a slightly lesser evil than clandestine abortion. However, complications were not prominent among the demonstrated adverse sequelae summarized in Hyde's (1970) review of Soviet medical publications. Abortion of the first pregnancy was not shown to have significant deleterious effects (Kornet, 1972). Repeat abortions were not deemed to be a major contributing factor to the rise in infant morality (Davis and Feshbach, 1980).

To combat clandestine abortion, the Ministry of Health on 16 March 1982 by Order No. 234 gave women the right to have the procedure performed in a hospital of their own choosing, provided their decision was confirmed by a committee of three local physicians. This allowed women to have abortions outside their area of residence, facilitating the use of early abortions in specialized institutions (Popov,

1994b). At the same time, the use of pills for oral contraception was further restricted.

On 31 December 1987 the Ministry of Health published Order No. 1342, entitled "About the statement of instruction on the order of permission for the operation of artificial interruption of pregnancy on nonmedical grounds." Physicians were authorized to provide legally induced abortion through the gestation age of 27 weeks on juridic, genetic, broad medical, sociomedical, and social grounds (when confirmed as reasonable by a medical commission).

Early abortions (mini-abortions) began to be performed outside the government health system in the mid-1980s, usually with imported equipment. At that time they were neither officially approved by the Ministry of Health nor registered (Popov, 1994b). It was not until 5 June 1987, in the Ministry of Health Order No. 757, entitled "Regarding the statement of instruction and the order for conducting the operation for artificially interrupting pregnancy in its early stages by the method of vacuum-aspiration" that official permission for early abortions was granted, legalizing them in any clinic, independent of the woman's place of residence. Officially, this type of induced abortion included those performed during the earliest gestation period up to 20 days by vacuum-aspiration after obligatory pregnancy testing. In the statistical reports of 1989 mini-abortion was reclassified as "*menstrual regulation* by vacuum-aspiration during the early gestation period" and reported separately but included in the total abortion statistics.

1988–97

On 29 March 1988 the Ministry of Health legalized commercial abortions. Fees could be charged by physicians who had a legal private practice, in private clinics, and in commercial sections of state hospitals (Russia/MOH, 1988). Such privatization constituted legalization of already existing illegal and paralegal commercial abortion services. Women with money often preferred clandestine procedures because they could avoid time-consuming laboratory tests, waiting delays, and overcrowded assembly line hospital conditions. They were also guaranteed personal attention and the availability of anesthesia, plus assurance of secrecy because the physicians involved provided another medical reason (other than abortion) for their absence from work. The average charge was 50 rubles for early abortion and 100 rubles for later abortion, representing 25–50 percent of a woman's average monthly income at the time (David, 1992a; Ruevekamp, 1994).

The state monopoly on the provision of abortion service was restored in a 29 December 1988 Decree of the USSR Council of Ministers, outlawing procedures performed in commercial centers, nonstate hospitals, or in private practice. However, commercial health centers continued to operate in the "gray" market and flowered after the 1991 collapse of the Soviet Union (Popov, 1994a).

By the time of the October 1990 International Conference "From Abortion to Contraception," convened in Tbilisi by the Zhordania Institute under UNFPA/WHO auspices, Soviet health officials acknowledged that abortion had become a major public health problem. Over 3,000 gynecologists were said to be performing abortions full-time and more than 20,000 women were absent from work each day because of abortion (David, 1992a). It has been estimated that approximately 50 percent of the MOH reproductive health budget was spent on abortion and associated complications (Bowers et al., 1995). In Georgia, estimates suggested that the total amount of money derived from unregistered abortions exceeded the entire state health budget (David, 1992a). In the Russian Federation, a list of medical indications for pregnancy interruption was published in 1993 (Russia/MOH, 1993a).

Morbidity

In 1993, up to 30 percent of Russia's 800 maternal deaths were caused by abortion complications (Zelinskaya cited by Bernard, 1994), an increase from about 26 percent in the early 1990s (Popov, 1995a). In St. Petersburg over 40 percent of all maternal deaths in 1990–92 were attributed to abortion. The maternal mortality ratio was over 73 times that of the United Kingdom (Stephenson and Porter, 1994).

The Abortion Culture

Objective reasons for abortion were gradually enveloped by psychological tolerance. Abortion had become a method of fertility regulation available to every woman, a symbol of personal freedom in a society in which the state tried to control all aspects of life. By the 1990s, abortion was "perceived as a routine, although certainly unpleasant, medical procedure, comparable, say, to the removal of a tooth" (Remennick, 1993: 53). Because the fetus was not considered a person, induced abortion was not deemed to be murder. This view was enhanced by an ultra-cautious attitude toward modern contraceptives. They were perceived as something "unnatural" and labeled harmful by physicians who never mentioned their advantages. Given that the abortion procedure was brief and easily available, and ethical objections were weak, few women persisted in following the difficult and expensive path of finding efficient and personally acceptable contraceptives (which were often in short supply).

In writing about the reliance on abortion for fertility regulation, Remennick (1991, 1993: 46) observed that "abortion has never been a matter of choice for Russian women, but rather a pressing necessity emanating from the lack of any alternative." Official figures showed that in 1988 contraceptive supplies met only 25 percent of demand (Khomassuridze, 1993). The "abortion culture" and the abortion industry, both legal and illegal, were firmly established.

In 1998 abortion was available at any stage of pregnancy after medical examination and with a physician's referral. Up to 12 weeks, the referral is given on request. Between 13 and 22 weeks, medical or social indications may be used to justify the referral. After 28 weeks, only medical indications are considered. Women 15–18 years of age do not require parental consent. While abortion remains free of charge in some regions of Russia, insurance coverage is provided in others only for mini-abortions and for medically prescribed abortions for social reasons (e.g., women refugees, women with three or more children, women prisoners, or women over age 45). Coverage depends on local financial capabilities. The cost of private abortion varies and in Moscow can equal the average monthly wage (Ballaeva, 1997). Many women try to self-abort despite the risks. The quality of public abortion services and the attitude of providers still leave much to be desired (Grebesheva, 1996).

ABORTION REGISTRATION AND SURVEYS

Beginnings

Through the end of the 1920s the USSR was the world leader in the study of abortion. The information base was unique with respect to magnitude of data collection, depth of data analysis, and the accessibility of secondary information (Popov, 1991; Avdeev et al., 1994). Although abortion statistics continued to be compiled, no official data were published by the USSR/MOH from 1929 to September 1988.

Somc information on abortion during the restrictive period (1935–55) and subsequent relegalization was published by Sadvokasova (1969) Working in the Central Scientific Office for Health Statistics in the Ministry of Health, Sadvokasova had responsibility for the collection, evaluation, and analysis of record cards completed by hospitals for women admitted for legal termination or for treatment of incomplete abortions. As shown in Table 11.1, Sadvokasova's data indicated a substantial rise in abortion after relegalization in 1955 and a continuing increase, exceeding live births. These observations were confirmed by Katkova (1968), Sokolova (1970), and others.

Registration Changes and Distortions

The statistical registration of induced abortion was virtually unchanged from 1956 to 1991. During this period, abortion was differentiated into the following categories: spontaneous abortion, induced therapeutic abortion, induced abortion on request, out-of-clinic (clandestine) abortion and, since 1988, early abortion (by vacuum aspiration). The officially published statistics consisted of the absolute number of induced abortions on request, the abortion rate per 1,000 women aged 15–49, the abortion ratio per 100 births (born alive or dead), the number of abortions to first time pregnant women, the number of abortions to women under ages 14 and 17, and the number of deaths resulting from induced abortion. The indices were calculated separately for legally induced abortions and for clandestine induced abortions USSR/MOH (1975–91).

Available data sets include information on USSR abortions for 1954–90 and the Russian Federation for 1957–92, on registered abortions 1971–90 by republics and some years for selected subunits (Avdeev et al., 1993, 1994; Popov, 1991). Although there are considerable differences between these data sets (not shown here), there is little doubt that the abortion rates and ratios were higher than anywhere else in the world.

In the 1980s and 1990s the statistical registration of abortions was revised twice: slightly in 1988 and fundamentally in 1991 and 1992. In 1988 the registration of early abortions was introduced. In 1991 and 1992 the Ministry of Health of the Russian Federation disseminated new statistical registration forms (Russia/MOH, 1992b, 1992c). Abortions were differentiated into the following categories: legal induced abortion on request, spontaneous abortion, therapeutic induced abortion, clandestine induced abortion, and abortion without clear stated grounds. The last category was not considered to be an induced abortion because of the vagueness of the cause. Early abortions were still registered separately from induced abortions because they were not officially considered abortions, but rather a method of menstrual cycle regulation. An age distribution system was also introduced.

Clandestine Abortions

Abortion obtained unofficially, or "on the left" in Russian idiom, had a particular appeal for women, especially unmarried women, willing to pay a qualified gynecologist for more personal care, privacy, and avoidance of long lines and demeaning conditions in women's health centers (Popov, 1980; Ryan, 1987). After 1991, the classification of clandestine abortions was limited to those performed by the woman herself (self-induced abortion); those performed by a licensed physician in a clinic but unregistered; and those performed by an untrained person, out-of-clinic, and not registered (a criminally induced abortion).

Before 1991 all clandestine abortions were included by the USSR Ministry of Health in one category—"out-of-clinic" abortion, defined as officially registered but performed outside an approved clinic. This was not a valid enumeration of clandestine abortion, but merely an indicator of complications and side effects following clandestine abortions that ultimately required professional medical treatment and hospitalization. Only serious complications resulted in the official registration of clandestine abortions. Until 1991 the authorities considered all out-of-clinic abortions resulting in death or serious bodily injury as criminal procedures.

The number of registered criminal abortions was low because confirmation of its criminal nature was needed in each case. The main criterion for criminal classification was direct evidence of interference (i.e., uterus perforation or sepsis). However, confirmation by the woman herself was also required before a criminal investigation could be initiated.

The main problem of the statistical system introduced in 1991 was its total incomparability with the previous one. Under the new classification most clandestine abortions were registered as abortions for unknown reasons. Only a small portion of clandestine abortions was registered as criminally induced. As a result, the number of registered criminal abortions remained low, because confirmation by the woman herself was very rare. When clandestine abortion was not detected or proved, the abortion was registered as one with an unestablished cause, or "spontaneous," and excluded from the total number of induced abortions. As a result, the 1991 decline in the reported number of induced abortions was a statistical artifact, the result of transferring clandestine-induced abortions into the category of "spontaneous abortion."

Comprehension of the importance of these changes in the induced abortion registration system is essential in explaining the decrease in the number of abortions in the Russian Federation from 4.1 million in 1990 to 3.4 million in 1992, and 2.5 million in 1997 (Table 11.2). Another reason for the decline in reported abortions may be the greater reliance on mini-abortions performed within 17–20 days of amenorrhea. Registration is known to be incomplete (Popov, 1994a, 1995a). Still another problem is the lack of registration of abortions performed in commercial health centers. The many newspaper advertisements for abortions in private clinics and by privately practicing physicians suggest that the number escaping registration is considerable (Popov, 1995a, 1995b).

Finally, MOH data may only be part of the national statistics. There is some debate about whether information always included data from independent medical institutions operated by large ministries and enterprises. In 1992 practically every large ministry had its own health facilities, including the Ministry of Defense, the Ministry of Railroad Transportation, the Ministry of the Interior, the Committee of State Security (former KGB), the Academy of Sciences, and others. Large industrial enterprises typically had their own hospitals providing abortion services. Registration was not compulsory. Monthly registration of induced abortions was introduced in 1992 in the Russian Federation (Russia/MOH, 1992b). Although present Goskomstat reports include abortions reported by the Ministry of Health and other institutions, it is not known whether coverage is complete (Zakharov, 1998b).

Reported Data

The officially reported total number of abortions in the Russian Federation peaked at 5.6 million in 1966, subsequently stabilized around 4.5 million between 1975 and 1985, and then began a gradual descent to 4.1 million in 1990 and 2.7 million in 1996. Some 2.5 million abortions were reported for 1997 with the comment that 70 percent of pregnancies were terminated (Russia/MOH, 1998). Official figures are widely believed to be underestimates (e.g., Khomassuridze, 1993). The reported statistics, including mini-abortions, suggest abortion rates of 125.6 per 1,000 women aged 15–44 in 1990, 103.7 in 1992, and 75.7 in 1997. During those years the legal abortion ratio per 1,000 live births was 2,063 in 1990, 2,165 in 1992, 2,320 in 1993, and 2,016 in 1997 (Table 11.2).

Official statistics indicate that most abortions are performed in older women. In 1995 only 12 percent of registered abortions were in teenagers, 68 percent in women aged 20–34, and 20 percent in women over age 35. Most women attending an abortion clinic are married and use the procedure to limit childbearing or for spacing after the first delivery (Khomassuridze, 1993). While some studies of registered abortions suggest that 85 percent of first pregnancies are delivered, others indicate that most terminations of first pregnancies go unrecorded (Popov, 1991; Remennick, 1993). Clandestine abortions are said to be performed especially among teenagers and unmarried women in rural areas (Grebesheva, 1992). While abortions among teens and primigravida, particularly among women under age 17, increased in the early 1990s, they declined in subsequent years (Grebesheva, 1994; Kulakov, 1995; Chervyakov, 1998).

Estimates of the number of abortions an average sexually active woman will have during her fertile years range from two to five (Grebesheva, 1992; Remennick, 1993; Popov et al., 1993; Avdeev et al., 1994). Some women are known to have had 10 or more abortions (David, 1974; Manulova, 1991b; Dorman, 1993; Bowers et al., 1995). The proportion of women who had been pregnant but never experienced an abortion varied in different surveys between 15 and 25 percent.

As also noted by others (Avdeev, 1994; Willekens and Scherbov, 1994; Popov, 1995a), it has become increasingly difficult to interpret the decreasing trend in the reported number of abortions. These difficulties result from changes in official rules and regulations on the provision of abortion services and changes in the statistical registration system, the decreasing proportion of women of reproductive age in the population, and an increasing use of modern contraceptives. Among the vexing problems are: the nonregistration of abortion services in commercial health centers, medical units associated with state ministries and large industrial enterprises; the incomplete registration of early abortions; the 1991 changes in the registration system, and the lack of accurate figures on clandestine abortions after 1991. The official statistics do not capture the dynamics of commercial medical services and are reported as improperly calculated aggregate indices. The present system can only be characterized as an incomplete reflection of actual trends that fails to inspire confidence.

Post-Soviet Surveys

Published estimates of the abortion rate per 1,000 women of fertile age (15–49 years) for 1992 range from 81 (Avdeev et al., 1994) to 98 (Popov, 1996). The Avdeev data are based solely on MOH reports and differ from Popov in the treatment of incomplete abortions. Estimates based on a single question of fertility related practices during the past year, part of the household questionnaire prepared for the Russian Longitudinal Monitoring Survey (RLMS), suggest an abortion rate of 56 per 1,000 women aged 15–49 in 1994 (Entwistle and Kozyreva, 1997). Adjusting the RLMS estimate for likely underreporting, using the results of Anderson et al. (1994), the estimate increases to 62 per 1,000 women. If the maximum point within the 95 percent confidence interval (68 per 1,000) were used and adjusted for underreporting, the estimate would further increase to 75 per 1,000 women, still lower than the 1994 MOH figures of 82, which is deemed an overestimate in the RLMS report.

A 1996 three-city survey by the All-Russian Centre of Public Opinion and Market Research (VCIOM), with technical assistance from the U.S. Centers for Disease Control and Prevention (CDC), focused principally on issues of abortion, contraception, and pregnancy (Russian Centre et al., 1997, 1998). About 2,000 women aged 15–44 from each site were interviewed. The surveys included a detailed reproductive history over the previous five years for each respondent, asking on a monthly basis whether the woman had been pregnant, obtained an abortion, or used a contraceptive. While not intended to reflect national rates, official abortion and fertility statistics for the three surveyed areas were quite similar to those of the Russian Federation. The abortion rate in the three Russian cities of the VCIOM/CDC study ranged from 77 to 99 per 1,000 women aged 15–44, substantially higher than the rate of 56 for women aged 15–49 initially suggested by RLMS (before making adjustments) and higher than the 72 per 1,000 and 75 per 1,000 rates reported by the MOH for two of the three cities. No comparable data were available for the third site where VCIOM/CDC had reported a 99 per 1,000 rate. While not directly comparable, the VCIOM/CDC findings tend to support the opinions expressed by other observers that the reported MOH figures are underestimates and do not confirm the RLMS assertions that official statistics overestimate the actual number of abortions (Goldberg, 1997).

In a 1996 study conducted by the Russian Center for Obstetrics, Gynecology, and Perinatology in cooperation with WHO/Europe, 352 women were interviewed immediately after abortion, using a standardized questionnaire. Of these women, 221 lived in Moscow and 121 in the rural area around Moscow; 90 percent were married or living in union (Vikhlayeva and Nikolaeva, 1996). Nearly 50 percent of the women married before their twentieth birthday; 90 percent of first pregnancies were carried to term. Of the subsequent pregnancies, only 32 percent resulted in deliveries. Among the major reasons given for the present abortion were unplanned pregnancy (96 percent) or difficult living and economic conditions. Only 15 percent of the urban and 13 percent of the rural women had received contraceptive counseling before sexual debut. After the first deliveries or abortions, this figure rose to 45 percent and 36 percent, respectively. About 76 percent experienced the abortion as "psychological torture" and 35 percent expressed moderate or severe pain. Nevertheless, 42 percent said they would resort to abortion again, 37 percent did not know, and about 20 percent preferred delivery. Most women would prefer to rely on a safe method of contraception if they had ready access to accurate information and regular supplies.

Given the increasing antiabortion activities in the Russian Federation, often supported with American funds (Hockstader, 1994), women may be more reluctant to report having had abortions, especially mini-abortions that have not been recorded. While there is general evidence that increasing availability of contraceptive supplies has reduced the incidence of abortions, there are also indications that unrecorded abortions have increased among teens and unmarried women. Thus, reported rates, whether based on officially registered data or on surveys, remain open to questions.

CONTRACEPTION/ABORTION

Restrictive Regulations

Throughout Soviet history the right to family planning was formally proclaimed *de jure*, following international conventions, but was never realized *de facto*. Family planning services were nonexistent or inaccessible, owing to the lack of reliable information, the absence of specialized medical units and qualified personnel, and the unavailability of modern contraceptives. Voluntary sterilization was prohibited following Stalin's recriminalization of abortion (USSR/PCH, 1939; Popov, 1994c). Pronatalist financial and social measures, reduced in 1948, were restored in 1981 to stimulate the births of second and third children (Popov, 1993).

After the relegalization of abortion in 1955, studies of aborting women reflected the widespread lack of reliance on any then available methods of contraception (Heer, 1965a, 1965b; Verbenko, 1968). IUDs, which had been described as harmful in a 1958 Soviet medical handbook, began to be produced in limited quantities in 1968 (Moffet, 1987) and became more widely available in the 1970s and 1980s. Diaphragms were manufactured in only one size. Russian produced condoms, made of thick dark latex and mockingly called "galoshes" (David, 1974), were unpleasant to use (Moffet, 1987; David, 1992a).

On 9 October 1970 a MOH letter recommended use of oral contraceptives on prescription. High dosage pills, imported from Central and Eastern European countries, induced numerous side effects (Moffet, 1987; Khomassuridze, 1993). On 1 August 1971, by MOH Special Order No. 16, use of the pill was restricted to specified therapeutic grounds and prohibited for contraceptive purposes. In 1974 the MOH published 30 contraindications for use of the pill. In 1975 a MOH letter suggested that physicians encourage pregnant women to carry to term. In the mid-1980s, the MOH repeated its anti-pill warnings (Polchanova, 1986; Popov et al., 1993; Visser et al., 1993). Erratic supply problems, especially in rural areas, and concerns about possible carcinogenic side effects continued to discourage use (Lubell, 1979). Strict prohibitions against vasectomy and female sterilization remained in effect (Stern and Stern, 1980).

At the beginning of the 1990s a few gynecologists and social demographers punctured the myth of the pill as a cancer promoting agent, noting that, statistically speaking, late abortion was the riskiest form of fertility regulation. Low dosage pills began to be imported from Germany and the Netherlands (Khomassuridze, 1993). These efforts had limited influence on the medical profession or on the media that associated the pill with declines in morality and a rise in promiscuity in the West (Remennick, 1991). The authorities and the medical profession did all they could to discredit new contraceptive methods and prevent their widespread use (Makusheva, 1983; Khomassuridze, 1985; Popov et al., 1993). In contrast to the problems supposedly associated with the pill and men's discomfort with condoms, abortion seemed comparatively harmless and convenient. Following the collapse of the Soviet Union, imported contraceptives became more available in the market place, albeit at high prices. International donor agencies provided a wide array of modern contraceptives at low or no cost in an effort to reduce reliance on abortion.

At the end of 1990 the ban on voluntary sterilization was lifted for women meeting very specific medical criteria such as having three or more children or being over age 30 with two children (USSR/MOH, 1990; Popov, 1994c). Vasectomy was not mentioned in the MOH order. On 28 December 1993 male and female sterilization were declared methods of family planning (Russia/MOH, 1993b; Popov, 1994a, c). The procedure could be performed in response to a written request by any person over age 35 who had at least two children. However, these criteria could be waived when there were medical indications and the person consented (Fluss, 1994). By 1994 commercial clinics began to advertise sterilization services. Official statistics did not differentiate between procedures provided for medical or contraceptive reasons and were not considered reliable (Popov, 1996).

In sum, from the beginning of the 1980s fertility regulation in the Soviet Union was determined by parallel processes: broad provision of early abortion; continuation of the fight against clandestine abortion; localized attempts to make modern contraceptives available; the continuation of official efforts to limit the widespread availability of modern contraceptives; and the commercialization of abortion services. By *de facto* prohibiting oral contraceptives while gradually extending the grounds for and availability of abortion, the Soviet public health system developed a model of fertility regulation which relied primarily on low cost abortions.

Kon (1995: 183) wonders why Soviet medicine took such a reactionary stance. He ascribes much to the "ignorance and lack of principle" among medical personnel who clung to outmoded information and freely accepted any position presented as "anti-Western." Leaders in the Soviet medical community were unwilling to renounce abortion services in favor of something new. Moreover, abortion clinics produced a tidy sum for the MOH and physicians were not averse to taking their share. With the low salaries paid to physicians, the profit to be made performing abortions still acts as a strong disincentive to providing contraceptive counseling (Schehl, 1996).

There were also more general reasons. The authorities feared that widespread availability of the pill might lead to a greater decline in the birth rate. Furthermore, any switch from an abortion strategy to one of effective contraception would mean a substantial extension of women's rights, replacing state control over reproductive behavior with self-determined control. In this scenario the role of the physician would be reduced to a technician-consultant. That amount of potential autonomy went against the paramount principles of the Soviet regime and past patriarchal experience. The situation improved in recent years.

Contraceptive Practice

Little is known about contraceptive practice between 1938 to the mid-1970s. Official statistics did not include contraception. Available demographic and social studies seldom met acceptable scientific standards (Popov, 1993). Informal nonrepresentative inquiries in major cities of European Russia in the 1970s suggested that most couples engaged in unprotected sex, relying on withdrawal and hope (David, 1974; Smith, 1977; Remennick, 1993). When Urlanis (1968, 1974) lamented the failure to conduct research on reproductive behavior and recommended that sample surveys be initiated on contraceptive practices and frequency of abortion, his priorities were not shared by the MOH or the Academy of Sciences.

Popov et al. (1993) summarized the results of five studies conducted during 1976–84, three in Moscow and one each in Saratov and Tartu, plus a national survey organized by Goskomstat in 1990. Rhythm, condoms, vaginal douches, and withdrawal were the main contraceptive methods. Only 1 to 3 percent of women interviewed preferred oral contraceptives and about 10 percent IUDs. The majority of respondents perceived the pill to be ineffective, inconvenient, unavailable, or harmful. The IUD was assessed somewhat more favorably. The condom was the most acceptable method despite its low rating for convenience.

Avdeev (1994) estimated a contraceptive prevalence rate (CPR) of 34.8 percent in 1979 and 31.4 percent in 1988. In subsequent nationally representative random sample surveys in St. Petersburg and Kaluga the CPR was estimated at 68 percent. However, of all women currently using a method, only 53 percent used modern contraceptives (Turner, 1992; Hutter, 1996). A trend toward more modern methods was reported subsequently (Lehert et al., 1992; Visser et al., 1992, 1993).

In 1992, the only two condom factories in the Russian Federation stopped production because they could not afford to import latex. The single IUD plant was closed because of complaints about the quality of its products (Jones, 1992; Turner, 1992). Although this situation was subsequently resolved, abortion remained the major method of fertility regulation.

In 1996 the baseline component of The Russian Women's Reproductive Health Survey (RWRHS) was conducted in Ivanovo Oblast and the cities of Perm and Yekaterinburg (Russian Centre et al., 1997, 1998). Ivanovo and Yekaterinburg were the sites of considerable international donor support for training and provision of contraceptive supplies in the public sector. Perm had not received similar assistance. The preliminary report showed a CPR among women in registered and unregistered marriages ranging from 69 percent in Perm to 78 percent in Ivanovo. Modern method use exceeded traditional methods by 3:1 or higher. The IUD accounted for more than half (27–35 percent) of all modern method use, followed by condoms (12–14 percent), and pills (5–11 percent). Female sterilization was only 1–2 percent. Reliance on traditional methods was 14–18 percent. Use of condoms, pills, and periodic abstinence increased with education. In all three sites slightly more than half of married nonusers of contraception cited inability to become pregnant, current pregnancy, desire to become pregnant, or lack of sexual activity as their excuse for nonuse. Reasons such as method cost or unavailability, lack of information, preference for abortion, or partner objections were not very common. Respondents held "extremely negative opinions" about conventional abortions and mini-abortions (Russian Centre et al., 1997: 10).

As of 1999, oral contraceptives are still not produced in the Russian Federation. However, international donors have extended their programs to more than 14 regions, working with women's consultation centers of maternity houses (Antarsh, 1998). Although the pill no longer requires a prescription (Borisov et al., 1997) and attitude toward low-dosage hormonal methods have become more positive, supplies in the private sector continue to be irregular and expensive. Women often have to change from one brand of pill to another. Pill use would probably be higher if supply and cost problems could be resolved (Russia/MOH, 1994; Grebesheva, 1996; Kamsiouk, 1998). IUDs are purchased in the marketplace for hospital insertions. In 1996 the production of condoms was about 370 million, not sufficient to cover the demand (WHO/EURO, 1997). Spermicides, popular among teens, cost 30,000 rubles per package, more than half the minimum monthly salary (Dmitrieva, 1996). Misperceptions, even among motivated physicians, suggest a continuing need for postgraduate courses and the integration of family planning into the medical school curriculum (Visser et al., 1993).

THE WOMAN QUESTION

The historian Laura Engelstein (1992) showed how deeply imbedded the position of women, gender ideology, and the perceptions of sexuality were in the social and political life of the last half century of Czarist rule. The institutional consolidation of professional expertise and the emergence of commercial culture along with the capitalist marketplace generated a contest over the authority to regulate sexual conduct, determine the boundaries of individual autonomy, and demarcate public from private life. Citing a wealth of resources, Engelstein demonstrated how imperial laws governing family relations and gender rights reflected patriarchal values underlying both the autocratic regime and the peasant community. Until the 1880s access to education was limited as were positions in state service (Stites, 1978; Malysheva, 1992).

As noted in Chapter 3, socialist views before and after the October 1917 Revolution influenced Soviet legislation, granting women equal rights in all social and private areas, including marriage and family relations (Lenin, 1976). Although the realities confronting the Bolsheviks forced a halt in the progress toward attaining genuine equality, millions of women from lower class backgrounds gained educational and employment opportunities beyond the dreams of their parents and grandparents (Engel and Posadskaya, 1998).

By the time severe restrictions were imposed on abortion in 1936, Soviet ideology had already long abandoned its earlier view that woman's primary function, like that of man, was social activism and that bearing children should be accessory to this function (Goldman, 1993). The status of the family changed from being a barely tolerated "survival of the past" to an indispensable primary cell in Soviet society (Stloukal, 1993). As noted previously, traditional sex roles in the family were reasserted. Penalties were imposed on couples failing to register their marriages.

With the suppression of sexuality in Soviet society and the double and triple burdens assumed by women in their families, the individual freedom of the early 1920s was replaced by social control, although the 1936 Constitution declared that women had "equal rights with men in all spheres of economic, state, cultural, public, and political life" (Orlova, 1973). The percentage of women working outside the home was not much different from that of men but the proportion of women with higher education was greater for all cohorts born after 1950 (Vishnevsky, 1996). Receiving little support in the working world and even less at home, women often felt deceived (Gray, 1989). Complaints mounted that Soviet women occupied an underprivileged position in the family and in public life. When a small group of women organized to protest in Leningrad, it was dissolved by the KGB (Mamonova, 1984).

In the early stages of *perestroika*, there were calls for "women to return to their natural predestination" (Gorbachev, 1987a: 117; 1987b), exacerbating the long-standing discussions about home and labor market responsibilities (Lapidus, 1978; Voronina, 1988). In 1987 Soviet women still constituted 51 percent of the labor force, more women than men had a university education, and women predominated in teaching and medicine (Maddock et al., 1994). However, women occupied the worst paid and least prestigious jobs and were poorly represented on the higher rungs of the vocational ladder. Men also dominated in the party and state hierarchies.

The women's movement did not represent a serious public force. Gorbachev's attempts to reanimate the Communist Party-directed Union of Soviet Women were not very popular. Small clusters of women began to organize in Moscow and Leningrad. In 1990 the Moscow Center for Gender Studies, attached to the Institute of Social-Economic Problems of Population of the Soviet Academy of Sciences, became the first facility of its kind in the Soviet Union and in Eastern Europe. It collected information, conducted research, and helped to defeat antiwomen's rights legislation (Posadskaya et al., 1995).

After 1991, with the transition to the market economy and the overall economic collapse, the position of women greatly deteriorated. Women were the first victims of economic reform. Countrywide figures show that the percentage of women among the registered unemployed rose from 61 percent in 1992 to 68 percent in 1993 before declining to 60 percent in 1996 (Russia/SCS, 1996). Whereas in earlier years women received about 70 percent of men's salaries, in 1993, it was closer to 40 percent. Laws against sexual harassment in the work place were seldom enforced (Stanley, 1994). Access to kindergartens and nurseries diminished as the result of closures and/or escalating prices. Poverty became strongly skewed toward women (Fong, 1993). The number of women deputies in the USSR Supreme Soviet declined from 33 percent in 1985 to 16 percent in 1989 and to only 5 percent in the regional and local Soviets. In the December 1993 elections, the gender-specific Women of Russia Party, formed from a nucleus of the Union of Russian Women's organizations, received 20 seats but lost them in 1995 when the party failed to gain

5 percent of the vote in the parliamentary elections. It subsequently splintered and lost political significance (Ballaeva, 1998).

In 1993 a Presidential Committee on Problems of Women, Family, and Demographics was established in the president's office but has been handicapped by a lack of funds to implement planned activities. The stress induced by economic crises and general lawlessness fueled increasing rates of domestic violence, rape, and murder of women. Reported incidence exceeds Western figures by four or five times (Horne, 1999). Sex trafficking flourishes (Specter, 1998). In 1995 there was not a single home for battered women in all of Russia (Hockstader, 1995; *The Economist*, 1995).

The collapse of the never realized revolutionary goals produced a backlash with large groups of women wanting to turn back the clock to the pre-Soviet days of patriarchal family life. *Robotnitsa*, a magazine for women workers, published an interview with a noted filmmaker who said "In my soul, I am an Asiatic and consider woman to be a lesser being" (cited by Randolph, 1996: 116). The comment provoked no public reaction from its millions of readers. At year end 1998 the head of the Liberal Democratic Party advocated the promotion of sex tourism as a means of earning millions in foreign currency (Zhirinovsky as cited by Williams, 1999).

The woman question remains unresolved (Jancar, 1974, 1978; Sacks, 1976; Lapidus, 1978, 1982; Heitlinger, 1979; Wolchik, 1981; Zakharova et al., 1989; Yedlin, 1980; Kon, 1993, 1995; Malysheva and Bertaux, 1996). However, an independent women's movement is gradually gaining strength, holding seminars, working with policymakers, and disseminating information about the lives of women abroad and in Russia with the goal of attaining actual rather than theoretical gender equality and securing women's rights (Posadskaya and Azghikhina, 1995; Dmitrieva, 1996; MCGS, 1996; Malysheva et al., 1998).

SEX AND SOCIETY

Historical Trends

In his extensive reviews of trends in Russian sexual beliefs and practices, Kon (1993, 1995) describes how the October 1917 Revolution produced a breakdown in the old normative cultural patterns and moralization about sexuality. In the 1920s premarital and extramarital relations were widespread among students and young workers. Freud's major works were translated into Russian, psychosexual development was widely discussed, and fairly sophisticated questionnaire surveys were conducted. Kon (1995) cites Golod's observation that the empirical research of the time demonstrated the destruction of the traditional norms and values rather than the emergence of new ones. The "sexual freedom" brought by the October Revolution was "freedom from," not "freedom for." For many, uncertainty prevailed.

Beginning in the 1930s, Stalin's sexophobia induced a counterrevolution aimed at total control. Sex surveys and sexological research disappeared together with sociology and social psychology. Access to Freud's works was restricted. Criminal punishment for male homosexuality was introduced in 1934, abortions were banned in 1936, and sex education vanished from the schools. A strict puritanical morality was proclaimed and traditional links between sexuality and marital family relations were restored. In an Orwellian sense, if a person was primarily a productive force working for the future of humankind, he or she should produce material commodities during work time and children at home at night. Everything else had to be eradicated. "For many decades Soviet society hypocritically portrayed itself as utterly asexual, even sexless" (Kon, 1993:15). In the 35 years between 1930 and 1965 the Soviet Union produced no scientific sexual literature of its own, books published abroad were very difficult to obtain, and no objective data on sexual behavior were available. When references were made to sex, it was taken for granted that it could only happen between adults of the opposite gender in multiples of two within marriage and for the purpose of procreation (Attwood, 1993).

Kon's own experience well reflects the development of sexology in Russia. His book on *Culture and Sexology* (Kon, 1981), commissioned by the Hungarian Communist Party publishers, appeared in 1981. A revised version, *Introduction to Sexology*, was published in the German Democratic Republic in 1985 and immediately sold out in both German states. It was not until 1988 that the Russian language text could be printed. Eventually, 550,000 copies were produced but not a single copy was available in bookstores. The entire print run was distributed among medical and scientific institutions. The average reader could obtain a copy only on the black market.

During the time of *perestroika* and *glaṣnost*, all forms of social control disintegrated. As Kon notes (1995: 112), "the long-ignored problem of differentiation between 'decent' eroticism and 'squalid' pornography, between sexual freedom and the exploitation of sex for commercial purposes, became uncommonly acute and urgent." The gradual liberalization and transformation of the Soviet regime caused its sexual stance to change "from one of brutal suppression to one of awkward taming" (Kon, 1995:267). Since the collapse of the Soviet regime in 1991 sexuality has been further de-romanticized, commercialized, and trivialized.

The integration of reproductive rights into the post-Soviet cultural context is hampered by the preservation of traditional puritanical values regarding sexuality. Reproductive behavior continues to be a derisive topic among large segments of the Russian population (Dmitrieva, 1996, 1999). In 1997, the Moscow city authorities discontinued an advertising campaign on TV, billboards, and buses that sought to arrest the spread of AIDS by promoting the use of condoms (*Segodnya*, 1997). Meanwhile the reported incidence of syphilis soared from 4.2 cases per 100,000 population in 1988 to 263 in 1996 (Tichonova et al., 1997). AIDS is becoming a major public health problem.

Sexuality Education

Disputes about whether Russian adolescents needed sexuality education, and if so, what it should involve, lasted until 1983 when a two-part course on preparation for marriage and family life was introduced into all Soviet schools. Twelve hours of "Hygienic and Sexual Education" were taught to15 year olds within a course on anatomy and physiology. "The Ethics and Psychology of Family Life" was a 34-hour course for 16 and 17 year olds that often provoked laughter in the classroom (Williams, 1994). While some elements of sexuality education were included, the course was designed primarily to serve the state's political objective of transforming society. The word "sexual" was rarely used in standard school syllabuses. When it was, it was often with the purpose of condemning the importance of sex (Muckle, 1987). The new Soviet man and woman were to embody the personal discipline demanded by communist morality, a code of behavior that strove to equal personal actions with the public interest and required personal sexual restraint (Field, 1996; Rivkin-Fish, 1996). The program, which had been largely ignored, was discontinued in the late 1980s because teachers were poorly prepared, textbooks failed to consider the diverse cultural backgrounds of the geographic regions, and the discussion was about everything else but sex (Popova, 1996; Randolph, 1996). The official silence on sexuality was such that a young Russian participating in one of the early 1988 live TV satellite linkups between the United States and the USSR responded to an American inquiry by declaring "We have no sex here." No one in the Moscow studio audience laughed (Vanden Heuvel, 1993; Attwood, 1993). At the same time, news reports on the activities of Russian adolescents demonstrated the opposite.

Perestroika did not signal a change in the Communist Party's official attitude toward sexuality. While the Ministry of Health acknowledged the problems of unwanted adolescent pregnancies and abortions, the Ministry of Education abolished the unsatisfactory course on ethics and family life without substituting anything for it. In effect, there was no sexuality education (Popova, 1992; Komyssova, 1992; Kon, 1992).

In 1989 a group of physicians and psychologists with support from the MOH organized the Moscow Center for Forming Sexual Culture. Lectures on sexuality were provided to schools and youth clubs. While the teenagers were very receptive, the center was forced to close due to lack of funds. Its activities were subsequently resumed by similar centers within teacher and medical training institutions and by similarly named centers in Yaroslavl and other major cities as well by the Russian Family Planning Association and the Russian Sexological Association (Popova, 1996). However, none of these endeavors, supported largely by nongovernmental organizations, was ever adopted for national use by the Ministry of Education. Sexuality education or sexology were not included in the curricula of medical students (Dmitrieva, 1996). Attempts at sexual enlightenment were frequently denounced as a "capitalist, Jewish plot." One slogan of the failed 1991 coup was "Fight Sex and Violence" (Kon, 1991, cited in David, 1992a).

Following a 1989 VCIOM survey, a 1991 VCIOM survey confirmed the low state of sexuality education in the country. Only 13 percent of the nationally representative respondents replied affirmatively to the question "Did your parents ever talk to you about sexual education?" While parents had spoken with 15 percent of the girls, they had done so with only 10 percent of the boys. Some 28 percent of those under age 25 replied affirmatively, but only 5 percent of those over age 25 had ever had any parental guidance. Urban parents talked about sex with their children more frequently than rural parents (Kon, 1992). A 1994 VCIOM survey reported that more than half the respondents would welcome the introduction of a course on sex education for 14–16 year olds; the approval rating was a bit lower for 10–13 year olds. Respondents with higher education and urban residence were most supportive. The 1994 research confirmed the results of the 1989 VCIOM survey where respondents expressed the opinion that the most effective method of educating young people on sexual matters would be through the implementation of special courses at the elementary school level (Bodrova, 1995, 1996). The majority of men (83 percent) and women (72 percent) agreed with the idea that information on contraceptives be made available to teenagers.

In 1996 the Ministry of Education launched a sexuality education pilot project with the support of the United Nations Fund for Population Activities (UNFPA) and under the auspices of UNESCO. It was expected to take three years to develop the curriculum, prepare the materials, and test them in the schools. The project was initiated with a survey of seventh to nineth grade students, their parents, and their teachers in 16 schools in eight regions of Russia that agreed to participate (Chervyakov, 1997). One of the findings was that 78 percent of teachers expressed the view that parents should assume primary responsibility for the sexuality education of their children. However, only 5 percent of teenagers considered it acceptable to discuss sexuality with their parents, who were also reluctant to initiate such conversations. More than half the parents never discussed sexuality with their children; another 25 percent did so only once or twice; and only 20 percent of mothers (fewer fathers) initiated such talks several times. The primary restraining factors were psychological and educational unreadiness. More than 75 percent of parents favored the publication of special textbooks explaining what should be said and how to their children. Nearly as many parents (66 percent) thought it useful to organize seminars for parents in the schools their children attended. Teachers sought retraining in pedagogical institutes.

Unfortunately, before the survey results were completely analyzed, the Ministry of Education dispatched to 30,000 schools a package of five alternative sex education programs it had developed on its own without consultation and unconnected to the UNFPA/UNESCO project. A storm of protest ensued, led by vocal pro-life activists who complained to the Duma's National Security Committee about Western ideological subversion of Russian children. Some Russian Orthodox priests told their parishioners that all bad things in Western life are rooted in sex education, that Western governments are trying to ban it, and that only the Russian Government, at the instigation of the "Sexological Industrial Complex," was acting against its own best interests (Kon, 1997). Strong opposition was also voiced by "Communist die-hards who still cling to the Soviet policy of vigorous suppression of public sexuality" (*Moscow Times*, 1997). The argument was made that Western capitalists introduced pornography and immorality into Russia and that sex education was a means of increasing both. Some mental health specialists attacked sex education as likely to lead to psychological pathologies (Williams, 1997). The result of all this agitation was that the Attorney-General's office declared the Ministry of Education survey "illegal." Another attempt to develop a national program of sex education was halted (*Itar-Tass*, 1997). Meanwhile the communist faction of the Duma succeeded in reducing the federal family planning budget by more than half (Gadasina, 1997).

A pilot project, initiated by the Center for Formation of Sexual Culture in Yaroslavl in consultation with Russian and American specialists, produced a framework for sexuality education, textbooks, and teaching materials. It is among the few resources still available to Russian youth along with the regional activities of the Russian Family Planning Association (Alesina, 1996; Mitlyng, 1997). As of early 1999 there is no national policy on sexuality education in the Russian Federation. Few autonomous local school districts offer any systematic sex education programs.

Adolescent Sexuality

In 1965 Golod surveyed 500 students at 10 Leningrad colleges. For political reasons, the findings were not accepted as a doctoral dissertation and not published until some years later (Golod, 1993; Kon, 1995). Asked to indicate their attitude to premarital sex, 45 percent of students approved, 33 percent took a neutral position ("uncertain"), and 22 percent opposed it. Seven years later, Golod conducted a repeat survey in the same colleges. This time 47 percent approved and 14 percent opposed. Additional studies were conducted in 1974 and 1978. Golod's data comprise the only methodologically sound sociological studies on sexual behavior and attitudes of Soviet youth in the 1960s and 1970s. They reflect a growing gap between generations in terms of sexual values and behavior. While publicly supporting the Communist Party's puritanical ideology, youths behaved quite differently in their private lives.

Perestroika and *glasnost* resulted in Russian adolescents becoming more sexually active at earlier ages (Bodrova and Mytil, 1992). Kon suggested that changes in adolescent sexual behavior, such as earlier sexual debut and acceptance of premarital sexuality, reflected universal generational changes made apparent but not generated by *glasnost*.

According to 1993 and 1995 surveys (Chervyakov, 1996), 25 percent of 16 year old girls and 38 percent of same age boys had experienced sexual debut by 1993; in 1995, the respective figures had risen to 33 percent and 50 percent. Adolescent sexuality was strongly related to social class and often violent. About 10 percent of teenage girls experienced sexual initiation under some degree of compulsion (Kon, 1997). The 1995 survey of 2,871 students from 16 to 19 years of age, residing in a large city, medium size city, and small town, reflected significant differences between the sexual experiences of city and small-town youth as well as between high school and vocational students. Among males, nearly half of Moscow (city) students had experienced sex while a third of Borisoglebsk (small town) students had done so. Among females, 30 percent of the Moscow students had sexual experience while just over 10 percent of the Borisoglebsk students did. Vocational students were 1.5 times more active than their high school peers.

One of the few published questionnaire studies of adolescent sexuality was reported from St. Petersburg (Lunin et al., 1995). A random sample of tenth graders was surveyed in 14 schools, stratified by socioeconomic district. The response rate was 94 percent. Among the respondents, 31 percent of males and 20 percent of females reported having experienced sexual debut. The responses reflected considerable misinformation, particularly about AIDS prevention. Most information sources were considered not very credible or not readily accessible. A subsequent study reported on a successful pilot program using the Internet (Lunin et al., 1997). It never went beyond St. Petersburg.

Russian Family Planning Association

The nongovernmental sector in the field of sexual and reproductive health did not exist until 13 January 1989 when the Soviet Family and Health Association (SFHA) was officially registered (Manuilova, 1991a). With the 1991 dissolution of the USSR, the SFHA became the independent International Association for Family and Health.

At the end of 1991 the Russian Family Planning Association (RFPA) was organized with support from the International Planned Parenthood Federation (IPPF) and the Russian Government (Grebesheva, 1992; Russia/MOH, 1992a). Accepted for IPPF membership in 1993, it has more than 50 branches throughout the country (Grebesheva, 1994, 1996). RFPA established a training center for medical professionals, organized seminars for physicians, nurses, and teachers, and developed three youth centers focusing on the needs of adolescents. With MOH and international donor support, RFPA publishes its own medical journal, translations of IPPF texts, and popular booklets. In working with the media and with parliamentarians, RFPA advocates family planning and reproductive rights, new ideas for many male Duma members. It seeks to facilitate the dissemination of information, better marketing of contraceptives, and improved quality of abortion services, while encouraging men to assume greater responsibility for pregnancy prevention.

Family planning programs were often met with public skepticism in the false belief that the purpose was to decrease the already low Russian birth rate. Once people understood that the actual purpose was to improve women's health and reduce reliance on abortion, they became more receptive to family planning (Schehl, 1996). However, despite the increasingly successful RFPA programs, the right to family planning in Russia remains restricted due to the lack of information outside the larger cities, lack of specialized medical personnel, and lack of wide access to modern contraceptives. In practice, only the right to abortion on request has been realized and that right is being challenged by the Russian Orthodox Church and its supporters.

FUTURE PERSPECTIVES

It is difficult to consider future perspectives in a country as large and diverse as the Russian Federation. Communism has been dismantled but nothing lasting has replaced it as yet. Moreover, communism is not dead and still exerts a powerful political influence. Hopes for achieving a better standard of living for most people have been repeatedly dashed by economic crises, including the 17 August 1998 devaluation of the ruble, default on debt, and soaring inflation. However, the impact of these events has been far greater in Moscow and other major cities than in the villages (*The Economist*, 1998e; Schmemann, 1998). The search continues for more solid ground on which to build a new democratic society.

As Talbott (1998: 54) noted, "the doom and gloom are no more justified now than was euphoria a few years ago. The drama of Russia's transformation is not over; its ending is neither imminent nor foreordained." A Russian political scientist (Pain, 1998) observed that people whose worldview was essentially formed post-*perestroika* are entering political life. Many do not remember Soviet ways. They openly demonstrate a new pattern of behavior in many spheres of life such as family formation. This includes increased use of modern contraceptives, more favorable attitudes toward family planning, and greater requests for improved educational programs organized by schools, public health facilities, and nongovernmental organizations, often supported by international donors and designed to reduce reliance on abortion. Women's issues are slowly becoming part of the political agenda (Azhgikhina, 1998). Gender equality is no longer ignored. Optimism lurks beneath smoldering discontent.

NOTES ON OTHER USSR SUCCESSOR STATES

On 8 December 1991 Belarus, the Russian Federation, and Ukraine signed the Minsk Agreement that announced that "the USSR has ceased to exist" (*Los Angeles Times*, 1991). The three countries formed the Commonwealth of Independent States (CIS). By October 1993 all the former republics except the three Baltic states had joined the CIS. Discussion of historical and current trends in reproductive behavior in the 14 republics (other than the Russian Federation) follows the geographical grouping suggested by the United Nations.

A major problem with the reported demographic data from the former USSR republics is that they were compiled by the All-Union territorial system with each republic MOH reporting to the USSR/MOH. A comparative summary was published by Popov (1991) with a more detailed presentation by Avdeev et al. (1994) Their papers include abortion rates for women aged 15–49. For comparative purposes the data presented here are for women age 15–44 years, usually only for the period 1990–96.

After the disintegration of the USSR, the independent successor states were free to decide which abortions to report. Mini-abortions were not always included. Data sources are cited in the tables. When choosing among different sources reporting different numbers, the highest official reports were selected. Caution in the use of these data is suggested. For the Central Asian Republics only data from 1990 onward are shown in the tables. Earlier figures cited in the text were provided by Zakharov (1998b). Abortion data are incomplete, in part because many procedures are performed by traditional health practitioners and not recorded (Kuzebaeva, 1998). Technical capability for data collection and utilization is still being developed (Heleniak, 1998).

Baltic States: Estonia, Latvia, and Lithuania

On 20 August 1991 Boris Yeltsin, then president of the Russian Soviet Federated Socialist Republic issued a decree recognizing the independence of Estonia, followed by Latvia and Lithuania. The three Baltic states had been annexed by the USSR in 1940 following the Molotov-Ribbentrop Pact. The effects of the transition on social and economic conditions, on reproductive health, and on the status of women are rather similar in all three countries as they try to move closer to their neighbors in Northern Europe. One concern is that 20–35 percent of maternal mortality continues to be due to induced abortion (Mogilevkina et al., 1996).

Lithuania has the largest population (3.7 million), followed by Latvia (2.5 million) and Estonia (1.5 million). Births, birth rates, and total fertility rates for all three countries are shown in Table 11.3. Abortions, abortion rates, and abortion ratios are presented in Table 11.4. Abortion has been legal in all three countries since 1955. Vacuum aspiration is the main method. The Catholic Church, which is strongest in Lithuania, is actively opposed to abortion, modern contraception, and sexuality education.

Official abortion statistics are very reliable in Estonia. Reporting of abortions is compulsory for all institutions including the private health sector. A new abortion register was introduced in 1993. In Latvia official abortion statistics underestimate the actual number; some procedures in private clinics may not be reported. Similarly, reported abortions in Lithuania are likely to be an underestimate. Since independence, the choice of effective contraceptives has increased but the cost is too expensive for many couples. Sterilization is not regulated by law in Estonia or Latvia. In Lithuania voluntary surgical sterilization is opposed by the Catholic Church. Hard data on contraceptive sales or on contraceptive practice are not readily available (Karro et al., 1997).

In *Estonia* the birth rate decreased from 16.6 per 1,000 population in 1960 to 15.8 in 1970, 15.0 in 1980, 14.2 in 1990, and 8.7 in 1997. The TFR declined from 2.16 per woman in 1970 to 2.02 in 1980, rose to 2.05 in 1990 and dropped to 1.24 in 1997. Termination of pregnancy is regulated by decrees from the Ministry of Social Affairs of 1992 and 1993. Abortion on request is allowed up to 12 weeks and abortion for medical reasons up to 20 weeks in state-owned and other licensed

Table 11.3
Live Births, Birth Rates, and Fertility Rates: Baltic States, 1960–97

	Estonia			Latvia			Lithuania		
Year	Live Births	Birth Rate	Total Fertility Rate	Live Births	Birth Rate	Total Fertility Rate	Live Births	Birth Rate	Total Fertility Rate
1960	20,200	16.6	1.97	35,500	16.7	1.94	62,500	22.5	2.57
1970	21,600	15.8	2.16	34,300	14.6	2.01	55,500	17.6	2.40
1980	22,200	15.0	2.02	35,500	14.1	1.86	51,800	15.1	2.03
1990	22,300	14.2	2.05	37,900	14.2	2.02	56,900	15.2	2.03
1991	19,300	12.3	1.79	34,600	13.0	1.86	56,200	15.0	2.01
1992	18,000	11.7	1.69	31,600	12.0	1.73	53,600	14.3	1.90
1993	15,200	10.0	1.45	26,800	10.3	1.51	46,700	12.5	1.67
1994	14,200	9.5	1.37	24,300	9.5	1.39	42,800	11.5	1.54
1995	13,600	9.1	1.32	21,600	8.6	1.25	41,200	11.1	1.49
1996	13,300	9.0	1.30	19,800	7.9	1.16	39,200	10.5	1.42
1997	12,600	8.7	1.24	18,800	7.5	1.11	37,800	10.2	1.39

Sources: Litinaviciene (1996), COE (1998), Katus and Zakharov (1997), and Zakharov (1998b).

Table 11.4
Legal Abortion Rates and Ratios: Baltic States, 1990–97

	Estonia			Latvia			Lithuania		
Year	Legally Induced Abortions	Legal Abortion Rate[a]	Legal Abortion Ratio[b]	Legally Induced Abortions	Legal Abortion Rate[a]	Legal Abortion Ratio[b]	Legally Induced Abortions	Legal Abortion Rate[a]	Legal Abortion Ratio[b]
1990	29,400	87.9	1,318	49,000	87.2	1,293	50,000	61.7	879
1991	29,400	87.6	1,523	38,800	69.1	1,121	40,800	50.2	726
1992	28,400	84.9	1,578	34,300	61.2	1,085	40,900	50.1	763
1993	25,600	78.6	1,684	31,300	57.2	1,168	35,200	43.0	754
1994	22,500	70.1	1,584	26,800	49.8	1,103	30,400	37.1	710
1995	20,500	64.5	1,507	25,900	48.9	1,199	31,300	38.2	760
1996	19,500	61.8	1,271	24,200	46.1	1,222	27,800	34.0	709
1997	19,200	61.4	1,524	21,800	41.6	1,159	22,700	27.7	600

[a]Per 1,000 women age 15–44. [b]Per 1,000 live births.

Sources: COE (1998), and Zakharov (1998b).

public and private health care facilities. Mini-abortions and abortions performed for medical reasons are free for women with health insurance. For other abortions provided for nonmedical reasons, insured women pay 50 percent of full cost, which in 1997 was equivalent to U.S.$50 (Karro et al., 1997). The abortion ratio decreased from 1,318 per 1,000 live births in 1990 to 1,271 in 1996, rising to 1,524 in 1997, highest among the Baltics. The abortion rate decreased from 87.9 per 1,000 women age 15-44 in 1990 to 61.4 in 1997. The use of modern contraceptives is increasing (Karro, 1997). Funds obtained from paid abortions are used as subsidies to make modern contraceptives more affordable for full time students, women in their first year after delivery, and for three months postabortion (Karro et al., 1997).

A 1992 survey of 360 women who had in-hospital abortions in Tallinn, Estonia, during 1991 found that ethnic Russian women were more likely than native Estonian women to have had a larger number of previous abortions, contributing disproportionately to the high incidence of abortion. For Estonian women abortion was more frequently a backup for contraceptive failure whereas it was more often the primary method of fertility regulation for Russian women. Estonians were more successful than non-Estonians in preventing unwanted pregnancies (Anderson et al., 1992). A related study showed that ethnic Estonians were somewhat less likely than Russian women to report their abortions. Unmarried women were also less likely than married women to report their abortions as were women who had delivered three or more children compared to those who had fewer than three children (Anderson et al., 1994).

Sexuality education is part of the school curricula for fourth, seventh, and tenth graders. Implementation has been hampered by lack of adequate teaching materials and/or insufficient teacher training. Pilot programs, some operated by the Estonian Family Planning Association, founded in 1994, are trying to facilitate the training of trainers.

In *Latvia* the birth rate declined from 16.7 per 1,000 population in 1960 to 14.6 in 1970 and 14.1 in 1980, rose to 14.2 in 1990, and fell to 7.5 in 1997. TFR increased from 1.94 per woman in 1960 to 2.01 in 1970, fell to1.86 in 1980, rose to 2.02 in 1990 and dropped to 1.11 in 1997. Abortions are provided in state medical institutions and in private clinics. The law requires parental consent for minors up to 18 years of age. A physician who knows about the pregnancy of a woman under age 16 is required to notify the state court system. However, these two requirements are not always strictly enforced (Karro et al., 1997). The cost of procedures ranges from U.S.$30 to U.S.$150. The legal abortion ratio fell from 1,293 per 1,000 live births in 1990 to 1,222 in 1996, declining to 1,159 in 1997. Imported contraceptives are in short supply and relatively expensive. Surgical contraception for women over age 35 with three or more children costs three months' salary (Lazdane, 1997). While sexuality education has not yet been incorporated into school curricula, it is being taught in about one-third of all Latvian schools. Information, education, and communication activities are conducted by the Latvian Family Planning Association, founded in 1994. In 1995 it published the first comprehensive Latvian booklet on contraception, translated IPPF guidelines, and established a youth group with a telephone hotline operated by peer counselors (Vjatere, 1995). In 1997 social marketing of contraceptives was initiated with IPPF support (Lazdane, 1998).

In *Lithuania,* despite the strong influence of the Catholic Church, small families have become traditional (Litinaviciene, 1996). Of the total number of families with children under 18 years of age in 1995, more than half (51 percent) had only one child, 40 percent had two children, 7 percent had three children, and 2 percent had four or more children. The birth rate declined from 22.5 per 1,000 population in 1960 to 17.6 in 1970, 15.1 in 1980, 15.2 in 1990, and to 10.2 in 1997. TFR decreased from 2.57 per woman in 1960 to 2.40 in 1970, 2.03 in 1980 and 1990, and to 1.39 in 1997.

Abortion was and is the main method of fertility regulation. Women now pay the equivalent of U.S.$15 to U.S.$25 per procedure. The provision of services has come under increasingly strong attack from U.S- funded antiabortion activists, who are plastering clinic walls with their propaganda (Lithuanian Frontline, 1994). The reported abortion ratio decreased from 879 per 1,000 live births in 1990 to 709 in 1996 and 600 in 1997. The abortion rate dropped from 61.7 per 1,000 women age 15–44 in 1990 to 34.0 in 1996 and 27.7 in 1997. Since not all abortions are recorded, UNFPA estimates that actual figures are much higher, approximating those in the other Baltic states (WHO/EURO, 1997).

Withdrawal remains the most popular contraceptive method, followed by condoms and IUDs. Contraceptives are imported and are expensive compared to abortion. Sterilization is not part of the national health policy discussion (Kuliesyte, 1998). In February 1998 the Lithuanian Family Planning Association, with IPPF support, launched a social marketing project featuring Cool and Cool Gold condoms (Open File/IPPF, 1998). Sexuality education is restricted by church influence, which often limits discussion to "saving the life and moral values of society" (Karro et al., 1997: 15). More practical information is promoted by the Lithuanian Family Planning Association. A helpline for teenagers was established by the Vilnius Family Planning Center in 1994 to provide contraceptive information (Bukute and Kabasinskiene, 1996).

Belarus, Ukraine, and Moldova

Births, birth rates, and total fertility rates for Belarus, Ukraine, and Moldova are shown in Table 11.5 while reported abortions, abortion rates, and abortion ratios are presented in Table 11.6. *Belarus* (formerly Bylorussia or White Russia) is perhaps closest to Russia in its culture and history. The 1986 nuclear accident at Chernobyl, just across the border in Ukraine, continues to influence social, health, and economic conditions. The birth rate declined from 24.4 per 1,000 population in 1960 to 16.2 in 1970, 16.0 in 1980, 13.9 in 1990, and 8.7 in 1997. The TFR decreased from 2.76 per woman in 1960 to 2.33 in 1970, 2.03 in 1980, 1.91 in 1990, and 1.23 in 1997. Only IUDs are domestically produced. Imported contraceptives are expensive and are insufficient to meet needs. The reported abortion ratio declined from 1,307 per 1,000 live births in 1980 to 1,825 in 1990, rose to 1,878 in

Table 11.5
Live Births, Birth Rates, and Fertility Rates: Belarus, Ukraine, and Moldova, 1960–97

	Belarus			Ukraine			Moldova		
Year	Live Births	Birth Rate	Total Fertility Rate	Live Births	Birth Rate	Total Fertility Rate	Live Births	Birth Rate	Total Fertility Rate
1960	200,200	24.4	2.76	878,800	20.5	2.23	87,900	29.3	3.36
1970	146,700	16.2	2.33	719,200	15.2	2.09	69,800	19.4	2.58
1980	154,400	16.0	2.03	742,500	14.8	1.95	79,600	19.8	2.36
1990	142,200	13.9	1.91	657,200	12.7	1.85	77,100	17.7	2.39
1991	132,000	12.9	1.80	630,800	12.1	1.81	72,000	16.5	2.26
1992	128,000	12.4	1.75	596,800	11.4	1.72	69,700	16.0	2.22
1993	117,400	11.3	1.61	557,500	10.7	1.55	66,200	15.2	2.10
1994	110,600	10.7	1.57	521,500	10.0	1.46	62,100	14.3	1.95
1995	101,100	9.8	1.39	492,900	9.6	1.40	56,400	13.0	1.76
1996	95,800	9.3	1.31	467,200	9.1	1.38	51,900	12.0	1.63
1997	89,600	8.7	1.23	445,500	8.9	1.30	45,600	10.6	1.51

Sources: U.N. (1995a), Haub (1995), COE (1998), USBOC (1998), Belarus/MOH (1998), Zakharov (1998b, 1999), Sarbanescu (1998b), and UNICEF/TDB (1998).

Table 11.6
Legal Abortion Rates and Ratios: Belarus, Ukraine, and Moldova, 1990–97

	Belarus			Ukraine			Moldova		
Year	Legally Induced Abortions[a]	Legal Abortion Rate[b]	Legal Abortion Ratio[c]	Legally Induced Abortions	Legal Abortion Rate[b]	Legal Abortion Ratio[c]	Legally Induced Abortions	Legal Abortion Rate[b]	Legal Abortion Ratio[c]
1990	259,500	116.9	1,825	1,045,300	96.5	1,590	81,800	83.1	1,061
1991	235,300	104.9	1,783	957,000	86.9	1,517	73,500	73.5	1,021
1992	234,300	103.6	1,830	932,300	84.0	1,562	71,400	70.9	1,024
1993	212,700	93.2	1,812	861,000	76.7	1,544	64,200	63.5	970
1994	207,700	90.5	1,878	798,500	71.1	1,531	58,800	58.0	947
1995	188,700	81.9	1,866	740,000	66.5	1,501	57,200	56.7	1,014
1996	169,900	73.6	1,753	687,000	62.3	1,470	46,000	45.8	886
1997	148,900	64.4	1,662	na	na	na	37,100	45.4	814

[a]Includes mini-abortions. [b]Per 1,000 women age 15–44. [c]Per 1,000 live births.

Sources: UN (1995a), Haub (1995), COE (1998), Belarus/MOH (1998), USBOC (1998), and Zakharov (1998b).

1994, and fell to 1,662 in 1997. The abortion rate decreased from 116.9 in 1990 to 81.9 in 1995 and 64.4 in 1997. Family and marriage centers treat infertility but seldom provide contraceptive information. Sexuality education is nonexistent. Family planning is not included in the medical school curriculum. There is no family planning association (Mikhnyuk, 1994).

Ukraine with its population of 52 million has almost as many people as France or Great Britain. The eastern provinces are dominated by ethnic Russians and the western ones by Ukrainians with many people of mixed parentage living in the center. More than 300 years of Russian governance reduced ethnic tensions. The Ukrainian birth rate per 1,000 population declined from 20.5 per 1,000 population in 1960 to 15.2 in 1970, 14.8 in 1980, 12.7 in 1990, and 8.9 in 1997. The TFR decreased from 2.23 per woman in 1960 to 2.09 in 1970, 1.95 in 1980, 1.85 in 1990, and 1.30 in 1997. There is considerable variation between urban and rural areas (Ukraine, 1994).

Nearly all urban and most rural young women know about modern contraceptives but access is very limited (Frejka, 1998). None are produced domestically. The reported legal abortion ratio ranged around 1,500 per 1,000 live births from 1970 to 1995, declining to 1,470 in 1996. The abortion rate dropped from 96.5 per 1,000 women aged 15–44 years in 1990 to 62.3 in 1996. Sexuality education is virtually nonexistent. While abstinence is widely recommended, as many as 83 percent of brides under age 20 were pregnant in the 1990s (Vornik and Govorlin, 1996). A Kiev survey of sexually active women aged 19 years or younger showed that 78 percent used only traditional methods of contraception (WHO/EURO, 1997). In the recent past the average Ukrainian woman was said to have had five abortions during her lifetime (Volk, 1995), but the situation is improving. The government is encouraging family planning and approved a national policy in 1996. A first National Family Planning Week was organized throughout the country in 1997. The Ukrainian Family Planning Association was established the same year (Antarsh, 1997). In 1998, training was initiated for designated trainers from family planning centers in every region (Antarsh, 1998).

In *Moldova* the birth rate declined from 29.3 per 1,000 population in 1960 to 19.4 in 1970, rose to 19.8 in 1980 and decreased to 17.7 in 1990 and 10.6 in 1997. The TFR declined from 3.36 per woman in 1960 to 2.58 in 1970, and 2.36 in 1980, and rose to 2.39 in 1990 before dropping to 1.76 in 1995 and 1.51 in 1997. Abortion remains legal and available under the same conditions as authorized in 1982 in the Soviet Union. The reported abortion ratio declined from 1,224 abortions per 1,000 live births in 1987 and 1,061 in 1990 to 814 in 1997. The abortion rate fell from 83.1 per 1,000 women age 15–44 in 1990 to 45.4 in 1997.

According to the 1997 Moldova Reproductive Health Survey (Serbanescu and Morris, 1998), the highest abortion rate per 1,000 women by age group occurred among women aged 25–29 years (81/1,000), followed by 20–24 year olds (74/1000). Although abortion rates decrease by age, they are higher than fertility rates for women over age 30. Moldovan women complete their desired family size at younger ages after which most pregnancies are unintended and intentionally terminated. Among women aged 15–44 years, married or living in union, 38 percent rely on the IUD, 22 percent on withdrawal, 6 percent on condoms, 7 percent on other methods, and 26 percent do not use contraception. Only 32 percent of sexually experienced unmarried young adults used contraception at sexual debut. Over half of those using a method relied on withdrawal and about one in three on condoms (Serbanescu et al., 1998).

In 1993 Moldova recognized and registered the Moldovan Family Planning Association. It is particularly active in sexuality education. A survey of 1,500 young people, aged 13 to 18 years showed that knowledge about sexuality is low. For most adolescents, pornographic videos and literature are the major sources of information (Moshin et al., 1996). To counter this trend, the Moldovan Family Planning Association is working closely with public and private media, providing consultation to a prime time TV sex education program (Hincu and Gilca, 1996). On 17 May 1994 the Ministry of Health signed a decree stipulating that each of Moldova's 44 districts will have one physician responsible for family planning services in a designated family planning room. International donors played a major role in providing clinical and counseling skills, management training, and contraceptive supplies (Antarsh, 1998).

The Caucasus: Armenia, Azerbaijan, and Georgia

The mountainous Caucasus region lies between the Black and Caspian seas, including the ancient lands of Armenia, Azerbaijan, and Georgia. Births, birth rates, and total fertility rates for all 3 countries are presented in Table 11.7 while reported abortions, abortion rates, and abortion ratios are shown in Table 11.8.

Predominantly Christian *Armenia* survived for 17 centuries in a largely Moslem region. The country is still recovering from the severe 1988 earthquake and the disastrous conflict with Azerbaijan over the ethnic Armenian enclave in Nagomo-Karabach. The birth rate declined from 40.1 per 1,000 population in 1960 to 22.1 in 1970, rose to 22.7 in 1980, and decreased to 22.5 in 1990 and 12.8 in 1996. TFR decreased from 4.69 per woman in 1960 to 3.17 in 1970 and 2.31 in 1980 before increasing to 2.84 in 1990 and declining to 1.60 in 1996 and to around 1.40 in 1997 (Khachikien, 1998). The reported abortion rate (not including mini-abortions), after falling in the 1970s and 1980s, rose to 31.1 per 1,000 women aged 15–44 years in 1990 and to 33.9 in 1996. The abortion ratio declined from 723 per 1,000 live births in 1975 to 422 in 1985 and 327 in 1990 before rising again to 650 in 1996. A 1994–95 study of 200 fairly well educated urban women attending an abortion clinic in Yeravan found that those younger than 20 years of age reported a median of one and those older than 40 years a median of eight induced abortions in their lifetimes. In the month of conception, 87 percent of the women had not used any form of contraception (Dolian et al., 1998) Oral contraceptives and IUDs have to be imported and are available primarily in specialized clinics (UNFPA/WHO/EURO, 1995). In the absence of sexuality education and contraceptive

Table 11.7
Live Births, Birth Rates, and Fertility Rates: Armenia, Azerbaijan, and Georgia, 1960–96

	Armenia			Azerbaijan			Georgia		
Year	Live Births	Birth Rate	Total Fertility Rate	Live Births	Birth Rate	Total Fertility Rate	Live Births	Birth Rate	Total Fertility Rate
1960	74,800	40.1	4.69	165,800	42.6	5.50	102,900	24.7	2.90
1970	55,700	22.1	3.17	151,000	29.2	4.66	90,200	19.2	2.68
1980	70,300	22.7	2.31	155,000	25.2	3.32	89,500	17.6	2.25
1990	79,900	22.5	2.84	183,000	26.3	2.74	92,800	17.0	2.21
1991	77,800	21.6	2.58	190,400	27.0	2.87	89,100	16.6	2.15
1992	70,600	19.2	2.35	181,400	25.2	2.76	72,600	14.9	1.79
1993	59,000	15.8	1.97	174,600	24.2	2.70	61,600	12.6	na
1994	51,100	13.5	1.70	159,800	21.6	2.30	51,900	10.7	na
1995	49,000	13.0	1.63	143,300	19.1	2.30	56,300	11.6	na
1996	48,100	12.8	1.60	129,200	17.1	2.30	53,700	11.1	na

Sources: COE (1998), Zakharov (1998b, 1999), Azerbaijan/SSC (1998), USBOC (1998), and Monnier (1998).

Table 11.8
Legal Abortion Rates and Ratios: Armenia, Azerbaijan, and Georgia, 1990–96

	Armenia			Azerbaijan			Georgia		
Year	Legally Induced Abortions	Legal Abortion Rate[a]	Legal Abortion Ratio[b]	Legally Induced Abortions	Legal Abortion Rate[a]	Legal Abortion Ratio[b]	Legally Induced Abortions	Legal Abortion Rate[a]	Legal Abortion Ratio[b]
1990	26,100	31.1	327	24,600	14.5	134	56,900	47.2	613
1991	27,300	31.8	351	37,900	22.0	199	49,600	40.9	557
1992	28,000	34.2	397	36,900	21.0	203	40,500	33.4	558
1993	27,900	33.8	473	39,800	22.5	228	36,700	30.0	596
1994	30,600	36.9	599	na	na	na	35,600	29.3	686
1995	30,700	33.6	626	na	na	na	na	na	na
1996	31,300	33.9	650	na	na	na	na	na	na

[a]Per 1,000 women age 15–44. [b]Per 1,000 live births.

Sources: COE (1998), Zakharov (1998b), Azerbaijan/SSC (1998), and USBOC (1998).

counseling before marriage, abortion continues to be the most common method of fertility regulation (Dolian, 1997; WHO/EURO, 1997).

In *Azerbaijan* the birth rate declined from 42.6 per 1,000 population in 1960 to 29.2 in 1970, 25.2 in 1980, 26.3 in 1990, and 17.1 in 1996. TFR decreased from 5.50 per woman in 1960 to 4.66 in 1970, 3.32 in 1980, 2.74 in 1990, and 2.30 in 1996. The absence of modern family planning methods is felt most in the rural areas. Fertility regulation is perceived to be the woman's responsibility. In 1992 less than 2 percent of women used contraceptives (UNFPA/WHO/EURO, 1995). The abortion ratio decreased from 375 per 1,000 live births in 1965 to 134 in 1990 before rising to 228 in 1993. The abortion rate remained at around 22.0 per 1,000 women age 15–44 between 1991 and 1993. The actual number of abortions is believed to be much higher (WHO/EURO, 1997). Sexuality education does not exist.

In *Georgia* the birth rate declined from 24.7 per 1,000 population in 1960 to 19.2 in 1970, 17.6 in 1980, 17.0 in 1990, and 11.1 in 1996. It was influenced in the 1990s by the civil war and continuing ethnic strife amidst deteriorating economic conditions. The TFR decreased from 2.90 per woman in 1960 to 2.68 in 1970, 2.25 in 1980, 2.21 in 1990, and 1.79 in 1992 (the latest date available). Under the leadership of the Zhordania Institute (Tbilisi) and with government and international support, considerable progress was made in the late 1980s in promoting modern methods of contraception, providing abortions on an anonymous outpatient basis, and training physicians in menstrual regulation techniques (David, 1992a). Although the number of established family planning clinics had to be curtailed due to economic constraints after the collapse of the Soviet Union, the use of oral contraceptives (nearly all donated) and IUDs rose from 5 percent to 19 percent of fertile age women in 1994. The reported abortion rate declined from 47.2 per 1,000 women 15–44 years old in 1990 to 29.3 in 1994, including mini-abortions The legal abortion ratio rose from 513 per 1,000 live births in 1970 to 953 in 1980 and 1,022 in 1981 before declining to 613 in 1990 and rising again to 686 in 1994. Since private abortions are usually performed outside of hospitals and not reported, abortion statistics are considered unreliable (Khomassuridze, 1993).

Central Asian Republics: Kazakstan, Kyrgyz Republic, Tajikistan, Turkmenistan, and Uzbekistan

The Central Asian republics comprise five countries: Kazakstan, Kyrgyz Republic, Tajikistan, Turkmenistan, and Uzbekistan. They share many common characteristics, including the organization of their health care system, established when they were part of the Soviet Union. The traditions of life under Islam, long suppressed, are being reclaimed by many men. The situation of women has deteriorated. They are the last to be hired and first to be fired. Most pregnancies are carried to term. While pronatalist allowances remain in effect, inflation has seriously eroded their value. Births, birth rates, and total fertility rates are shown in

Table 11.9
Live Births, Birth Rates, and Fertility Rates: Kazakstan and the Kyrgyz Republic, 1990–96

	Kazakstan			Kyrgyz Republic		
Year	Live Births	Birth Rate	Total Fertility Rate	Live Births	Birth Rate	Total Fertility Rate
1990	362,200	21.7	2.72	128,000	29.3	3.69
1991	354,100	21.0	2.80	129,500	29.2	3.67
1992	338,400	19.9	2.65	128,300	28.6	3.62
1993	316,200	18.6	2.55	116,800	25.9	3.28
1994	306,500	18.2	2.45	110,100	24.7	3.14
1995	277,000	16.6	2.40	117,300	26.2	3.31
1996	253,100	15.3	2.27	108,000	23.8	3.00

Sources: CIS/ISC (1998), Kyrgyz Republic/NSS (1997), Heleniak (1998), Zakharov (1998b, 1999), and USBOC (1998).

tables 11.9 and 11.11 while abortions, abortion rates, and abortion ratios are presented in tables 11.10 and 11.12.

None of the Central Asian republics has any legal prohibition against using, distributing, or advertising modern contraceptives, or providing sterilization (Darsky and Dworak, 1993). International donor assistance has greatly improved the supply situation in Kazakstan, Kyrgyz Republic, and Uzbekistan. Since 1996, UNFPA has concentrated its support in two regions in each of the five countries, working through international nongovernmental organizations (Antarsh, 1998). Most gynecologists are women. Only physicians may prescribe pills and insert IUDs. International assistance seems to have produced significant changes in attitudes (Babamuradova et al., 1995). While social marketing programs are making contraceptives more acceptable, local production of contraceptives has largely ceased due to a shortage of raw materials and equipment previously provided by the USSR (WHO/EURO, 1997).

Surveys conducted between 1995 and 1997 in *Kazakstan, Uzbekistan* and the *Kyrgyz Republic* suggest that little fertility regulation is practiced before the first birth and that abortion seldom involves a first or premarital pregnancy (Westoff et al., 1998). Contraception, both modern and traditional, is practiced by between 56 and 59 percent of married couples. The IUD is most popular, accounting for between 66 and 80 percent of use in all three countries. According to Ministry of Health data, contraceptive practice has increased considerably since the beginning of the 1990s, accompanied by a decline in abortion. However, as previously noted, the registered abortion data may be underestimates as unreported procedures performed outside government hospitals grow more popular. Also, since ethnic Russian women resorted to abortion more frequently than other nationalities, their

Table 11.10
Legal Abortion Rates and Ratios: Kazakstan and the Kyrgyz Republic 1990–96

	Kazakstan			Kyrgyz Republic		
Year	Legally Induced Abortions	Legal Abortion Rate[a]	Legal Abortion Ratio[b]	Legally Induced Abortions	Legal Abortion Rate[a]	Legal Abortion Ratio[b]
1990	355,200	93.0	978	73,800	77.2	573
1991	321,500	82.9	908	66,400	67.7	513
1992	346,400	89.1	1,023	59,400	59.4	463
1993	290,700	72.6	919	52,700	52.6	451
1994	261,800	65.3	854	49,300	48.0	448
1995	224,000	56.0	809	42,500	41.4	362
1996	194,200	51.9	767	34,100	32.6	318

Sources: CIS/ISC (1998), Kyrgyz Republic/NSS (1997), Heleniak (1998), Zakharov (1998b, 1999), and USBOC (1998).

Table 11.11

Live Births, Birth Rates, and Fertility Rates: Tajikistan, Turkmenistan, and Uzbekistan, 1990–96

	Tajikistan			Turkmenistan			Uzbekistan		
Year	Live Births	Birth Rate	Total Fertility Rate	Live Births	Birth Rate	Total Fertility Rate	Live Births	Birth Rate	Total Fertility Rate
1990	203,600	38.8	5.05	123,500	34.2	4.17	684,900	33.7	4.07
1991	208,400	38.9	5.01	124,800	33.6	4.09	714,400	34.5	4.10
1992	179,400	32.2	4.13	129,500	34.0	4.09	702,000	33.1	3.90
1993	184,400	33.1	4.35	140,800	34.4	3.93	683,600	31.5	3.80
1994	160,900	28.2	3.57	139,600	34.8	3.77	652,400	29.4	3.50
1995	164,900	28.6	3.70	125,000	28.1	3.62	672,400	29.8	3.60
1996	129,400	22.0	na	109,600	24.0	3.47	628,100	27.3	3.07

Sources: CIS/ISC (1998), Uzbekistan/Goskomstat (1996), Heleniak (1998), Zakharov (1998b, 1999), Henshaw (1999), and USBOC (1998).

Table 11.12
Legal Abortion Rates and Ratios: Tajikistan, Turkmenistan, and Uzbekistan, 1990–96

	Tajikistan			Turkmenistan			Uzbekistan		
Year	Legally Induced Abortions	Legal Abortion Rate[a]	Legal Abortion Ratio[b]	Legally Induced Abortions	Legal Abortion Rate[a]	Legal Abortion Ratio[b]	Legally Induced Abortions	Legal Abortion Rate[a]	Legal Abortion Ratio[b]
1990	55,500	49.0	270	37,200	45.8	297	192,500	42.4	281
1991	54,900	48.1	258	na	na	na	189,000	42.4	265
1992	47,000	39.3	262	37,000	42.1	282	168,300	36.6	240
1993	na	na	na	na	na	na	148,300	30.4	217
1994	na	na	na	na	na	na	120,400	23.6	184
1995	19,900	15.7	121	33,800	32.3	266	104,400	20.0	155
1996	17,900	13.9	138	na	na	na	104,600	18.2	166

Sources: CIS/ISC (1998), Uzbekistan/Goskomstat (1996), Heleniak (1998), Zakharov (1998b, 1999), Henshaw (1999), and USBOC (1998).

out-migration may be a factor in the decline of abortions (Westoff et al., 1998). Findings reported by others confirm that abortion has become less prevalent in Central Asia in recent years (e.g., Djusubalieva, 1994; Mahler, 1997) in part because younger women practice contraception more vigilantly after the birth of the first child and supplies are more readily available.

In *Kazakstan*, birth rates declined from 37.2 per 1,000 population in 1960 to 23.9 in 1980, 21.7 in 1990, and 15.3 in 1996. TFR decreased from 4.5 per woman in 1960 to 3.35 in 1970, 2.93 in 1980, 2.72 in 1990, and 2.27 in 1996. Some 15 percent of Kazak women have seven or more children (WHO/EURO, 1992a–1992f). According to the 1995 Kazakstan Demographic and Health Survey (1996); 59 percent of married women say they currently use a contraceptive method. Of these, 40 percent rely on the IUD. Experience with induced abortion was common; 41 percent of respondents had terminated one or more pregnancies. Among all women who had an abortion, 70 percent had more than one. Urban women were more likely than rural women to have terminated an unwanted pregnancy (47 percent versus 29 percent). Ethnic Russian women were more than twice as likely as Kazak women to have done so (58 percent versus 23 percent). Kazaks interviewed in Russian were more likely to terminate their pregnancies than Kazak women speaking only in Kazak (Agadjanian and Qian, 1997). Abortions are believed to have declined considerably, especially in Almaty (Djusubalieva, 1996; Nukusheva, 1996; Kaupova, 1996). The reported abortion rate in the country decreased from 93.0 per 1,000 women aged 15–44 years in 1990 to 51.9 in 1996. The abortion ratio declined from 978 per 1,000 live births in 1990 to 767 in 1996.

In the *Kyrgyz Republic* the birth rate dropped from 36.9 per 1,000 population in 1960 to 30.5 in 1970, 29.6 in 1980, 29.3 in 1990, and 23.8 in 1996. TFR decreased from 4.1 per woman in 1985 to 3.69 in 1990 and 3.00 in 1996 (Kyrgyz Republic/NSC, 1997). The government actively encourages family planning, a policy publicly supported by the republic's highest religious official, the Kadi. Contraceptives are supplied through international assistance programs. In Bishkek abortions started to decline as more women sought to prevent unwanted pregnancies (Doolotova and Ahmed, 1996; Hata, 1996). The reported abortion ratio declined from 759 per 1,000 live births in 1970 to 605 in 1980, and 573 in 1990, falling further to 318 in 1996. In 1997 more than 27,000 abortions were reported with an abortion rate of 19.4 per 1,000 women of reproductive age. Maternal mortality was 74 per 100,000 live births with 21 percent of the total due to abortion (ESCAP, 1998).

The preliminary report of the 1997 Kyrgyz Republic Demographic and Health Survey (Kyrgyz Republic/MOH, 1997) indicates that recourse to abortion was highest among Russian women (2.2 abortions per lifetime), somewhat lower for Uzbek women (1.9), and lowest for Kyrgyz women (1.3). Of currently married women surveyed, 60 percent reported using a method of contraception. The IUD was most popular (38 percent). Among modern method users, 97 percent obtained their supplies from the public sector.

Tajikistan, the republic most remote from Moscow, traditionally has had the highest birth rate in Central Asia and the youngest population. Over half the families have three or more children and around 9 percent are reported to have more than 10 children. The birth rate rose from 33.5 per 1,000 population in 1960 to 34.8 in 1970, 37.0 in 1980 and 38.8 in 1990, before falling to 22.0 in 1996. TFR decreased from 5.93 per woman in 1970 to 5.05 in 1990 and 3.57 in 1994, before rising to 3.70 in 1995. Reported abortion rates declined from 53 per 1,000 women 15–44 years old in 1975 to 39.5 in 1985 and 39.3 in 1992, subsequently falling to 13.9 in 1996. The abortion ratio decreased from 403 per 1,000 live births in 1960 to 358 in 1970, 274 in 1980, 270 in 1990, 262 in 1992, and 138 in 1996. Reported data are not considered reliable. The sharp decline in the birth rate and TFR may reflect the civil strife and the imposition of a birth registration fee equivalent to $3 (Heleniak, 1998). Condoms are unpopular and modern methods of contraception are in short supply. The civil war made it difficult to provide international assistance but training outside the country is being arranged for key personnel.

In *Turkmenistan* the birth rate declined from 42.4 per 1,000 population in 1960 to 35.2 in 1970, 34.3 in 1980, 34.2 in 1990, and 24.0 in 1996. TFR decreased from 4.70 per woman in 1985 to 4.17 in 1990, and 3.47 in 1996. Fertility regulation is perceived to be a woman's responsibility provided the husband approves. There is, however, little partner communication about family planning (Storey et al., 1997). Most men traditionally want large families and object to condoms as being against local customs. The reported abortion rate declined from 45.8 per 1,000 women aged 15–44 years in 1990 to 32.3 in 1995. The abortion ratio decreased from 423 per 1,000 live births in 1960 to 408 in 1970, 344 in 1980, 297 in 1990, and 266 in 1995. Unreported clandestine abortions are rising among younger women and teenagers in rural areas who lack adequate information about contraception (UNFPA/WHO/EURO, 1995).

In *Uzbekistan,* the most populous of the Central Asian republics, the birth rate declined from 33.9 per 1,000 population in 1960 to 33.6 in 1970, increased to 33.9 in 1980 and decreased to 33.7 in 1990 and 27.3 in 1996. TFR dropped from 5.97 per woman in 1970 to 5.15 in 1980, 4.07 in 1990, and 3.07 in 1996. The reported abortion ratio declined from 370 per 1,000 live births in 1970 to 298 in 1980, 281 in 1990, and 166 in 1996. The abortion rate decreased from 42.4 per 1,000 women aged 15–44 years in 1990 to 18.2 in 1996. A governmental family policy was initiated in 1992 (Kuzibaeva, 1998). According to the 1996 Uzbekistan Demographic and Health Survey (Uzbekistan Institute of Ob/Gyn, 1997), 16 percent of respondents had had an abortion. In the three years before the survey, 14 percent of all pregnancies ended in abortion. The prevalence of abortion increased with age, number of births, and level of education. Urban women were more than twice as likely to report an abortion than rural women. Uzbek women were about half as likely to do so as women of Russian ethnicity.

Throughout Islamic and still largely agricultural Central Asia sexuality education is practically nonexistent, family planning associations remain to be organized, and fertility regulation continues to be perceived as largely a woman's responsibility. However, economic realities are having an impact on the traditional preference for large families. For many couples contraceptive practice is impeded by cultural values and kinship traditions, religious beliefs, lack of consistent access to modern methods and services, at times unresponsive health services, and limited partner communication or indifference toward women's reproductive health. Condoms are unpopular and misconceptions about the potential health effects of oral contraceptives persist. IUDs are the method of choice and the most widely available. Vasectomy remains unacceptable. Some women perceive abortion as the only practical solution for an unwanted pregnancy while others reject it on moral and health grounds (Storey et al., 1997). Government family planning initiatives, supported by international donor assistance, are having an impact. Contraceptive practice is improving, especially among younger rural women after the birth of the second child. Abortions are declining. However, considerably more effort is needed to strengthen women's organizations, reproductive health, and reproductive rights in the Central Asian republics.

chapter 12

Former Yugoslavia and Successor States

Nila Kapor-Stanulovic and Henry P. David

EDITOR'S INTRODUCTION

Since coming into existence on 4 December 1918, the area constituting the former Yugoslavia has had a turbulent history. It emerged after World War I as a Serbian Kingdom forged from the union of Serbia, Bosnia-Herzegovina, Croatia, Slovenia, Montenegro, and the southern Slav territories wrested from the dissolved Austro-Hungarian empire. The largest country in the Balkans, Yugoslavia bordered on Italy, Austria, Hungary, Romania, Bulgaria, Greece, Albania, and the Adriatic Sea. Its population was largely Slav and Christian. Belgrade, the capital of Serbia, also served as the capital of the country. The principal language was Serbo-Croat.

Serbia, the largest republic, began as a collection of principalities, formed a kingdom in the Middle Ages, and, after its defeat at the Battle of Kosovo in 1389, became a vassal of the Ottoman Turkish empire. The Serbs in mountainous Montenegro were never completely conquered. Yugoslav Macedonia was part of an ancient domain contested by Serbia and Bulgaria. The people living in these areas adopted the Orthodox Christian faith and the Cyrillic alphabet. The early Croatian kingdom became a vassal of Hungary and later a restive province of the Austro-Hungarian empire, as did Slovenia. Croats and Slovenians embraced Roman Catholicism and the Latin alphabet.

Ethnically mixed Bosnia was a mountainous kingdom that annexed Herzegovina from Serbia in the fourteenth century before falling to the Turks. The 1878 Berlin Congress, which divided the Balkans after the defeat of Turkey in the Russo-Turkish war, assigned Bosnia and Herzegovina to the Austro-Hungarian empire. The assassination of Austrian Archduke Ferdinand in the Bosnian capital of Sarajevo ignited World War I. At the time of joining the former Yugoslavia in 1918, Bosnia's population was 43 percent Muslims, 37 percent Orthodox Serbs, and 17 percent Croats.

The decade after the establishment of Yugoslavia in 1918 was marked by continuing political struggles between the forces of Serbian centralism and Croat federalism. Some cohesion was achieved by the declaration of a royal dictatorship on 6 January 1929. The name of the country was formally changed to Yugoslavia (meaning South Slavs). Following continuing political struggles and shifts from pro-German to pro-Allied governments, German and Italian forces invaded on 6 April 1941. Control of the country was divided. Croatian fascists (Ustashis) were allowed to establish a puppet state, including most of Bosnia-Herzegovina. "Ethnic cleansing" emerged as Croatian Ustashis killed Bosnian Serbs, and Serbian nationalists (Chetniks) killed Croatians and Muslims. Meanwhile, the Croatian-born leader of the communist partisans, Josip Broz, known as Tito, formed the National Liberation Army. He opposed ethnic nationalism while fighting Chetniks, Ustashis, and Germans; developed a civil administration in liberated territories; and eventually prevailed in his struggle against the royalist government-in-exile (West, 1996). Following the end of World War II, a Federal Republic was proclaimed on 29 November 1945. It consisted of six constituent republics: Bosnia-Herzegovina, Croatia, Macedonia, Montenegro, Serbia, and Slovenia, plus two autonomous provinces within Serbia—Kosovo-Metohija and Vojvodina.

The constituent republics were subordinated in all matters to the Tito-led central government in Belgrade. In June 1948 the Moscow dominated Cominform denounced Tito's independent socialist policies and expelled Yugoslavia. In the early 1950s, after Tito had become president of the republic, Yugoslav socialism shifted from central decision making toward a policy of economic self-management, organized around workers' councils and decentralized local governments. In April 1963 the country's name was officially changed to Socialist Federal Republic of Yugoslavia (SFRY). Until his death in 1980, Tito's personal authority succeeded in keeping internal rivalries submerged (West, 1996).

Among the SFRY politicians who understood the implications of Tito's death was Slobodan Milosovic, a high Communist Party official who was elected president of Serbia in 1987. His policies encouraged the reemergence of a Greater Serbian ideology and a step-by-step descent into war (Silber and Little, 1997). Serbian expansionism, Croatian nationalism, and Slovenian secessionism combined to destroy the Yugoslavia Tito had created. Dissatisfaction with a sharply deteriorating economy, ethnic disharmony, and the long-lingering cultural and religious differences also contributed to the breakup of the SFRY and civil war (Magas, 1993). Multiparty elections in Slovenia and Croatia in 1990 brought former communist leaders with new ideologies to power in those republics, while Milosovic retained the presidency of Serbia.

The dissolution of the SFRY began in June 1991 when Slovenia and Croatia declared their independence. This secession was opposed by Serbia and Montenegro, and by the Serbs living in Croatia. A war with Slovenia lasted only a few days. In Croatia, fighting continued until a cease-fire was brokered by the United Nations

(U.N.) in January 1992. In May 1992 the international community recognized Slovenia and Croatia as independent states.

In Bosnia-Herzegovina events were precipitated by a declaration of independence. After the European Union recognized Bosnia-Herzegovina, war broke out on 6 April 1992. The Republic of Macedonia proclaimed its sovereignty on 8 September 1992. When Greece vetoed European Union recognition of the name Macedonia (which was the name of its own northern province) the U.N. recognized the republic in April 1993 as the Former Yugoslav Republic of Macedonia (FYROM). Although fighting did not spread to Serbia, Montenegro, and Macedonia, the war at the borders and events in Bosnia-Herzegovina had a dramatic effect on vulnerable groups in the Yugoslav population. Sanctions imposed by the U.N. Security Council in May 1992 on Serbia and Montenegro affected almost every aspect of daily life in these two republics. Over three million refugees and displaced persons throughout the whole territory of the Former Yugoslavia imposed a heavy strain on the social service infrastructure (Pine and Mertus, 1993). The vast majority of refugees were housed with host families rather than in collective centers. After much diplomatic maneuvering, an American-brokered peace agreement was initialed in Dayton, Ohio on 21 November 1995 (US/ Department of State, 1995) and signed in Paris on 14 December 1995. It ended four years of fighting and preserved Bosnia as a unified state, divided into two almost equal parts: the Muslim-Croat Federation and the Serb Republic . NATO troops continue to maintain the peace but the mentality of war lingers, blocking national reconstruction and reconciliation (Cohen, 1998).

In April 1992 the Federal Assembly in Belgrade adopted a constitution for the newly established Federal Republic of Yugoslavia (FRY). It consisted of Serbia (including the previously autonomous provinces of Kosovo and Vojvodina) and Montenegro. Milosovic was reelected president of Serbia and in July 1997 elected FRY president. The FRY experienced political conflicts in Montenegro, military clashes in Kosovo, and hyperinflation. Although the signers of the 1995 Dayton Accords agreed to resolve peacefully the future status of Kosovo (90 percent populated by Albanians), there have been no realistic negotiations toward restoring autonomy or eventual independence (Neier, 1997). Clashes continue between separatist ethnic Albanians and Serbian government forces (Finn, 1999).

The economic and social crises that followed the 1991 dissolution of the SFRY produced a sharp drop in living standards and the quality of communal services amidst a rise in pauperization and social problems, especially in urban areas. The hardest hit were families with children and the elderly (UNICEF, 1996b). Unemployment reached nearly 50 percent. Political turbulence and ethnic strife continue. Only Slovenia reaffirmed its Habsburg heritage and reduced its Balkan affiliation (*The Economist*, 1998a).

As will be noted in subsequent sections, the former Yugoslavia was a trendsetter in its official support for women's rights and voluntary fertility regulation. It became the first country to proclaim in its constitution that "it is a human right to decide freely on childbearing" (Yugoslavia, 1974). It was a Yugoslav obstetrician/gynecologist who introduced the vacuum aspiration method of abortion to the West (Novak, 1968). Groups of dedicated women spearheaded the drive for access to legal abortion and modern contraceptives, and later fought to maintain their rights in the separate republics.

This chapter will present trends from the SFRY, the FRY, and the successor states of Croatia and Slovenia. Little information is available from Bosnia and Macedonia. After a presentation of demographic patterns, there is a review of abortion legislation, abortion trends, contraception, and psychological aspects of the abortion/contraception relationship. There follows an overview of population policies/family allowances, the woman question, and sex and society, including sexuality education and adolescents.

We are most grateful to Dr. Mirjana Rasevic, Director of the Center for Demographic Research, Belgrade, who provided valuable data that are cited extensively in the chapter.—H.P.D.

DEMOGRAPHIC TRENDS

Former Yugoslavia (SFRY)

A multinational community, the SFRY was a demographic mosaic. The considerable ethnic and cultural diversities and regional differences in levels of economic and social development, history, social structures, religion, and customs greatly influenced population trends. Population growth rates ranged from low to among the highest in Europe. Literacy rates and other socioeconomic indicators as well as the quality and availability of health care, and social services varied over wide ranges. The country's five official nationalities and over 20 ethnic minorities displayed the diverse and often contradictory imprints of three major religions (Orthodox, Roman Catholic, and Muslim) and three cultural heritages (Central European, Ottoman, and Veneto-Mediterranean).

Between 1860 and 1900, the Serbian birth rate hovered around 40 per 1,000 population. A somewhat similar situation prevailed for Croatia (Mitchell, 1975). After the establishment of Yugoslavia as a national entity in 1918, the birth rate decreased from 36.7 in 1921 to 26.6 in 1947, before rising to 30.2 in 1950. As shown in Table 12.1, the birth rate declined to 26.9 in 1955, dropping further with the liberalization of abortion in February 1960 to 19.6 in 1967. The last available birth rate for the SFRY was 14.0 in 1990. Sources for demographic data are indicated in the tables.

On the basis of 1953–91 data, majority ethnic groups, that is, those numbering over 100,000 inhabitants in 1991, could be divided into two different populations. The first consisted of those persons who declared themselves as Serbs, Montenegrins, Yugoslavs, Hungarians, and Croats. They displayed low fertility rates, high mortality, low or negative natural increases in population, advanced aging, and a high female/male ratio. The second group was composed of ethnic Albanian, Muslim, and Roma (gypsy) populations, characterized by high fertility rates, low general mortality, and a relatively young age structure with a high male/female ratio (Penev, 1995a). In 1993 the total fertility rate of ethnic Albanian women in Kosovo was the highest in Europe, with an average family size of 6.5 children per woman. Some 60 percent of the population was under 18 years of age (UNICEF, 1996d). The low fertility zone of Serbia stabilized at about 20 percent below the population replacement level in the course of the last two decades (Rasevic, 1995b). Comparisons of total fertility rates, birth rates, and legal abortion ratios in the republics of the SFRY, 1960–90 are shown in Figures 12.1, 12.2, and 12.3.

Federal Republic of Yugoslavia (FRY)

The 1991 FRY census, boycotted by Albanians for political reasons, showed a total population of 10.3 million for Serbia and Montenegro of which over 9.7 million lived in Serbia (Yugoslavia/FSO, 1997). In late 1994, the FRY had 10.5 million inhabitants, of whom 9.9 million lived in Serbia and 0.6 million in Montenegro (FSO data cited in UNICEF, 1996b). The majority of individuals in Serbia and Montenegro are Slavs and Orthodox Christians, using both the Cyrillic and Latin alphabets. In the Kosovo-Metohija province of Serbia, over 80 percent of the population are Albanians who are mostly Muslim. Hungarians in Vojvodina are largely Roman Catholic. Muslims of Slav origin live in parts of Montenegro and southwest Serbia.

As shown in Table 12.2, the crude birth rate for the population residing in the FRY declined from 21.6 per 1,000 population in 1960 to 17.7 in 1970 and 1980 to 14.7 in 1990 and 13.0 in 1996. The total fertility rate (TFR) decreased from 2.59 per woman in 1960 and 2.28 in1970 to 2.08 in 1990, and to 1.86 in 1996 . Data for Serbia are presented in Table 12.3 but do not fully reflect the diverse demographic trends between the different ethnic groups (Penev, 1995a, 1995b). For example, in Kosovo-Metohija, the average household size in 1981 was 8.5, as compared to the Serbian average of 3.9 for the same year.

Croatia

According to the 1991 census, the population of Croatia was 4.8 million, consisting of 78 percent Croats, 12 percent Serbs, 4 percent other nationalities, and 6 percent undeclared. As shown in Table 12.4, the birth rate fell from 18.4 per 1,000 population in 1960 to 11.6 in 1990 and 10.8 in 1995 before rising to 11.7 in 1996. The TFR declined from 2.20 per woman in 1960 to 1.63 in 1990 and 1.47 in 1994, before rising again to 1.67 in 1996 with the decline in abortions and increase in births.

Slovenia

At year end 1997 Slovenia had nearly two million inhabitants, 64 percent of whom lived in the capital, Ljubljana. In the 1991 census 85 percent of the

Table 12.1
Birth, Abortion, and Fertility Rates and Ratios: Socialist Federal Republic of Yugoslavia, 1950–90

Year	Live Births	Legally Induced Abortions	Other Abortions	Total Registered Abortions	Birth Rate	Legal Abortion Rate[d]	Legal Abortion Ratio[e]	Total Fertility Rate
1950	494,200	na	na	na	30.2	na	na	3.74
1955	471,400	na	na	na	26.9	na	na	3.20
1960[a]	432,600	76,700	56,600	133,300	23.5	18.3	177	2.82
1961	422,200	104,700	59,800	164,500	22.7	35.9	248	2.77
1962	413,100	150,000[b]	50,000[b]	200,000[b]	21.9	34.9	363	2.74
1963	407,400	146,800	68,200	215,000	21.4	33.3	360	2.68
1964	401,100	158,600	69,000	227,600	20.8	35.1	395	2.67
1965	408,200	182,400	63,000	245,400	21.0	39.8	447	2.69
1966	399,800	195,500	69,700	265,200	20.4	42.2	489	2.63
1967	389,600	210,700	65,500	276,200	19.6	44.9	541	2.55
1968	382,500	na	na	245,800	19.1	52.5	na	2.47
1969[c]	382,800	207,300	46,200	253,500	18.9	53.6	542	2.43
1970	363,300	205,000	49,400	254,400	17.8	53.4	564	2.28
1971	375,800	210,100	51,800	261,900	18.2	43.6	559	2.36
1972	380,700	215,800	51,900	267,700	18.3	44.6	567	2.35

1973	379,100	na	na	na	18.1	na	na	2.30
1974	382,900	na	na	na	18.1	na	na	2.27
1975	388,000	238,700	54,500	293,200	18.2	48.5	615	2.26
1976	392,400	na	na	na	18.2	na	na	2.25
1977	384,600	na	na	na	17.7	na	na	2.18
1978	381,400	268,700	56,500	325,200	17.4	53.9	704	2.15
1979	378,800	na	na	na	17.1	na	na	2.12
1980	382,100	297,300	59,100	356,400	17.1	59.5	778	2.12
1981	369,000	302,700	59,400	362,100	16.4	61.0	820	2.08
1982	378,800	na	na	na	16.7	na	na	2.12
1983	374,600	341,200	31,400	372,600	16.4	68.2	911	2.10
1984	377,400	358,300	26,600	384,900	16.4	71.3	949	2.11
1985	366,600	366,000	23,000	388,900	15.9	72.9	998	2.05
1986	359,600	359,100	29,700	388,900	15.5	70.6	999	2.00
1987	359,300	365,700	23,200	388,900	15.3	71.6	1,018	2.00
1988	356,300	343,300	24,800	367,800	15.2	66.7	964	1.98
1989	336,400	na	na	na	14.2	na	na	1.88
1990	333,700	na	na	na	14.0	na	na	1.88

[a]Liberalization of abortion law (February 1960). [b]Estimate. [c]Further liberalization of abortion (April 1969). [d]Per 1,000 women age 15–44. [e]Per 1,000 live births.

Sources: David and McIntyre (1981), SFRY/FSO (1991), U.N. (1995a), Stloukal (1995), and Rasevic (1998).

Figure 12.1
Total Fertility Rates: Republics of the Former SFRY, 1960–95

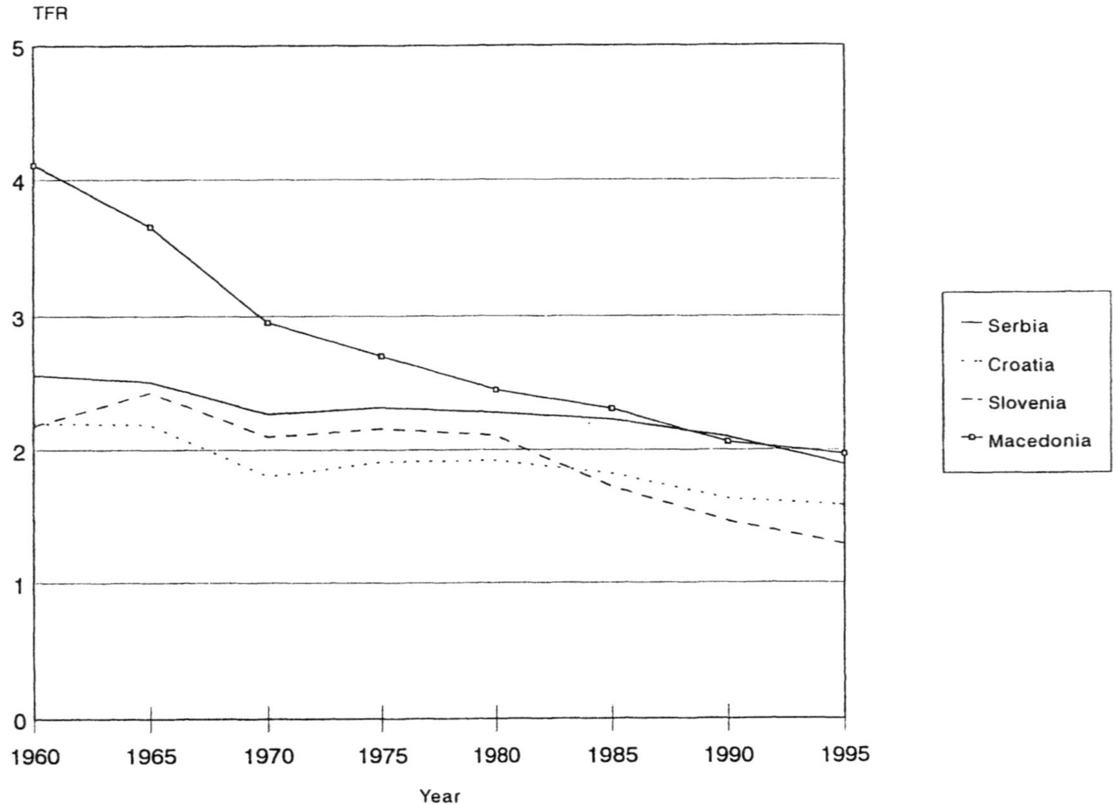

Replacement level fertility is 2.1.

Sources: See data from tables 12.3–12.6.

Figure 12.2
Birth Rates: Republics of the Former SFRY, 1960–95

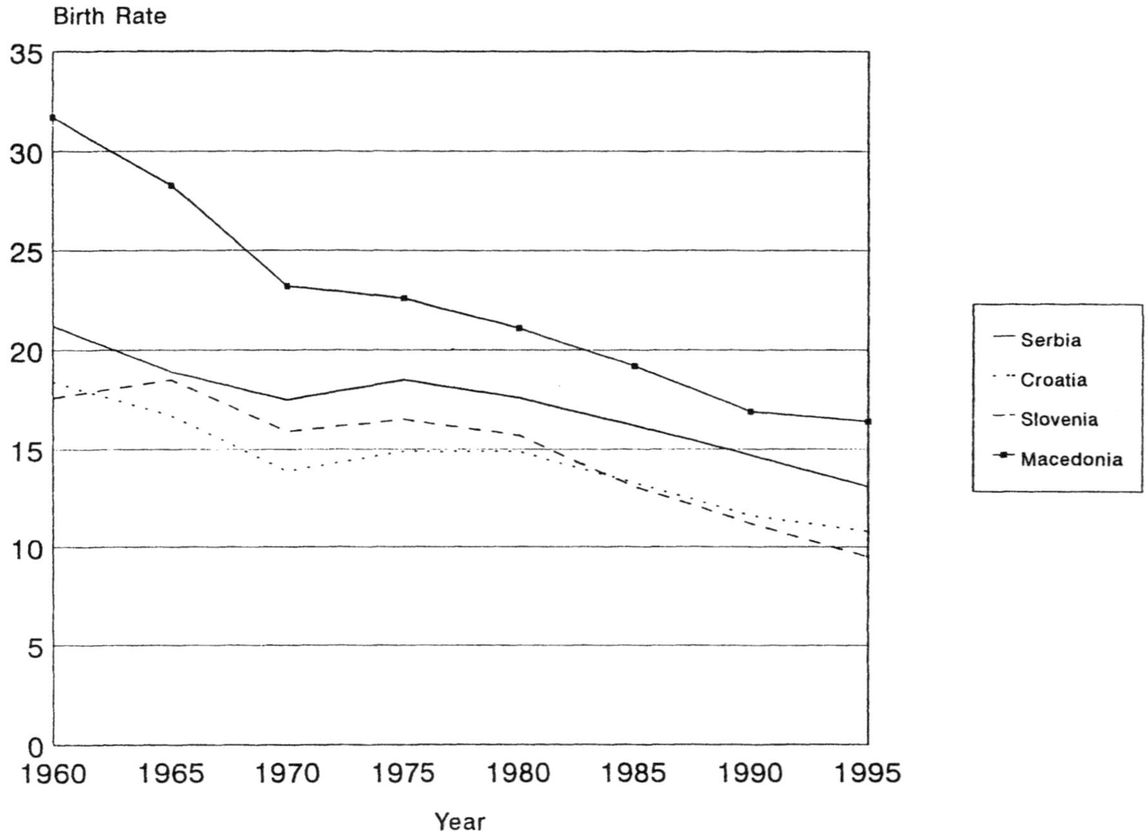

Sources: See data from tables 12.3–12.6.

Figure 12.3
Legal Abortion Ratios: Republics of the Former SFRY, 1960–95

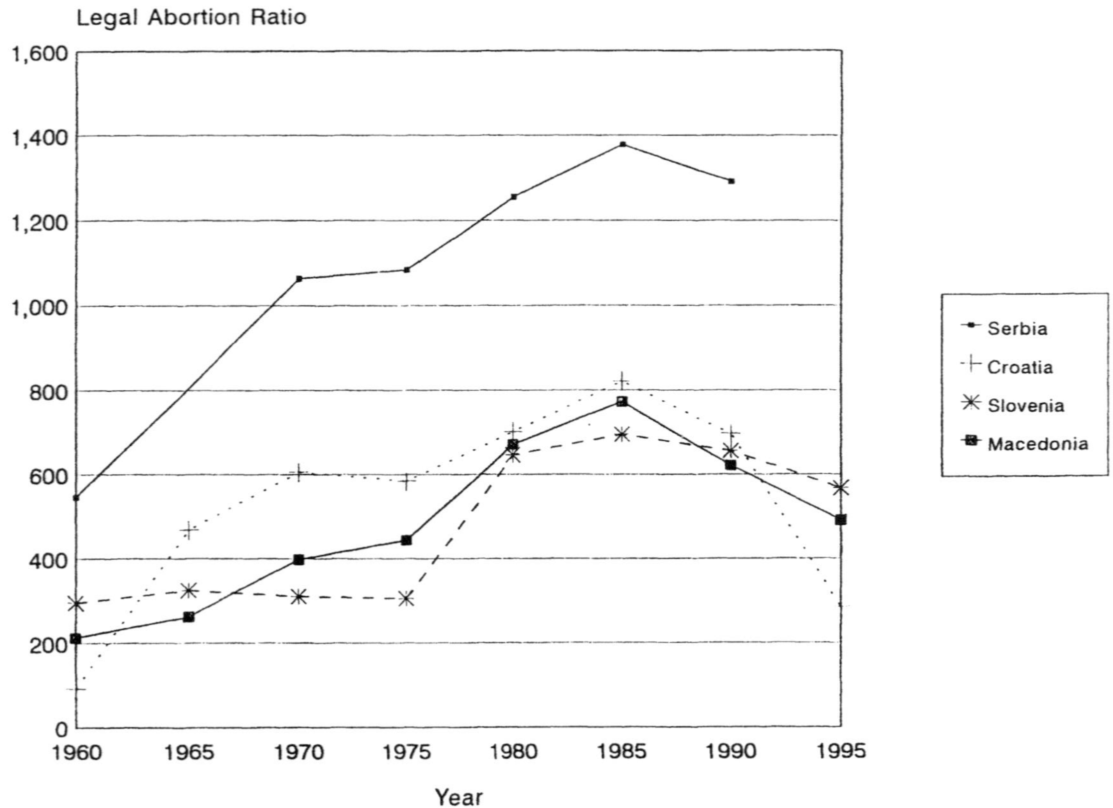

Legal abortions per 1,000 live births.

Sources: See data from tables 12.3–12.6.

Table 12.2
Birth, Abortion, and Fertility Rates and Ratios: Federal Republic of Yugoslavia, 1950–96

Year	Live Births	Legally Induced Abortions	Birth Rate	Legal Abortion Rate[a]	Legal Abortion Ratio[b]	Total Fertility Rate
1950	210,400	na	29.5	na	na	3.57
1955	190,700	na	24.9	na	na	2.90
1960	174,100	na	21.6	na	na	2.59
1965	162,600	na	19.2	na	na	2.53
1970	157,600	na	17.7	na	na	2.28
1975	172,800	182,000	18.5	na	1,053	2.31
1980	173,300	210,000	17.7	na	1,212	2.26
1981	161,900	na	16.3	na	na	2.11
1982	170,000	na	17.0	na	na	2.22
1983	168,300	215,000	16.7	na	1,277	2.21
1984	172,800	222,100	17.0	na	1,285	2.29
1985	166,600	222,600	16.3	na	1,336	2.21
1986	164,400	220,300	16.0	na	1,340	2.22
1987	165,100	213,400	16.0	na	1,292	2.20
1988	163,900	201,600	15.7	na	1,230	2.19
1989	154,600	201,700	14.8	90.5	1,305	2.06
1990	155,000	195,700	14.7	na	1,263	2.08
1991	152,300	157,500	14.6	70.7	1,034	2.08
1992	140,800	142,300	13.5	62.4	1,011	1.92
1993	141,000	119,300	13.4	51.4	846	1.91
1994	137,600	119,200	13.1	52.8	866	1.85
1995	140,500	96,900	13.3	44.9	690	1.87
1996	137,700	na	13.0	na	na	1.86

[a]Per 1,000 women age 15–44. [b]Per 1,000 live births.

Sources: FRY/FSO (1997), COE (1998), and USBOC (1998).

Table 12.3
Birth, Abortion, and Fertility Rates and Ratios: Serbia, 1950–96

Year	Live Births	Legally Induced Abortions	Birth Rate	Legal Abortion Rate[a]	Legal Abortion Ratio[b]	Total Fertility Rate
1950	198,500	na	29.5	na	na	3.54
1960	160,900	88,000	21.2	na	547	2.56
1965	150,300	na	18.9	na	na	2.51
1966	146,200	122,900	18.6	na	841	2.41
1967	147,500	na	18.2	na	na	2.41
1968	148,200	154,700	18.0	na	1,044	2.39
1969	152,000	na	18.3	na	na	2.40
1970	146,900	156,100	17.5	na	1,063	2.27
1971	151,100	na	17.9	na	na	2.30
1972	154,800	161,800	18.1	na	1,045	2.32
1973	156,100	na	18.1	na	na	2.30
1974	160,300	na	18.4	na	na	2.33
1975	162,300	176,000	18.5	na	1,084	2.32
1976	165,400	na	18.7	na	na	2.35
1977	161,400	na	18.0	na	na	2.30

1978	159,600	201,900	17.7	na	1,265	2.27
1979	158,100	na	17.4	na	na	2.24
1980	162,700	204,100	17.6	na	1,254	2.28
1981	151,500	na	16.2	na	na	2.14
1982	159,400	na	17.0	na	na	2.26
1983	157,600	207,900	16.7	na	1,319	2.24
1984	162,300	214,400	17.0	na	1,321	2.31
1985	155,900	214,800	16.2	na	1,378	2.23
1986	153,900	212,400	15.9	na	1,380	2.21
1987	154,500	205,300	15.9	na	1,329	2.22
1988	153,800	193,600	15.7	na	1,259	2.21
1989	144,900	193,800	14.7	na	1,338	2.09
1990	145,600	188,000	14.7	na	1,291	2.10
1991	139,600	150,500	14.6	72.1	1,078	2.08
1992	131,300	136,000	13.5	63.7	1,036	1.91
1993	132,100	na	13.5	na	na	1.91
1994	128,700	na	13.0	na	na	1.85
1995	131,000	na	13.1	na	na	1.89
1996	128,600	na	12.9	na	na	1.83

[a]Per 1,000 women age 15–44. [b]Per 1,000 live births.

Sources: FRY/FSO (1995) and Serbia/BOS (1997).

Table 12.4
Birth, Abortion, and Fertility Rates and Ratios: Croatia, 1950–96

Year	Live Births	Legally Induced Abortions	Other Abortions	Total Registered Abortions	Birth Rate	Legal Abortion Rate[a]	Legal Abortion Ratio[b]	Total Fertility Rate
1950	na	na	na	na	24.9	na	na	2.94
1960	76,200	7,000	11,300	18,300	18.4	na	92	2.20
1961	74,200	12,300	11,300	23,600	17.8	na	166	2.14
1962	72,300	18,900	10,500	29,400	17.2	na	261	2.11
1963	69,900	24,000	10,800	34,800	16.5	na	343	2.07
1964	68,900	29,000	10,700	39,700	16.2	na	421	2.07
1965	71,200	33,300	10,400	43,700	16.7	na	468	2.19
1966	71,300	35,200	8,100	43,300	16.6	na	494	2.14
1967	67,100	37,200	5,200	42,400	15.5	na	554	2.06
1968	65,400	38,100	4,900	43,000	15.0	na	583	1.99
1969	63,600	35,600	6,400	42,000	14.5	na	560	1.91
1970	61,100	37,000	6,400	43,400	13.9	na	606	1.80
1971	64,900	36,200	4,900	41,100	14.6	na	558	1.97
1972	66,000	36,800	5,200	42,000	14.9	na	558	1.96
1973	67,400	37,200	5,400	42,600	15.1	na	552	1.96
1974	67,300	39,000	4,500	43,500	15.0	na	580	1.92
1975	67,000	39,900	5,400	45,300	14.9	na	585	1.91
1976	67,100	40,300	4,800	45,100	14.9	na	601	1.88

1977	68,000	na	na	na	15.0	na	na	1.89
1978	68,700	na	na	44,600	15.1	na	na	1.91
1979	69,200	na	na	na	15.2	na	na	1.99
1980	68,200	47,800	na	51,300	14.9	na	701	1.92
1981	67,500	55,900	na	na	14.7	na	828	1.92
1982	66,700	46,000	na	na	14.5	na	690	1.90
1983	65,600	48,000	6,700	54,700	14.2	na	732	1.88
1984	64,900	48,800	6,900	55,700	14.0	na	752	1.87
1985	62,700	51,500	4,100	55,600	13.3	na	821	1.82
1986	60,200	50,100	4,600	54,700	12.9	na	832	1.76
1987	59,200	48,600	8,500	57,100	12.7	na	821	1.74
1988	58,500	42,900	8,100	51,000	12.5	na	733	1.79
1989	55,700	43,200	na	na	11.9	na	776	1.63
1990	55,400	38,600	na	na	11.6	na	697	1.63
1991	51,800	33,400	na	na	10.8	na	778	1.53
1992	47,000	26,200	na	na	9.8	25.8	557	1.48
1993	48,500	25,200	6,000	31,200	10.2	25.0	520	1.52
1994	48,600	19,700	na	na	10.3	19.1	405	1.47
1995	50,200	14,300	na	na	10.8	13.9	285	1.58
1996	53,800	12,300	na	na	11.7	12.4	229	1.67

[a]Per 1,000 women age 15–44. [b]Per 1,000 live births.

Sources: Stampar (1971–76), David and McIntyre (1981), Croatia/CBS (1996), COE (1998), and USBOC (1998).

Table 12.5
Birth, Abortion, and Fertility Rates and Ratios: Slovenia, 1950–97

Year	Live Births	Legally Induced Abortions	Birth Rate	Legal Abortion Rate[a]	Legal Abortion Ratio[b]	Total Fertility Rate
1950	36,000	na	24.5	na	na	3.01
1955	32,100	400	20.9	na	12	2.58
1960	27,800	8,200	17.6	na	295	2.18
1965	30,600	9,900	18.5	na	324	2.43
1970	27,400	8,500	15.9	na	310	2.10
1975	29,800	9,100	16.5	na	305	2.16
1980	29,900	19,300	15.7	na	646	2.11
1981	29,200	na	15.2	na	na	1.94
1982	28,900	na	15.0	na	na	1.93
1983	27,200	na	14.1	na	na	1.82
1984	26,300	18,600	13.5	na	707	1.75
1985	25,900	18,000	13.1	na	695	1.72
1986	25,600	18,000	12.9	na	703	1.65
1987	25,600	16,800	12.9	na	660	1.64
1988	25,200	16,500	12.6	na	655	1.63
1989	23,400	15,900	11.7	na	680	1.52
1990	22,400	14,700	11.2	na	656	1.46
1991	21,600	14,000	10.8	na	648	1.42
1992	20,000	13,300	10.0	na	650	1.34
1993	19,800	12,200	9.9	27.4	616	1.34
1994	19,500	11,300	9.8	25.1	580	1.32
1995	19,000	10,800	9.5	23.9	568	1.29
1996	18,800	10,200	9.4	22.7	543	1.28
1997	18,200	9,700	9.1	21.6	533	1.25

[a]Per 1,000 women age 15–44. [b]Per 1,000 live births.

Sources: Andolsek (1988), Slovenia/MOL (1994), Slovenia/SO (1995), U.N. (1995a), COE (1998), and USBOC (1998).

Table 12.6

Birth, Abortion, and Fertility Rates and Ratios: Former Yugoslav Republic of Macedonia, 1950–96

Year	Live Births	Legally Induced Abortions	Birth Rate	Legal Abortion Rate[a]	Legal Abortion Ratio[b]	Total Fertility Rate
1950	49,600	na	40.3	na	na	5.70
1955	49,100	na	36.2	na	na	4.91
1960	44,100	9,400	31.7	na	213	4.11
1965	42,400	11,100	28.3	na	262	3.66
1970	37,900	15,100	23.2	na	398	2.95
1975	39,600	17,600	22.6	na	444	2.70
1980	39,800	26,700	21.1	na	671	2.45
1981	39,500	na	20.6	na	na	2.46
1982	39,800	na	20.5	na	na	2.46
1983	39,200	27,400	19.9	na	699	2.40
1984	38,900	29,700	19.5	na	764	2.34
1985	38,700	29,900	19.2	na	773	2.31
1986	38,200	27,000	18.7	na	707	2.26
1987	38,600	30,700	18.7	na	795	2.27
1988	37,900	30,900	18.1	na	815	2.21
1989	35,900	30,400	17.0	na	847	2.09
1990	35,400	22,000	16.9	na	622	2.06
1991	34,800	23,200	17.1	na	667	2.30
1992	33,200	19,900	16.2	42.9	599	2.18
1993	32,400	18,800	15.7	40.4	580	na
1994	33,500	16,500	17.2	37.5	492	2.08
1995	32,200	15,800	16.4	35.3	491	1.97
1996	31,400	14,200	15.7	31.6	452	1.90

[a]Per 1,000 women age 15–44. [b]Per 1,000 live births.

Sources: U.N. (1995a), Republic of Macedonia/SO (1997), COE (1998), and USBOC (1998).

inhabitants declared themselves Slovenians, 3 percent Serbs, 3 percent Croats, and less than 1 percent each Hungarians or Italians. About 7 percent were undeclared. As shown in Table 12.5, the birth rate decreased continually since 1950 when it was 24.5 per 1,000 population to 17.6 in 1960, 15.9 in 1970, 15.7 in 1980, 11.2 in 1990, and 9.1 in 1997. The TFR declined from 3.01 per woman in 1950 to 2.18 in 1960, 1.46 in 1990, and 1.25 in 1997. It is unusual for Slovenian families to have more than two children.

Former Yugoslav Republic of Macedonia (FYROM)

According to the 1994 census, the total FYROM population was 2.8 million. Of these, 69 percent were Macedonian, 22 percent Albanian, 3 percent Turks, and 2 percent each Serbs and Romanis. As shown in Table 12.6, the birth rate declined from 40.3 per 1,000 population in 1950 to 23.2 in 1970, 21.1 in 1980, 16.9 in 1990, and 15.7 in 1996. Meanwhile, TFR decreased from 5.70 per woman in 1950 and 4.11 in 1960 to 2.06 in 1990 and 1.90 in 1996.

Bosnia and Herzegovina

In 1991 the population of Bosnia-Herzegovina was 4.4 million. More than one million residents left their homes during 1992–95, becoming displaced persons or refugees. The birth rate declined from 21.4 per 1,000 population in 1970 to 13.8 in 1991 whie TFR decreased from 2.67 per woman in 1970 to 1.70 in 1990 (COE, 1997). Little data are available for subsequent strife-torn years.

ABORTION

Legislation

Prior to the enactment of the 1929 Penal Code, abortion was considered homicide and punished severely except in cases that involved unmarried women. The 1929 Penal Code permitted abortion only for strictly defined medical reasons to save the woman's life (Beric and Stevanov, 1970; Andolsek, 1988). Based on the number of reported cases of abortion complications, the number of illegal procedures was estimated to be sizeable (Rasevic, 1990c). By Decree No. 27 of 11 January 1951, sociomedical indications became acceptable grounds for termination of unwanted pregnancy (Yugoslavia, 1952). This action represented official recognition of the growing resort to illegal abortion and was an attempt to reduce associated health risks. However, the circumstances where abortion would be permitted were not clearly specified and the use of abortion did not rise to very high levels. Decree No. 33 of 16 February 1960 broadened the range of social conditions justifying pregnancy termination. Abortion was to be authorized on request if birth of the child "would result in a serious personal, familial, or economic situation for the pregnant woman which cannot be averted in any other way." Public Health

Instruction No. 06-1164 of 28 October 1960 provided details regarding the initiation of abortion requests, the work of district abortion commissions and regional appeals commissions, forms to be utilized, reporting procedures, confidentiality safeguards, and so on., but did not define the actual social and economic circumstances that might prove to be "serious." Abortions increased rapidly (Simoneti et al., 1976).

After considerable public debate a new General Law on the Interruption of Pregnancy was promulgated by the Federal Assembly on 26 April 1969 (Yugoslavia, 1969). It reaffirmed in Article 4 that "the interruption of pregnancy will be done on the demand of the woman if during pregnancy and after birth she could fall into serious personal, family, material and other troubles." Previous articles delineated health, genetic, and juridical reasons. Beyond three months of gestation, abortions could be performed only if the interruption of pregnancy "will not cause serious damage to the health" of the woman, as decided by district abortion commissions and/or regional appeals commissions on the basis of the 1960 criteria. Termination was permitted at any time if continuation of the pregnancy posed a serious threat to the woman's health. Abortions could be performed only by gynecologists in authorized medical facilities under conditions safeguarding the woman's privacy and assuring confidentiality of records. Registration was mandatory. The law further stipulated that the woman and her partner must be warned of the potential dangers of abortion and be instructed about methods for preventing future undesired pregnancies. Moreover, the law required "that social and educational institutions and other organizations which deal with the problems of the security of mother and child, within its duties, acquaint women and youth with the harmful consequences of the interruption of pregnancy and with the advantages of the application of means and methods of contraception" (Yugoslavia, 1969).

The former Yugoslavia became the first country to proclaim that "it is a human right to decide freely on childbirth"—a statement included in Paragraph 191 of the Federal Constitution adopted on 21 February 1974 (Yugoslavia, 1974). This right "can be limited only for medical reasons." Implementation of the Federal Constitution was subsequently accomplished through legislation in the constituent republics (except Macedonia) and autonomous provinces, with each giving due regard to local conditions (Simoneti et al., 1976). For example, on 20 April 1977 the Slovenian Assembly legislated that an abortion must be performed at the request of the woman, provided the pregnancy was of less than 10 weeks duration. After that time, termination was permitted "only if the procedure entails a risk to the woman's life, health, or future motherhood that is less than the risk to the woman or the child associated with continuation of the pregnancy or childbearing," as determined by a district abortion commission. Appeal of abortion denial to a regional abortion commission had to be decided within seven days. Mature minors did not require parental consent. Parents were informed about the abortion (requested or performed) only if such action was deemed to be in the minor's best interest (Yugoslavia/Slovenia, 1977). Croatia followed on 21 April 1978 but required parental consent for minors under age 16 (Yugoslavia/Croatia, 1979). The law of Bosnia-Herzegovina stipulated that a minor above age 16 could request abortion if she was employed. The statutes of the other constituent republics and autonomous provinces indicated that abortions could be performed for minors without parental consent. There was no requirement, however, to obtain the partner's consent at any age (Zupancic, 1978). In Macedonia an abortion could be performed only if more than one year had elapsed since the previous termination.

After approving the 1969 Law on the Interruption of Pregnancy, the SFRY Federal Assembly (1969) passed a Resolution on Family Planning. It espoused a liberal tradition whose central aim was the humanization of all aspects of reproductive processes with no limitation placed on the availability of safe means of fertility regulation, including abortion. Moreover, every effort was to be made to ensure not only that the means of fertility regulation were available but also that the population was fully informed (Besemeres, 1980). Since pronatalist or antinatalist policies were perceived as undesirable intrusions on individual freedom of choice (Tomsic, 1973), expressions of official views, such as those promulgated by the National Assembly in 1975, tended to reflect humanistic perspectives of socialist society rather than demographic goals (Macura, 1974a; Simoneti et al., 1976). While objectives were shared, specific measures, which by necessity had to be differential, were left to the discretion of the constituent republics, autonomous provinces, and communes. The cost of abortion and contraceptive counseling and supplies were covered by compulsory health insurance (Kapor-Stanulovic, 1989a).

The decentralization of population policymaking was quite unique among the socialist countries of Central and Eastern Europe. As time passed, considerable differences emerged among federal units on the legal bases of family planning. The right to family planning articulated in the 1974 Federal Constitution was interpreted almost exclusively as the right to abortion (Malacic, 1990; Rasevic, 1990 a, 1990b, 1990c; Stampar and Kapor-Stanulovic, 1990; Avramov, 1991).

Current Status

The laws promulgated in the SFRY republics still apply in the present-day Federal Republic of Yugoslavia (FRY) as well as in the successor republics. The FRY reconfirmed the SFRY regulations in its new Law on the Interruption of Pregnancy (Serbia, 1995). After Croatia declared its independence in June 1991, a church supported vocal antiabortion minority steadily urged that restrictions be placed on access to abortion. While the law was not changed, women found it more difficult to obtain procedures. Some hospitals declined to perform abortions until the Ministry of Health ordered them to do so. However, individual physicians were permitted to refuse their services. Some who did so in hospitals performed abortions privately at high fees (*Dan*, 1995). During the war abortions were more widely available, especially for refugees who claimed to be victims of rape (Pine and Mertus, 1993). A 1992 public opinion poll indicated that nearly 70 percent of the population opposed abortion restrictions and 79 percent objected to restrictions on the sale of contraceptives (Pine and Mertus, 1993). With the support of the National

Program for Demographic Development, the Croatian Population Movement, headed by a Catholic fundamentalist, distributed posters admonishing women to "Have One More Child." The poster showed a smiling preschooler in army fatigues flashing the two-finger victory sign (Hockenos, 1996).

In Slovenia the Roman Catholic Church took a strong stand against abortion in 1990. However, a comparison of results from 1969 and 1990 opinion polls using identically worded questions showed that the percentage of Slovenians who opposed the right to abortion declined from 43 percent to 27 percent while the percentage in favor increased from 50 percent to 66 percent. Women were more liberal in their views than men (Meznaric and Ule, 1994). In December 1991 the Constitutional Court declared that abortion was a basic human right and would continue to be available in Slovenia on the same basis (until the tenth week of pregnancy) as before independence (WGNRR, 1992; WHO/EURO, 1997).

Procedure

In the SFRY the procedure for obtaining an abortion began with the pregnant woman's request. Formal commission approval was not necessary up to 10 weeks of gestation, although, depending on local legislation and custom, parental consent was required for minors under age 16. After 10 weeks the request was directed to a First Level (District) Commission. A decision had to be made promptly, usually within a week. If the woman's request was denied, she could appeal to a Second Level (Regional) Commission, which was obligated to render a decision within three to seven days (depending on the republic or province). The vast majority of abortion requests was approved; for example, 97 percent in Croatia during 1967–70 (Stampar, 1973b), 95 percent in Slovenia in 1972 (Simoneti et al., 1976), and over 90 percent in Vojvodina in 1972 (Kapor-Stanulovic and Beric, 1973). Nearly 95 percent of approvals were based on social indication, about 3–4 percent on medical grounds, and 1 percent on juridical reasons (Simoneti et al., 1976).

Abortions were performed by gynecologists in designated hospitals and outpatient clinics. In small towns abortions were performed by visiting medical teams. A working woman was entitled to paid leave during the time she was in the hospital. No fee was charged for an abortion induced for medical reasons. According to Stampar (1971d), the total cost of an abortion performed on request for social reasons was equal to the cost of a five-year supply of oral contraceptives, including medical supervision and cervical cancer screening. There was no additional compensation for the gynecologist, who was usually rotated through the hospital abortion service for a six-month period and performed about six abortions per day.

After the breakup of the SFRY abortions became relatively more difficult to arrange. With the decline of the health system, lack of medicines and anesthetics, women are often asked to participate in covering expenses. The amounts vary from center to center. Following the reemergence of private practice an increasing proportion of women seek abortion in the private sector, preferring the confidentiality of the physician's office and the higher quality of services offered. The fee varies between U.S.$70 and U.S.$120, which is approximately half of an average monthly salary.

Clandestine Abortions

Traditionally there have been two kinds of illegal abortions, those induced by nonphysicians and those provided by physicians. The Penal Code imposed a relatively mild punishment of three months' imprisonment for the provider of an illegal abortion with higher penalties in more serious cases. Women inducing their own abortions or cooperating in illegal procedures were not punished (Maggs, 1977).

Most women preferring privately practicing physicians did so to avoid probing questions from abortion commissions, the perceived embarrassment of waiting in abortion clinics, and the possible violation of the confidentiality of records (Benson, 1973). Following the 1969 legislation eliminating the necessity of abortion commission approval before 10 weeks' gestation, the incidence of illegal abortion declined greatly (Jurukovski and Sukarov, 1971; Rasevic, 1993; U.N./PD, 1995).

Maternal Mortality

Registered data on mortality attributed to abortion are not completely reliable. One explanation may be that such deaths are often placed in the category "cause unknown." Nearly all reported abortion related deaths since 1970 were the consequence of illegal termination and represented about one-third of total maternal mortality (Stampar, 1973a, 1975–76; Yugoslavia/FSO, 1991; Rasevic, 1990c, 1994). A major contributing factor to the improved safety of legal abortions was the introduction of electric vacuum suction curettage, developed in China and subsequently introduced to Yugoslavia (Novak, 1968; Beric and Kupresanin, 1971; Andolsek, 1974).

Trends

The findings presented in Table 12.1 were derived from official sources believed to be accurate (Mojic, 1969; Stloukal, 1995; COE, 1997; Rasevic, 1998). The ratio of legal abortions rose from 177 per 1,000 live births in 1960 to 564 in 1970, to 778 in 1980, and 1,019 in 1987. The abortion rate increased from 18.3 abortions per 1,000 women aged 15 to 44 years in 1960 to 53.4 in 1970, 59.5 in 1980, and 71.6 in 1987. Throughout the SFRY, except in Kosovo, almost half of all pregnancies ended in abortion. Nearly 91 percent of the women were married and over 70 percent had had a prior procedure (Rasevic, 1990a).

A comparison of 1967 regional figures shows that total registered abortions equaled 21 percent of live births in Macedonia, 30 percent in Bosnia-Herzegovina, 48 percent in Slovenia, 60 percent in Montenegro, 66 percent in Croatia, 113 percent in Serbia, and 167 percent in the autonomous province of Vojvodina. Abortion ratios were highest in the two largest republics and Vojvodina (Petric, 1969).

A 1970 representative sample survey of married women aged 15–49 years confirmed the persistence of the traditional reliance on abortion, especially among women 25–34 years of age, when the wanted family size had been reached or when it was desirable to increase the interval between deliveries (Lalovic, 1971). Some 32 percent of the 6,246 respondents acknowledged having had abortions; most of these women reported one or two abortions, but 14 percent mentioned 20 or more prior terminations (Ginic, 1971). Within the constituent republics and autonomous provinces, the percentage of women who said they had an abortion ranged from 66 percent in Vojvodina and 62 percent in Serbia to 21 percent in Kosovo (Lalovic, 1971). However, abortion data gathered in interviews tended to be underestimates. For example, Stampar (1973–74) noted that Croatian women only admitted about two-thirds of legal abortions performed during the immediately preceding year.

Regional Data

One of the most complete regional abortion data sets was reported from Croatia by Stampar (1971a–d, 1972a, 1973a, 1973b, 1976). As noted in Table 12.4, births and birth rates declined since 1959, accompanied by an increase in abortions. Repeat abortions represented 70 percent in 1990, declining to 62 percent by 1993 (Stampar, 1976, 1995).

The largest number of abortions per 1,000 women of fertile age were performed in the Autonomous Province of Vojvodina (Zarkovic, 1972, 1973). The number of births declined while the number of legal abortions and the abortion ratio rose (Beric, 1972; Kapor-Stanulovic and Beric, 1973; Breznik, 1981). By 1986 the number of abortions declined to 38,700 and the abortion ratio to 1,521 (Rasevic, 1990c). In a study of 2,500 women of fertile age in Vojvodina, 1,694 (67.8 percent) had 4,786 induced abortions, an average of 2.82 abortions per woman who had abortion experience and an average of 1.91 for the entire sample group (Ferencovic-Nikasinovic et al., 1975–76/1978). This compared with a Yugoslav countrywide average of 0.69 abortions for women of fertile age (Zarkovic, 1972, 1973). It was not unusual to find women in Vojvodina who had experienced more than 10 abortions in their reproductive years (Kapor-Stanulovic and Beric, 1979).

In the Autonomous Province of Kosovo, bordering on Albania, the number of births rose from 41,600 in 1960 to 43,400 in 1965, while abortions increased during these years from 1,300 to 4,400. By 1986 the number of abortions increased to 12,300 and the abortion ratio stood at 225 per 1,000 live births (Rasevic, 1990c).

In Serbia Proper (without Vojvodina and Kosovo), the number of births declined from 86,500 in 1960 to 78,000 in 1965, while abortions increased from 25,000 to 73,800 during those years. The abortion ratio rose from 289 per 1,000 live births in 1960 to 946 in 1965, 1,518 in 1969, and 2,182 in 1986 (Rasevic, 1990c).

In Slovenia the abortion ratio rose from 310 per 1,000 live births in 1970 to 703 in 1986. It subsequently declined to 650 in 1992 and 533 in 1997. In Macedonia the ratio increased from 398 in 1970 to 847 in 1989, dropping to 452 in 1996. In Croatia the ratio peaked at 832 in 1986, decreasing to 229 in 1996 (See respective tables and Rasevic, 1990c). In Bosnia-Herzegovina, family planning services became a low priority in the midst of armed conflict. In Tuzla, the abortion ratio increased from 77 abortions per 100 deliveries in 1991 and 1992 to 97 in 1993. In Sarajevo, abortions outnumbered deliveries by two to one. When contraceptives became freely available again in 1996, abortions declined to one per three deliveries (Omeragic, 1998).

By 1988 nearly 80 percent of all abortions for Serbian women were repeat procedures (Yugoslavia/FSO, 1991). Approximately every fifth Serbian woman requesting an abortion had already experienced four or more prior procedures. However, in Kosovo nearly half (45 percent) of all procedures were first time abortions. In Vojvodina 23 percent were first-time procedures and 6 percent were the fourth occurrence. In Montenegro 34 percent were first abortions in 1988 while for 11 percent of women it was the fourth time.

Historically, in the areas now constituting the FRY, the number of registered abortions exceeded the number of live births from 1975 to 1992. Abortions increased from 182,000 in 1975 to 222,600 in 1985 and then decreased to 202,000 in 1989 and 96,900 in 1995. The 1989 abortion rate was 90.5 per 1,000 women between 15 and 44 years of age. Almost 10 percent of women who terminated their pregnancies in 1989 had not carried a prior pregnancy to term and almost 20 percent had only one child. For 22 percent of the women, the procedure was their fourth or higher order abortion (Rasevic, 1994; COE, 1998).

In Bosnia, the abortion rate declined considerably since 1995. Abortions outnumbered births by 3:1 during the war years. Subsequently the proportions were reversed (*The Times* (London), 1996). As shown in Table 12.6, the number of legal abortions in Macedonia fell from 29,900 in 1985 to 14,200 in 1996. During the same period of time the abortion ratio dropped from 773 per 1,000 live births to 452.

Abortion Denied

A follow-up study of women denied abortion during the period of 1 July 1970 to 30 June 1971 was reported from Zagreb, the capital of Croatia (Stampar 1972b, 1973b). Of the 269 women who had not appealed denial of their initial request for abortion or whose appeal had been refused, 50 women could not be traced and were lost to follow-up. Of the remaining 219 women, 30 had carried to term. No delivery records were found for the other 189 women. Thus, 86 percent of the Zagreb women whose addresses could be traced appear to have terminated their unwanted pregnancies after permission for legal abortion had been denied (Stampar, 1973b).

Subsequently, studies of denied abortions were conducted in Varazdin (northern Croatia) and in Zadar (southern Croatia). As in Zagreb, these areas were selected because they had well-organized health services and precise birth registration systems. In both areas about 83 percent of the women who did not appeal the denial of their initial request or lost their appeal managed to avoid term delivery, similar to the 86 percent figure in Zagreb (Stampar, 1973b).

CONTRACEPTION

Availability

The traditional method of fertility regulation was *coitus interruptus*, backed by abortion. A 1963 study of 1,960 Slovenian women indicated that 44 percent used *coitus interruptus* (withdrawal), 22 percent preferred diaphragms, 5 percent condoms, and 10 percent other combinations; 19 percent did not practice contraception. About 30 percent of the women said that legal abortion was an acceptable method of fertility regulation (Andolsek, 1964). In the late 1960s contraceptive practice was least accepted in the Kosovo and Metohija region, where only 25 percent of 5,000 women aged 15-49 years acknowledged use of any form of contraception (Breznik and Rasevic, 1970).

Following the 1969 Federal Assembly Resolution on Family Planning, modern contraceptives gradually became more readily available through health centers, medical practitioners, and pharmacies. Contraceptive consultations were offered free of charge throughout the entire country, but constituent republics and autonomous provinces could charge different prices for various contraceptives. In some republics all means of contraception were free; in others women paid a small participatory fee for contraceptives as they did for other prescriptions (Tomsic, 1973). While the public health nurse was a key figure in the delivery of family planning services, responsibility for the prescription of modern contraceptives was traditionally in the hands of specialists in obstetrics and gynecology who were reluctant to yield their prerogative to general practitioners despite recommendations of the Federal Council for Family Planning (Benson, 1973).

Following an extensive pilot study, the Demographic Research Center of the Institute of Social Science in Belgrade conducted a nationally representative survey of fertility-regulating behavior in October/November 1970 among married women age 15–49 (Breznik, 1971). Among the 3,216 women (51 percent) then practicing contraception 68 percent reported reliance on withdrawal. Pills were preferred by 8.8 percent of women 20–24 years old, 5.8 percent of those aged 25–29 years, and 4.3 percent of those 30–34 years of age. Of the 6,246 total respondents, 2,016 (32 percent) acknowledged having had one or more induced abortions (Ginic, 1971). Further analysis showed that use of more effective contraceptive methods increased proportionally with higher education and participation in the nonagricultural labor force (Rasevic, 1971). Regional surveys were reported from Macedonia (Jurukovski, Antonovski, and Ljatkova, 1972), from Croatia (Zaplatic, 1972; Stampar, 1976), and Vojvodina (Beric et al., 1978).

Sterilization was not considered a family planning method in the SFRY. It was usually performed only on strict medical indications, along guidelines developed by the Yugoslav Association of Gynecologists and Obstetricians (Andolsek, 1971).

Voluntary sterilization was available in Slovenia to persons over 35 years of age, provided six months had elapsed following medical authorization (Yugoslavia/Slovenia, 1977).

In 1990 oral contraceptives and IUDs were available only sporadically in the FRYand then at very high cost compared to average income. *Coitus interruptus* remained the preferred method. It was considered natural, easy to use, and not injurious to health. Well suited to a culture in which traditional beliefs and attitudes remained firmly interwoven, it supported the concept of an active male role and the passive-submissive role of the woman in intimate relationships (Rasevic, 1993, 1994).

In Slovenia the use of modern contraceptives increased. In Macedonia oral contraceptives were preferred by 75 percent of women using contraceptives in 1994. Condoms were never widely accepted; reliance on them declined considerably (UNICEF, 1996c).

ABORTION/CONTRACEPTION

The national effort in the SFRY to reduce reliance on abortion and encourage more effective contraceptive practice in a country where legal abortion and modern contraception were easily available, equally accessible, and virtually cost-free stimulated a unique research program during 1973–74 at the Clinic of Obstetrics and Gynecology of the University of Novi Sad Medical School. The objectives were to explore determinants of choice behavior in fertility regulation, reasons for the predominance of abortion over contraception, and promising approaches to more successful use of modern contraceptives (Kapor-Stanulovic and Friedman, 1978).

The first study included 48 married couples. Both partners consented to be interviewed separately but simultaneously shortly after the woman had had an abortion but before onset of her next ovulation. Most commonly, the abortion decision was reached after discussion between the spouses, usually immediately upon discovery of pregnancy. Conversations on topics dealing with fertility regulation were relatively infrequent and short. Although both partners had already thought about plans for preventing future pregnancies and each tended to believe that agreement had been reached jointly, in actual fact there was very little concordance on both the extent to which a decision had been made and the specific method of contraception. In 87.5 percent of couples, husbands had decided that a certain contraceptive, not previously used, should be tried. However, only 37.5 percent of the wives had decided to start using a particular contraceptive; 48 percent had not reached a decision on how to regulate future fertility at a time when she would once again be at risk for conception. Of the 41 wives available for a follow-up interview 14 months later, 46 percent had started using a modern contraceptive, but half had discontinued its use. About 63 percent relied on withdrawal or did nothing to prevent another pregnancy. Six of these women had already had another abortion and one had given birth.

The general conclusion was that prospects for changing the manner of regulating reproductive behavior were slight. Indecision and resistance to modern contraceptives were considerable among the wives. The actions of most of the couples demonstrated their lack of readiness to regulate future fertility by methods other than those that had proven inadequate in their prior experience. Although abortion was perceived by both men and women as psychologically more costly than contraception, great value was placed on its effectiveness at a time when the desired timing (and number) of childbirths in the family was threatened (Kapor-Stanulovic and Friedman, 1978).

The second Novi Sad study attempted to understand why a majority of women continued to choose abortion over modern contraception. To examine differences between aborting and contracepting women, 72 women were randomly selected from among those waiting for the procedure at the abortion clinic and 72 women who had come to the contraception clinic (Kapor-Stanulovic and Friedman, 1978). The extensive findings reflected a pattern of differences between the two groups, suggesting a greater degree of control, planning, and husband-wife agreement in the group of contracepting women (Group C) as compared to the aborting women (Group A). The heavier reliance on abortion in Group A was not so much the result of preference for abortion as failure to contracept successfully. The major method of Group A women had been withdrawal, which was not practiced by any of the Group C women.

Although there was a larger number of rural and less well-educated women in the Group A sample, their behavior was not significantly different from urban women. Significant differences did emerge, however, in the perception of Group A and Group C women of the extent of their personal influence over their own fates. This difference may explain the choice of ineffective contraceptive methods through which Group A women tried to prevent pregnancy. Having a low sense of personal efficacy may be incompatible with the use of modern contraceptives, which require a woman's initiative and continuing vigilance. In such a context, contraception was perceived as having more disadvantages and a higher psychological cost. By the same token, having a higher sense of personal efficacy increased the probability of Group C women to accept modern contraceptives and regulate reproduction much earlier in their fertility careers (Kapor-Stanulovic and Friedman, 1978). Subsequent Novi Sad studies suggest that postabortion contraceptive counseling and health education programs have selective effects (Beric et al., 1978). Women generally experienced the greatest stress just before the abortion with relief following afterward (Kapor-Stanulovic, 1985).

In a 1988 study of 473 women who had an abortion at the University Department of Obstetrics and Gynecology in Ljubljana, Slovenia, including 102 repeaters, 96 percent of sampled women were practicing some form fertility regulation at time of conception. However, the percentage using an effective method fell from nearly half to 16 percent a few months prior to conception. The study found no evidence that women using a less effective method of contraception had a more liberal view of abortion or viewed it as a fallback method. Women's continuity of contraceptive practice and perceptions of rhythm, withdrawal, or barrier methods as effective means of fertility control indicated that they had not anticipated having to rely on abortion (Obersnel-Kveder, 1992). Easy access to abortion did not reduce motivation for contraception.

In 1990 a study was conducted of 400 women selected at random (every third woman) coming for pregnancy termination to the Institute for Mother and Child Health Care in Belgrade (Rasevic, 1993, 1993–94, 1994). The sample represented 1 percent of the total number of women having abortions in Belgrade; 94 percent were between 20 and 44 years of age and 64 percent were married. For 35 percent it was the first abortion while 20 percent had had four or more prior procedures. The majority of the women (63 percent) had one or two living children and 61 percent had secondary level schooling. Individual interviews were conducted on the day the appointment for the abortion was made and again two months after the procedure.

The woman's responses showed that reliance on induced abortion was the consequence of a discordance between a felt need for fertility regulation and the unacceptability of modern contraceptive methods. This discordance resulted from several factors, including insufficient and inaccurate knowledge about reproductive physiology and contraception, a belief that modern contraceptives were harmful to health, and barriers in partner relations. The situation was further confounded by the few organized efforts to promote sexuality education.

In an exercise in which women were asked to rate the relative psychological cost of abortion and contraception, contraception was shown to have a higher cost than abortion for these women. Less than one-third had tried the pill, condoms, or the IUD. Any negative experiences with contraceptive methods, whether their own or somebody else's, had been generalized to the point where contraceptives were perceived as complicated, complex to use, and having bad health consequences whereas abortions were rated as an unpleasant experience with a moral burden. Withdrawal had been used by 54 percent of the Belgrade women when they became pregnant. Their preference for this method was ascribed to its freedom from health effects and its avoidance of interaction with health service providers. Nearly half the women had had an unpleasant experience with a provider and expressed a felt need for more open and human discussion of contraception in relation to their individual problems, fears, attitudes, and values (Rasevic, 1993, 1993–94, 1994).

A survey of 437 Belgrade physicians in 11 public health institutes found that condoms were used by medical staff and recommended to their patients. About one-third of the female staff never used contraception, two-thirds had abortions, and one-third had two or more terminations (Ristic, 1994).

POPULATION POLICIES AND FAMILY ALLOWANCES

Family allowances have a long history in the former Yugoslavia. After 1945 it was believed that allowances available to all would encourage low income families to have children (Simoneti et al., 1976). However, both pronatalist and antinatalist measures were gradually perceived as undesirable intrusions on freedom of choice and not proper matters of federal legislation. The 1974 constitutional reforms confirmed the evolving policy that questions of family allowances be left to the discretion of the constituent republics, autonomous provinces, and local communities. Accordingly, there were no pronatalist incentives in Yugoslavia in the sense in which that term was usually understood in the other socialist countries of Central and Eastern Europe.

In the late 1970s and in the 1980s, it was increasingly acknowledged that promoting family planning was insufficient to solve emerging demographic imbalances (Rasevic, 1990c). Vojvodina passed the "Family Planning Resolution" in 1977, Serbia adopted the "Population Replacement Resolution" in 1983, and Macedonia approved a "Population Policy Resolution" in 1987. The SFRY Federal Assembly passed a "Resolution on Population Development and Family Planning" in 1989. Balanced population renewal in all the regions was declared the main objective of demographic policy. Beyond birth grants and family allowances for low income families, the SFRY provided universally free and accessible health care and education, comprehensive material benefits, birth grants, pensions, workmen's compensation, unemployment insurance, and social assistance for the poor.

A woman worker had the right to uninterrupted maternity leave of at least 105 days to cover her pregnancy and confinement. Maternity leave could begin 45 days before the expected delivery date. On return to work the mother had the right to work only four hours a day until the child was eight months old (Simoneti et al., 1976). In 1973 Croatia extended maternity leave until the child was one year old. A woman worker could stay at home for 105 days after delivery and then work four hours daily until the baby was eight months old, or she could take 180 days full leave and then begin working full time. In either case she received full pay. Maternal leave could be extended to 12 months on an unpaid basis but still counted as full time work for pension purposes (*Narodny Novine*, 1973). Preschool child care facilities were often inadequate.

While the FRY is attempting to maintain much of the system, the economic collapse eroded the real value of allowances. The effectiveness of social services diminshed amidst demands placed on a failing infrastructure. Earlier recommendations for reform (Milosavljevic, 1988; Kapor-Stanulovic, 1989b; Macura, 1989) could not be implemented.

Much of the FRY's present legislation includes efforts to resolve demographic imbalances. Most important is the Labor Relations Act (Serbia, 1991), which stipulates eligibility criteria and duration of maternity leave. Mothers are given 12 months' leave for the first and second child; 24 months for the third child; and 9 months for the fourth child and all subsequent children. On expiration of leave, a woman can work half-time until her child is three years old and if it is in poor health or handicapped.

The Law on Pensions and Disability Insurance (Serbia, 1992) rewards mothers who have given birth to their third child with two extra years of pension entitlement. The Law on Child Welfare (Serbia, 1994) guarantees that the salary for employed mothers continues to be paid in full during maternity leave for the first, second, and third child, and at 80 percent for every subsequent child. In municipalities with negative natural increase, the allowance is 100 percent for the fourth child also. For nonemployed mothers, an allowance amounting to 30 percent of the republic's average net wage is provided up to one year after the birth of the first, second, or third child. In areas with a negative natural increase this right is also extended to the fourth child. A one-time birth grant, equaling the republic's average net wage, is awarded to the mother for each of her first three children (and for the fourth child in low natality areas). Monthly child allowances are given for the first three children up to age 19. Free services are provided at preschool institutions for the third child. Similar benefits exist in Croatia (Croatia/National Committee, 1994), Slovenia (Slovenia/MOL, 1994), and in Macedonia/SO (1995).

The effectiveness of population policies has long been a subject of debate (Kapor-Stanulovic, 1988a, 1988b, 1988c; Rasevic, 1995a, 1995b; Rasevic and Petrovic, 1995; Petrovic, 1995). Theoretically, the Serbian measures should encourage parents to have three children. In areas with negative population growth, the birth of a fourth or fifth child is encouraged, while in areas with high population growth birth of a fourth or fifth child is discouraged. These measures are perceived as discriminatory by ethnic Albanians, who usually have more than three children. Current demographic trends in Serbia and Montenegro suggest that the population policy measures have had little, if any, impact (UNICEF, 1996b). Belgrade surveys indicate that Serbian women resort to abortion regardless of age, marital status, education, occupation, or other social, psychological, or cultural characteristics. Their attitudes are not affected by age or education. Since they gave priority to their perceived individual and family needs, it seems unlikely that pronatalist policy measures would persuade them to have more children (Rasevic, 1995a, 1995b).

THE WOMAN QUESTION

Historical Development

Yugoslavia's resistance to German occupation during World War II was the springboard for women's emancipation from their traditional patriarchal subordination in the family. The civil, political, and social rights that women acquired after World War II had been earned by participation in partisan activities. However, despite the legal rights and equality promulgated in the communist system, women remained subordinated and segregated in all walks of life (Milic, 1993).

Consideration of "the woman question" was complicated by the multinational, multireligious, and multilingual configuration of the SFRY, compounded by increasing differences in levels of socioeconomic development between regions. Most women did not have full-time jobs. They worked primarily in traditional private sectors (e.g., agricultural handicrafts, private catering establishments) and predominantly in auxiliary labor (Wertheimer-Baletic, 1979). During the period 1961–71, the number of economically active women employed outside of agriculture increased by 52 percent, while the proportion in agriculture declined by 12 percent.

In the 1970s the number of employed women increased faster than that of employed men, resulting in a rise in the proportion of women in the total labor force from 29 percent in 1964 to 34 percent in 1974 (Wertheimer-Baletic, 1979). Differences in the proportion of women in total employment by constituent republics and autonomous provinces were related to economic development. The lowest proportion existed in predominantly Muslim Kosovo, but rose from 16 percent in 1971 to 21 percent in 1975. The highest proportion remained stable, around 43 percent in industrial Slovenia. Croatia, Vojvodina, and Serbia Proper also were above the national average in women's participation in the labor force (Sokolov, 1974; Wertheimer-Baletic, 1979).

The considerable sociocultural differences in the perceptions of women's roles and responsibilities in the constituent republics and autonomous provinces slowed the progress of emancipation. Despite the shifts in urban-rural population distribution in the postwar years, the SFRY remained one of the least urbanized areas in Europe (Rusinow, 1972, 1973). Illiteracy among women living in the developing regions was proportionally high, which, together with their low status, effectively prevented them from exercising their rights (Tomsic, 1973).

Although the economic position of women improved considerably, some discrimination persisted in spite of court warnings to adhere to legally recognized principles (Simoneti et al., 1976). Legal provisions could not overcome deeply rooted traditions, especially in rural areas and in the professions. People changed more in theory than in practice. Women's organizations were, however, able to exert considerable influence on the development and implementation of family planning policies (Besemeres, 1980).

By voicing alarm about dependence on abortion to regulate fertility and launching public campaigns to foster more effective contraception, the Conference for Social Activity of Women created an atmosphere within which family planning was linked with the status of women, progress for working women, and human rights (Tomsic, 1973). Party, government, and public health functionaries shared women's concerns about the rising tide of clandestine abortion and associated health hazards. While the decrees of 1952 and 1960 were designed, in part, to move termination of unwanted pregnancy into medical facilities, thus reducing health risks, there was a concomitant emphasis on encouraging the replacement of abortion, whether legal or illegal, with modern contraception (Rusinow, 1972). This comprehensive approach fostered an ethical rationale within which family planning was perceived as the basis for ascertaining that every child be a wanted child. The principle of self-management, and with it the decentralized responsibility for socioeconomic development, was extended to include responsibility for fertility behavior and preparation for responsible parenthood (Tomsic, 1973).

While statistics are lacking, it can generally be said that over the years Yugoslav women made considerable gains in social and economic equality. Examination of trends in education indicate that the proportion of girls in school enrollment gradually increased to within 1 or 2 percent of the proportion of females in the total population (with the exception of Kosovo). The proportion of women in the total work force also increased, from 23 percent in 1952 to 38 percent in 1992. In Kosovo women were much more disadvantaged in terms of health, education, and status within the community and family. The lowest percentage of employed women was in Kosovo (23.8 in 1992).

As the status of women improved, the average age of marriage increased. In 1980, 26.1 percent of brides were under age 20, compared with 23.4 percent in 1990 and 24.1 in the FRY in 1996 (COE, 1998). The greatest number of women married between ages 20 and 24, while most men tended to marry between ages 25 and 29. Kosovo had the highest rates of marriage and the lowest rates of divorce; the reverse situation prevailed in Vojvodina.

The negative image of women's organizations during the communist regime persisted until the end of the 1970s when a new generation of women was ready to challenge the gap between Marxist theory and social reality. By 1979 small groups of women organized in major towns but never evolved into a mass movement. In 1991 Serbian, Croatian, and Slovenian mothers together with Muslim mothers from Bosnia-Herzegovina and Macedonia joined in demonstrations to demand the discharge of their sons kept in the Federal army long after their military service was over. After Milosovic called on Serbian women to boycott the protests, the women's groups lost their spontaneity and failed to unite opponents to the war (Drakulic, 1993).

Current Trends

The political representation of women has been greatly reduced. Whereas in the single-party communist system women accounted for 19 percent of the membership in all republic assemblies and at the federal level, their share declined to 4.8 percent by 1992 when there were only three women ministers in all the republics of the former Yugoslavia (Milic, 1993). After the first free elections in 1990, the percentage of women parliamentarians was 13 percent in Slovenia, 4.5 percent in Croatia, 4 percent in Montenegro, 3.3 percent in Macedonia, 2.9 percent in Bosnia-Herzegovina, and 1.6 percent in Serbia (Papic, 1993). The low political participation reflected the strong influence of the conservative nationalist parties' propaganda, promoting a very traditional image of women as childbearers and housewives (Drakulic, 1993). After 50 years of "socialist emancipation" women were ill prepared to participate actively in the process of transition to democracy.

Employment conditions for women have seriously deteriorated. The average income of women is 20–40 percent lower than that of men. The economic situation varies from republic to republic. In 1989 Slovenia had only 3 percent of its population below the poverty level compared to 82 percent in Kosovo. The largest percentage of the unemployed, between 50 and 60 percent, have been women (Drakulic, 1993). However in Slovenia, women accounted for a smaller share of the unemployed (44 percent) in 1992 than men; but, in the cohort of Slovenians aged 18 to 30 years, more women than men were without jobs (Meznaric and Ule, 1994).

At the end of 1998, the position of women in the former Yugoslavia reflects cultural heritage, "communist" experience, and "transition," mixed with the consequences of war (which affected the population in the successor states in different degrees, but nearly always for the worse). The adverse effects of U.N-imposed sanctions on Serbia and Montenegro further worsened living conditions. "The economy of survival" required women to use all their resourcefulness to make ends meet. They returned to more traditional and less expensive means of providing the essentials for the family.

Some argue that in actuality women were not pushed into returning to the private domain; they willingly chose it of their own accord. In a society where the public domain is destroyed and where anomy, apathy, and chaos rule, the private domain is preferred. The focus of this escape to privacy is parenthood. This does not imply having more children. The mother's role is more demanding now even when there are only one or two children.

SEX AND SOCIETY

Family Planning Association

The family planning movement in Yugoslavia developed from the initiatives to liberalize abortion and increase the practice of effective contraception. A Coordinating Committee for Family Planning was formed within the Conference for Social Activity of Women in Yugoslavia in 1963. In June 1967 it was upgraded to the status of a Federal Council for Family Planning and became an Associate Member of the International Planned Parenthood Federation, attaining full membership status in 1971. The Federal Council acted as an umbrella organization for republic and province councils for family planning, each of which was constituted of representatives delegated by relevant organizations, institutions, and government agencies. In Serbia, for example, there were council representatives from the Institute for Health Protection of Young People, the Serbian Medical Association, the Serbian Red Cross, the Conference for Social Activity of Women in Serbia, Serbian Youth Federation, Serbian Student Federation, Serbian Trade Union Federation, ministries of Health and of Education and Culture, the Federation of Pharmaceutical Societies, and 15 other institutions, including the Serbian Association of Journalists (Rusinow, 1972).

During the Ninth Communist Party Congress in March 1969, the League of Yugoslav Communists passed a resolution in support of family planning. Subsequently, on 25 April 1969, the Yugoslav Federal Assembly (1969) approved a Resolution on Family Planning. It began by affirming that "it is one of the basic human rights and duties for parents to be able to plan the size of their families and the spacing between births." It continued that "for this to be feasible society should make it possible for married couples to get information about modern methods of birth control and provide them with adequate means to plan their families so that they can decide how many children they will have and at what intervals." The resolution acknowledged that abortion "is still the most widespread means of birth control and contraceptive measures are slow in expanding," urged wider dissemination of contraceptive information, and noted that the realization of family planning objectives "requires full involvement of educational, welfare and health institutions, sociopolitical and other organizations and associations, administrative agencies and other social factors" (Yugoslavia/Federal Assembly, 1969).

The institutional commitment to family planning by a wide array of sponsors produced a rationale emphasizing the importance of extending human freedom and independence, emancipating women, and enhancing individual responsibility. Individuals were encouraged to acquire knowledge of the means of preventing an unwanted pregnancy "in order to eliminate spontaneity in this delicate sphere of human life and to achieve, through family planning, the goal that each newborn child should be a wanted child" (Petric, 1969). Guidelines issued by Serbia (1995) reaffirm the wider promotion and use of contraceptives.

Sexuality Education

Sexuality education was introduced into elementary schools in 1960 (Andolsek, 1964). By 1964, it had been officially incorporated into the curricula of Bosnia-Herzegovina, Croatia, Macedonia, Serbia, Slovenia, and parts of Montenegro. The position of the Federal Council for Family Planning that sex education should be an integral part of all education, beginning with kindergarten and extending through all grades and the armed forces, was reaffirmed by the Resolution of the Yugoslav Federal Assembly (1969). Although legally compulsory, implementation was mixed, restrained by lack of teacher training and local resistance (Novak-Reiss, 1978).

At the beginning of the 1990s it became apparent that, despite being compulsory, sexuality education had not been effective. Teachers in secondary schools revealed that they felt uncomfortable in discussing sexuality issues and would prefer a physician or nurse to assume these responsibilities. However, medical school curricula provided little information on family planning and even less on the sexual and reproductive health of young people (Kapor-Stanulovic, 1991).

Adolescent Sexuality

For the entire SFRY, the proportion of legal abortions per 1,000 young women aged 15–19 years increased from 14 in 1969 to 50 in 1986. In Bosnia-Herzegovina, the proportion rose from 4 to 156, in Montenegro 3 to 5, Croatia 13 to 16, and in Serbia Proper 21 to 25. In Macedonia, the proportion remained stable at 13, and in Slovenia, it declined from 26 to 24. In Kosovo it also remained stable at 3 while Vojvodina saw a rise from 21 to 28 (Rasevic, 1990c).

During 1972–73 a quota sample survey was conducted with interviews of 1,086 girls and 1,231 boys attending three kinds of secondary schools in four regions of Croatia (Trenc and Beluhan, 1971). Among 15–19-year-old girls, 17.7 percent had experienced intercourse, ranging from 11.1 percent for the 16 year olds to 37.2 percent for those age 19 years. Among the boys, 19.6 percent had experienced intercourse by age 16 and 80.5 percent by age 19. Although there was considerable knowledge about contraceptives, actual use was limited. About 10 percent of teenage girls having intercourse relied on the pill; the majority did not use any method. Of the 19-year-old adolescent women having intercourse, 14.3 percent had already had an abortion (Beluhan et al., 1972–73). A similar reliance on abortion to terminate unwanted pregnancy was found among Belgrade high school students (Petric, 1969). In a review of marriage license applications involving 207 young women under age 17 during 1974–76 in Zagreb, it was noted that 30 percent of the brides were more than three months pregnant (Stampar, 1976). However, a 1990 study showed that only 4.2 percent of all abortions were performed in teenagers. By 1993 it was a similarly low 4.5 percent. One explanation was that young women in Croatia became sexually active at a later age (Pine and Mertus, 1993; Stampar, 1995).

In an interview survey of 5,234 adolescents under age 20 attending secondary schools in six towns in Croatia in 1990, 22 percent of girls and 49 percent of boys acknowledged having had sexual intercourse. Most girls experienced sexual debut when 17–18 years of age; boys began one year earlier. The majority of respondents with sexual experience did not use contraception. Some 4.5 percent of the girls had an unwanted pregnancy (Stampar and Beluhan, 1991).

Kapor-Stanulovic (1980) studied a representative sample of 130 adolescent boys and 120 adolescent girls drawn form the school-age population of Novi Sad (Vojvodina). A segment of the structured questionnaire was similar to one administered in 1964 to a similarly drawn sample of 560 adolescent girls. Results indicated that in 1979 teenage girls were significantly better informed about pregnancy prevention; 90 percent of the 1979 sample claimed having knowledge of contraception compared to 70 percent in 1964. While only 11 of the 120 adolescent women admitted having had sexual experience, this proportion was significantly higher (at the 0.05 level) than that found in 1964 (13 of the 560 women). Sexual activity began around 14–16 years of age, much as it did in 1964, with peer boys as partners. Almost 77 percent of the 1979 sample requested the provision of additional sex education in the schools (Kapor-Stanulovic, 1980). In a 1986 Serbian sample, 11 percent of all abortions occurred in the 15–19-year-old age group (Rasevic, 1990a).

In a more recent survey on sexuality-related knowledge and attitudes of adolescents in urban Vojvodina (Kapor-Stanulovic, 1996) a sample of 102 teenagers of both sexes, aged 13–18 years, showed that peers are the most frequent source of information about sexuality for both boys and girls. Girls are more likely than boys to obtain information from their mothers. Although 51 percent of boys and 49 percent of girls consider lack of knowledge about contraceptives the main reason for the relatively large number of teenage pregnancies, only 51 percent of boys and 53 percent of girls admit they know what is contraception. Withdrawal is still considered a safe method of preventing unwanted pregnancies (42 percent boys and 51 percent girls). Over 80 percent of boys and girls consider that both partners should take responsibility for pregnancy prevention.

Data from Slovenia indicate that adolescents terminated 18 percent of unwanted pregnancies in 1965 compared to 39 percent in 1980. In 1965 adolescent abortions represented 3.4 percent of all abortions compared to 9.2 percent in 1980 (Andolsek, 1988). The most rapid increase in the incidence of abortion was observed among 15–17-year-old girls.

According to a 1990 study in Croatia, around 22.1 percent of girls and around 30 percent of boys between the ages of 14 and 19 years were sexually experienced. However, only 43 percent of these students used "some form" of contraceptives. Over half of them had multiple partners, exposing themselves to sexually transmitted diseases and unwanted pregnancies (UNICEF, 1996d).

In addition to the poor quality of sexuality education in the schools, reproductive health services for adolescents in the early 1990s were either nonexistent in many parts of the former Yugoslavia, or else inadequate, with very little provision for confidentiality, privacy, or comfort for young clients. While well-meaning people were concerned that adolescents did not take advantage of free access to contraceptive services and supplies, thus risking unwanted pregnancies and likely abortions, more needs to be done than stating officially that young people have the right to free access to information and services (Kapor-Stanulovic, 1991). In Bosnia-Herzegovina reproductive health teams were organized under UNFPA/WHO auspices, providing supplies and disseminating information through lectures and the media (Omeragic, 1996).

FUTURE PERSPECTIVES

Future demographic trends within each of the successor states to the former Yugoslavia are likely to be affected in diverse measure by changes in population structure and personal values. These include the integration of refugees, emigration of young people, and the increased public role of the church. The continuing sociopolitical and economic crisis has drastically lowered the quality of life of the majority of the population and especially of women. Shrinking household budgets have burdened women by pushing them into resuming outdated housekeeping chores and finding extra jobs, mostly in the service sector where they are deprived of equal rights and more exposed to abuse and assault. After declining for 30 years, maternal mortality rates have increased. Between 1992 and 1993 maternal mortality rose from 4.3 to 10.3 maternal deaths per 100,000 live births in Croatia and from 9 to 16 maternal deaths per 100,000 in FRY (UNICEF, 1996a, 1996b). Life expectancy at birth for women is expected to become more unfavorable (Kapor-Stanulovic and Shabanah, 1991).

At a time when maintaining a satisfactory quality of life is a daily struggle, having one child seems enough to satisfy the basic desire for experiencing motherhood. There is little motivation for having additional children as long as political strife, ethnic struggles, and deteriorating living conditions persist.

References

To facilitate publication, titles appear in English translation with the language of origin indicated in parentheses. Diacritical marks have been omitted.

Adamik, M. (1991). Supporting parenting and child rearing: Policy innovation in Eastern Europe. In S. Kamerman and A. Kahn (Eds.), *Childcare, parental leave and the under 3: Policy innovation in Europe*. Westport, CT: Auburn House, 115–144.

Adamik, M. (1993). Feminism and Hungary. In N. Funk and M. Mueller (Eds.), *Gender politics and post-communism: Reflections from Eastern Europe and the former Soviet Union*. New York: Routledge, 207–212.

Adamik, M. (1997). How can Hungarian women lose what they never had? In M. Feischmidt, E. Magyari-Vincze, and V. Zentai (Eds.), *Women and men in East European transition*. Cluj: Editura Fundatiei Studu Europene, 42–61.

Adler, M. A. (1997). Social change and declines in marriage and fertility in Eastern Germany. *Journal of Marriage and the Family*, 59, 37–49.

Adler, N. E. (1975). Emotional responses of women following therapeutic abortion. *American Journal of Orthopsychiatry*, 45, 446–454.

Adler, N. E., David, H. P., Major, B., Roth, S. H., Russo, N. F., and Wyatt, G. (1990). Psychological responses after abortion. *Science*, 248, 41–44.

Adler, N. E., David, H. P., Major, B., Roth, S. H., Russo, N. F., and Wyatt, G. (1992). Psychological factors in abortion: A review. *American Psychologist*, 47, 1194–1204.

Agadjanian, V., and Qian, Z. (1997). Ethnocultural identity and induced abortion in Kazakstan. *Studies in Family Planning*, 28, 317–329.

Akkermann, S., and Mehlan, K. H. (1978). Law and fertility in the German Democratic Republic. In M. Kirk, M. Livi-Bacci, and E. Szabady (Eds.), *Law and fertility in Europe*. Vol. 1. Dolhain: Ordina, 274–297.

Albania. (1977). Penal Code, 15 June, Section 95.

Albania. (1988). Penal Code, Act. 5591, as amended by Decree No. 7251 of 15 September (In Albanian). Translated in *Annual Review of Population Law*, 1989, 16, 25.

Albania. (1992). Council of Ministers. Decision No. 226 of 27 May.

Albania/Institute of Statistics. (1991). *Statistical yearbook of Albania* (In Albanian). Tirana: Institute of Statistics.

Albania/Ministry of Health and Environment and UNICEF. (1993a). *Children and women in Albania.* Tirana: Ministry of Health and Environment.

Albania/Ministry of Health and Environment. (1993b). Women's health profile; Albania country report. Copenhagen: World Health Organization European Regional Office.

Albania/Ministry of Health and Environment. (1997). Personal communication, 23 January.

Alesina, I. (1996). Adolescent sexual health initiative in Russia. *Choices,* 25(2), 16–17.

Amendt, G., and Schwarz, M. (1991). *The life of unwanted children* (In German). Bremen: University of Bremen.

Anderson, B. A. (1987). The life course of Soviet women born 1905–1960. In J. R. Millar (Ed.), *Politics, work, and daily life in the USSR: A survey of former Soviet citizens.* Cambridge: Cambridge University Press, 203–240.

Anderson, B. A. (1991). *The role of abortion in fertility decisions in the Soviet Union: Results from analysis of data from Soviet emigres.* Research Report No. 91-231, Population Studies Center, Ann Arbor: University of Michigan.

Anderson, B. A., Katus, K. Puur, A., and Silver, B. D. (1992). *Characteristics of women having abortions in Estonia.* Research Report 92-254. Ann Arbor: Population Studies Center, University of Michigan.

Anderson, B. A., Katus, K., Puur, A., and Silver, B. D. (1994). The validity of survey responses on abortion: Evidence from Estonia. *Demography,* 31, 115–132.

Anderson, B. A., and Silver, B. (1986). Infant mortality in the Soviet Union: Regional differences and measurement issues. *Population and Development Review,* 12, 705–738.

Andolsek, L. (1964). Current status of abortion and family planning in Yugoslavia. Unpublished paper presented at the 4th Conference of the International Planned Parenthood Federation/Western Hemisphere Region, San Juan, Puerto Rico.

Andolsek, L. (1971). Sterilization in Slovenia, 1958–61. In *Proceedings of the 1st Yugoslav Symposium on Problems of Fertility and Sterility.* Zagreb.

Andolsek, L. (Ed.) (1974). *The Ljubljana abortion study, 1971–1973.* Bethesda: Center for Population Research, NICHD.

Andolsek, L. (1988). Yugoslavia. In P. Sachdev (Ed.), *International handbook on abortion.* Westport, CT: Greenwood.

Andorka, A. (1993). Population policies and fertility trends in Hungary. In H. P. David and D. Pierotti (Eds.), *Demographic and social effects of population policies in Europe.* Copenhagen: World Health Organization, 19–21.

Andorka, R. (1979). Birth control in the eighteenth and nineteenth centuries in some Hungarian villages. *Local Population Studies,* (22), 38–43.

Andorka, R., and Vukovich, G. (1985). The impact of population policy on fertility in Hungary, 1960–1980. In *International Population Conference, Florence, Solicited Papers,* Vol. 3. Liege: International Union for the Scientific Study of Population, 403–412.

Andrews, M. C., and Ovalioglu, G. (1994). *Albania and the World Bank.* Washington, DC: World Bank.

Antarsh, L. (1997). Family planning week in Ukraine. *AVSC News,* 35(3), 3.

Antarsh, L. (1998). Personal communication, 30 June.

Aresin, L. (1977). Family planning and marital counseling in the GDR. In M. Kozakiewicz (Ed.), *Family planning and sex education in socialist countries.* Warsaw: Polish Family Planning Association, 71–75.

Aresin, L. (1979). Counseling in the German Democratic Republic. *IPPF Europe Regional Information Bulletin,* 8(3), 1.

Aresin, L. (1987). Organized sexuality and family planning counseling in the GDR (In German). *Pro Familia Magazin,* 15(6), 19–20.

Aresin, L. (1993). East Germany: Two years after unification. *Planned Parenthood in Europe,* 22(1), 11–12.

Aresin, L., and Müller-Hegemann, A. (1978). *Young couples* (In German). Leipzig: Bibliographisches Institut.

Aresin, N., and Aresin, L. (1969). Development of family planning in the German Democratic Republic. *IPPF Medical Bulletin,* 3(4), 3–4.

Asbot-Thorma, J. (1998). Hungary. Report prepared for the N.I.S. and Eastern Europe Consultation, Hanover, Germany, 12–16 October. Summary from Population Institute, Washington, DC.

Ash, T. G. (1996a). Neo-pagan Poland. *The New York Review,* 11 January.

Ash, T. G. (1996b). Hungary's revolution: Forty years on. *New York Review of Books,* 14 November.

Atkinson, R. (1995). German unification lays heavy burden on eastern working women. *The Washington Post,* 29 March.

Atwood, M. (1986). *The handmaid's tale.* Boston: Houghton Mifflin.

Attwood, L. (1987). Gender and Soviet pedagogy. In G. Avis (Ed.), *The making of the Soviet citizen: Character formation and civic training in Soviet education.* London: Croom Helm, 107–135.

Attwood, L. (1993). Sex and the cinema. In I. Kon and J. Riordan (Eds.), *Sex and Russian society.* Bloomington: Indiana University Press, 64–88.

Avdeev, A. A. (1989). Abortion and fertility (In Russian). *Sotsiologicheskie Issledovania,* 15(3), 54–63.

Avdeev, A. A. (1994). Contraception and abortions: Trends in the USSR and prospects for the 1990s. In W. Lutz, S. Scherbov, and A. Volkov (Eds.), *Demographic trends and patterns in the Soviet Union before 1991.* London: Routledge, 131–146.

Avdeev, A. A. (1998). Personal communication to Stanley Henshaw, 22 February.

Avdeev, A.A., Blum, A., and Troitskaja, I. (1993). Abortion and contraception in Russia and the former Soviet Union: Past and present (In French). *Institut National D'Etudes Demographiques Dossiers et Recherches,* (41), 1–72.

Avdeev, A., Blum, A., and Troitskaja, I. (1994). The history of abortion statistics in Russia and the USSR up to 1991 (In French). *Population,* 49, 903–933; *Population: An English selection,* 7, 39–66.

Avramov, D. (1991). Yugoslavia. In J. J. Railu and A. Blum (Eds.), *European population, Vol. 1: Country analysis.* Paris: INED, 325–345.

Azerbaijan/State Statistical Committee. (1998). Communication to Population Reference Bureau, 25 February.

Azhgikhina, N. (1996). *Green roots: 16 interviews* (In Russian). Moscow: Aslan.

Azhgikhina, N. (1998). Personal communication, 4 June.

Babamuradova, M., Black, S., Black, S., Landry, E., and Margolis, A. *Evaluation of AVSC-supported programs in Kazakhstan, Kyrgystan, Turkmenistan, and Uzbekistan.* New York: AVSC International.

Baban, A. (1994). Sexual myths of Romanian women. Unpublished report.

Baban, A., and David, H. P. (1994). *Voices of Romanian women: Perceptions of sexuality, reproductive behavior, and partner relations during the Ceausescu era*. Bethesda: Transnational Family Research Institute.

Baban, A., and David, H.P. (1996). *Voices of Romanian women: Perceptions of sexuality, reproductive behavior, and partner relations during the Ceausescu era*. In S. Zeidenstein and K. Moore (Eds.), *Learning about sexuality: A practical beginning*. New York: The Population Council, 19–32.

Baban, A., and David, H. P. (1997). The impact of body politics on womens' bodies. In M. Feischmidt, E. Magyari-Vincze, and V. Zentai (Eds.), *Women and men in East European transition*. Cluj-Napoca: Editura Fundatiei Pentru Studii Europene, 156–170.

Bach, K. R. (1974). *Sex education in the socialist secondary school* (In German). Berlin: Deutscher Verlag der Wissenschaften.

Bach, K. R. (1987). Development, present situation, and tendencies of sexuality education in the GDR (In German). *Pro Familia Magazin*, 15(6), 8–10.

Bach, K. R. (1991). Basic position and goals of sexuality education in the GDR (In German). In J. S. Hohmann (Ed.), *Sexology in the GDR*. Berlin: Dietz, 239–261.

Badea, M., Stativa, E., and Stephenson, P. (1993). *Preventing maternal death in Romania: An analysis of maternity care and maternal mortality with recommendations for a safe Motherhood Programme*. New York: UNICEF.

Ballaeva, E. (1997). Personal communication, 26 December.

Ballaeva, E. (1998). Personal communication, 26 March.

Balogh, A., and Lampe, L. (1994). Hungary. In B. Rolston and A. Eggert (Eds.), *Abortion in the new Europe: A comparative handbook*. Westport, CT: Greenwood, 139–156.

Barbanov, A. A. (1991). Maternal and child health problems in the USSR. *Archives of Disease in Childhood*, 66, 542–545.

Barber, I. (1997). Women without democracy. *Open Society News*, Fall/Winter, 10–15.

Barr, D. A., and Field, M. E. (1996). The current state of health care in the former Soviet Union: Implications for health care policy and reform. *American Journal of Public Health*, 86, 307–312.

Bartonova, D. (1998). Abortion (In Czech). In Z. Pavlik (Ed.), *Population change in the Czech Republic, 1997* (In Czech). Prague: Department of Demography and Geodemography, Charles University, 45–51.

Batar, I. (1993). New abortion law in Hungary. *Entre Nous*, (22/23), 17.

Batt, J. (1991). *East Central Europe from reform to transformation*. London: Pinter.

Battiata, M. (1990). A Ceausescu legacy: Warehouses for children. *Washington Post*, 7 June.

Bebel, A. (1971). *Women under socialism*. New York: Schocken. (Originally published in German in 1875 under the title *Women in the past, present and future*, and later retitled *Woman and Socialism*.)

Beck, E. (1995). Bulgaria is slowing its backslide to chaos. *The Wall Street Journal*, 4 October.

Beck, E. (1996). Bulgarians protest as reforms kick in. *The Wall Street Journal*, 21 May.

Beck-Gernsheim, E. (1997). Declining birth rates and the wish to have children: Experience in East Germany (In German). *Zeitschrift für Bevölkerungswissenschaft*, 22, 59–71.

Beer, U. (1991). *Gender, structure, history* (In German). Frankfurt: Campus.

Belarus/Ministry of Health. (1998). Communication to Population Reference Bureau, April.

Belova, V. A., and Darsky, L. E. (1968). Opinions of women regarding future childbearing intentions (In Russian). *Vestnik Statistiki,* (8).

Belova, V. A., and Darsky, L. E. (1975/76). Opinion statistics in the study of the birthrate, I–IV. *Soviet Sociology,* 1975, 13(4), 3–41; 1975, 14(1), 33–61; 1975, 14(2), 13–50; 1976, 14(4), 60–109.

Beluhan, A., Benz, M., Stampar, D., and Trenz, P. (1972–73). Knowledge and use of contraception in adolescents attending secondary schools in Croatia (English summary). *Stanovnistvo,* 10/11, 204–217.

Benson, E. (1973). Family planning services in Yugoslavia. *International Nursing Review,* 20, 142–146.

Berelson, B. (1979). Romania's 1966 anti-abortion decree: The demographic experience of the first decade. *Population Studies,* 33, 209–222.

Berent, J. (1970). Causes of fertility decline in Eastern Europe and the Soviet Union. *Population Studies,* 24, 35–58; 247–292.

Bergmann, A. (1992). *Preventive sexuality: The beginnings of modern birth control* (In German). Hamburg: Rasch and Röhring.

Beric, B. (1972). Live births and legal abortion data for Vojvodina, 1960–1971. Unpublished summary.

Beric, B., Bregun-Dragic, N., and Breznik, D. (1978). *Low fertility variables and indirect abortion: The population of Vojvodina* (English summary). Novi Sad: University Gynecological Clinic and Belgrade, Demographic Research Center.

Beric, B., and Kupresanin, M. (1971). Vacuum aspiration using pericervical block for legal abortion as an outpatient procedure up to the 12th week of pregnancy. *The Lancet,* 2, 619–621.

Beric, B., and Stevanov, M. (1970). Epidemiological-legal aspects of the problems of abortion in Yugoslavia (English summary). *Stanovnistvo,* 8, 190–199.

Besemeres, J. F. (1975). Population politics in the USSR. *Soviet Union,* 2, 50–80; 117–144.

Besemeres, J. F. (1980). *Socialist population politics: The political implications of demographic trends in the USSR and Eastern Europe.* Armonk. NY: Sharpe.

Biberaj, E. (1990). *Albania: A socialist maverick.* Boulder, CO: Westview Press.

Blayo, C. (1989). Abortion in Europe (In French). *Espace, Populations, Societes,* 7, 225–238.

Blayo, C. (1991). Methods of birth prevention in Eastern Europe (In French). *Population,* 46, 527–546.

Blayo, C. (1992). Two routes, same destination. *Entre Nous,* (20), 8.

Blayo, C. (1993). The role of abortion in the countries of Central and Eastern Europe (In French). In *International Population Conference Montreal 1993, Solicited Papers, Vol. 1.* Liege: International Union for the Scientific Study of Population, 235–252.

Blayo, C., and Leitao, A. E. (1993). The access to contraception and sterilization. *Entre Nous,* (22/23), 4.

Bleuel, H. P. (1973). *Strength through joy: Sex and society in Nazi Germany.* London: Secker and Warberg.

Bock, G. (1984). Racism and sexism in Nazi Germany: Motherhood, compulsory sterilization, and the state. In R. Bridenthal, A. Grossmann, and M. Kaplan (Eds.), *When biology became destiny: Women in Weimar and Nazi Germany.* New York: Monthly Review Press, 271–296.

Bodrova, V. (1994). Reproductive behavior of Russia's population in the transition period: Tendencies towards depopulation. Unpublished report. Moscow: Russian Centre for Public Opinion Research (VCIOM).

Bodrova, V. (1995). People's notion about sexual behavior and sexual education in Russia. Unpublished paper. Moscow: VCIOM, May.

Bodrova, V. (1996). Russian attitudes on sex and youth. *Choices,* 25(1), 9–14.

Bodrova, V., and Mytil, A. (1992). Public opinion and sex education in the USSR (In French). *Population,* 47, 204–211.

Bogue, D. (1969). *Principles of demography.* New York: John Wesley.

Boh, K. (1993). Current family structures in Eastern and Central Europe. In *Changing families in changing societies, Proceedings of the International Conference in Brussels 1992.* Wiesbaden: Bundesinstitut fr Bevölkerungsforschung, 28–35.

Bohac, J. (1990). *The heritage of T. G. Masaryk* (In Czech). Prague: Prace.

Böhm, T. (1993). The women's question as a democratic question: In search of a civil society. In N. Funk and M. Mueller (Eds.), *Gender politics and post-communism: Reflections from Eastern Europe and the former Soviet Union.* New York: Routledge, 151–159.

Bollobas, E. (1993). "Totalitarian Lib": The legacy of communism for Hungarian women. In N. Funk and M. Mueller (Eds.), *Gender politics and post-communism: Reflections from Eastern Europe and the former Soviet Union.* New York: Routledge, 201–206.

Bolton, P., Corciova, M., and Köö, B. (1997). Experience with family planning and perceptions about voluntary sterilization among women seeking reproductive health services at four Romanian hospitals. Unpublished report. New York: AVSC International.

Borisov, S. B. (1989). Erotic texts as a source of sexual self-education (In Russian). *Sotsiologicheskie issledovaniya,* 16, 81–84.

Borisov, V., Sinelnikov, A., and Arkhangelsky, V. (1997). Expert opinions on abortion in Russia. *Choices,* 26(2), 23–26.

Bostandjiev, R. (1992). Sexual education in transition in Bulgaria. *Planned Parenthood in Europe* (21), 13.

Bostandjiev, T., and Vassilev, D. (1989). Sexuality of adolescents and youth in Bulgaria (In Bulgarian). Report presented at the Conference on Sexuality of Adolescents, Michajlevgrad.

Bowers, G., Wear, D., and Margolis, A. (1995). *The Russia reproductive health program strategy and action plan.* Arlington: Population Technical Assistance Project Report No. 94-033-016.

Brähler, E., and Richter, H. E. (1995). States of mind in East-West comparison (In German). Summary in *This Week in Germany,* 10 March.

Brandrup-Lukanow, A., and Mangiaterra, V. (1998). Mother and child health: Where is Eastern Europe going? *Safe Motherhood,* 26(2), 6.

Breznik, D. (1971). Survey of fertility of married women and family planning (English summary). *Stanovnistvo,* 9, 169–205.

Breznik, D. (Ed.) (1981). *Fertility and family planning in Yugoslavia.* Belgrade: Demographic Research Center.

Breznik, D., and Rasevic, M. (1970). Demographic development of the population of Yugoslavia with particular reference to fertility, reproduction, and birth control. Unpublished paper. Belgrade: Demographic Research Center.

British Medical Journal. (1992). 30 May and 13 June.

Bronfenbrenner, U. (1968). The changing Soviet family. In D. R. Brown (Ed.), *The role and status of women in the Soviet Union.* New York: Teachers College Press, 98–124.

Brown, A. (1979). Eastern Europe: 1968, 1978. *Daedalus*, 108, 151–174.

Brückner, H. (1968). *Sexual knowledge of our youth* (In German). Berlin: Deutscher Verlag der Wissenschaften.

Brückner, H. (1976). *Are you already thinking about love?* (In German). Berlin: Kinderbuchverlag.

Bukute, A., and Kabasinskiene, R. (1996). Lithuanian phone helpline reaching out to youth. *Choices,* 25(1), 16–17.

Bulgaria. (1956). Law of 27 April 1956. *International Digest of Health Legislation,* 8, 605–607.

Bulgaria. (1967). Decree No. 61 of the Central Committee of the Bulgarian Communist Party and Council of Ministers of 28 December 1967. *International Digest of Health Legislation,* 1968, 19, 589–602.

Bulgaria. (1968). Law of 16 February 1968. *International Digest of Health Legislation,* 1968, 19, 589–602.

Bulgaria. (1973). Instruction No. 0-27 of 20 April 1973 of the Ministry of Public Health to regulate the artificial interruption of pregnancy. *Dhurzhaven Vestnik,* 20 April. *International Digest of Health Legislation,* 24, 730–738.

Bulgaria. (1974). Amendment of 22 February 1974 to Instruction No. 0-27 of the Ministry of Public Health. *Dhurzhaven Vestnik,* 22 February, 7. *International Digest of Health Legislation,* 25, 541–548.

Bulgaria. (1989). *Public health statistics annual*, 1989. Sofia: PHSA.

Bulgaria. (1994). The demographic situation in Bulgaria. Report prepared for the International Conference on Population and Development, Cairo, 5–13 September.

Bulgaria/Central Statistical Office. (1988). *Statistical yearbook, 1988* and earlier years Sofia: Central Statistical Office.

Bulgaria/Ministry of Health. (1990). Decree No. 2 concerning the conditions and order for pregnancy termination. *Darzhaven Vestnick,* No. 12, 9 February.

Bulgaria/National Statistical Institute. (1997). *Statistical reference book of the Republic of Bulgaria 1995* and earlier years. Sofia: Statistical Publishing and Printing House.

Buresova, A. (1991a). Czechoslovakia 1991: Abortion and contraception. *Planned Parenthood in Europe,* 20(2), 6–7.

Buresova, A. (1991b). Sex education in Czechoslovakia. *Planned Parenthood in Europe,* 20(1), 12–13.

Burger, E. J., Fried, M. E., and Twigg, J. (1998). From assurance to insurance in Russian health care: The problematic transition. *American Journal of Public Health,* 88, 755–758.

Büttner, T., and Lutz, W. (1990). Estimating fertility responses to policy measures in the German Democratic Republic. *Population and Development Review,* 16, 539–555.

Callahan, D. (1970). *Abortion: Law, choice, and morality.* New York: Macmillan.

Camay, M. (1989). Foreword to the *Kanun.* In *Code of Leke Dukagjini.*

Campbell, R. (1974). *The Soviet-type economics* (3rd ed). Boston: Houghton Mifflin.

Campeanu, D. (1994). Voices of Romanian women: The status of Romanian women between 1980 and 1994. Unpublished report. Bucharest: Society for Interbalkan Cooperation of Romanian Women.

Caramelea, V. V. (1968). Romanian social and cultural anthropology and the study of man in industrial society (In French). *Annuaire Romain D'Anthropologie,* 5, 99–102.

Carlson, B. A. (1994). *Women and gender in countries in transition: A UNICEF perspective.* New York: UNICEF.

Carlson, E. D. (1998). Personal communication, 3 April.

Carlson, E. D., and Omori, M. (1998). Fertility regulation in a declining state socialist economy: Bulgaria, 1976–1995. *International Family Planning Perspectives*, 24, 184–187.

Carlson, E., and Tsvetarsky, S. (1996). Rising Bulgarian infant mortality: Fact or artifact? In G. Masuy-Stroobant and C. Gourbin (Eds.), *Sante et Mortalite des Enfants en Europe.* Louvain: Academia Bruylant/L'Harmattan, 469–473.

Carter, F. W., and Zagar, M. (1977). Postwar internal migration in Southeastern Europe. In L. A. Kosinski (Ed.), *Demographic developments in Eastern Europe.* New York: Praeger, 208–244.

Cassileth, B. R., Vlassov, V. V., and Chapman, C. C. (1995). Health care, medical practice, and medical ethics in Russia today. *Journal of the American Medical Association,* 273, 1569–1573.

Ceausescu, N. (1985). Decree No. 411, 26.XII. In *Official Bulletin* No. 76, 26 December.

Ceausescu, N. (1986). *Der Spiegel* (In German), 20 October, 217–218.

Center for Formation of Sexual Culture. (1977). *Framework for sexuality education for Russian youth* (In English and Russian). Yaroslavl: CFSC.

Center for the Advancement of Women. (1995). Directory of Women's Organizations and Institutions in Poland. Warsaw: Author.

Centers for Disease Control, Czech Statistical Office, Factum non Fabula, and World Health Organization Collaborating Center for Perinatal Medicine/Institute for the Care of Mother and Child. (1995). *1993 Czech Republic reproductive health survey.* Atlanta: Centers for Disease Control and Prevention.

Ceterchi, I., Zlatescu, V. D., Copil, I. M., and Anca, P. (1974). *The law and population growth in Romania* (In French). Bucharest: National Commission of Demography.

Ceterchi, I., Zlatescu, V. D., Copil, I. M., and Anca, P. (1975). *Romanian legislation on the population growth and its demographic effects.* Bucharest: National Demographic Commission, Law and Population Books Series, No. 12.

Chalidze, V. (1977). *Criminal Russia: Essays in crime in the Soviet Union.* New York: Random House.

Chao, P. (1977). *Women under communism.* New York: General Hall.

Chaze, S. (1992). Promotion of family planning in Romania (In French). Unpublished report. Tirgu-Mures: SECS.

Chazov, E. I. (1988). *Current Digest of the Soviet Press*, 40(27), 6–9.

Chazov, E. I. (1992). *Health and power: Remembrances of a "Kremlin physician"* (In Russian). Moscow: Izdatel'Stvo "Novosti."

Chen, L. C., Wittgenstein, F., and McKeon, E. (1996). The upsurge of mortality in Russia. *Population and Development Review,* 22, 517–530.

Chernev, T., and Stamencova, R. (1994). Bridging the information gap. *Entre Nous*, (26/27), 7–8.

Chervyakov, V. A. (1996). Survey supports arguments to start sexuality education in Russia. *SIECUS Report,* 25(2), 8–9.

Chervyakov, V. A. (1997). In-school sex education for teenagers in the Russian Federation: A report on the results of a sociological survey of the sexuality of secondary school students in 8 regions of Russia. Moscow: Unpublished paper.

Chervyakov, V. A. (1998). Personal communication, 9 March.

Chiochirca, M. (1994). Cited in *Evenimentul Zilei,* 19 March.

Choices. (1996). Abortion law in Bavaria restricted. *Choices,* 25(2), 30.

Cichocka, E. (1993). Do they deprave or do they waste money? (In Polish). *Gazeta Wyborcza,* July, (176).

Cichocka, E. (1994a). Double meaning of law (In Polish). *Gazeta Wyborcza,* (75).

Cichocka, E. (1994b). Abortion still forbidden (In Polish). *Gazeta Wyborcza,* (205).

Cichomski, B., and Sawinski, Z. (1993). *Polish General Social Survey; Structure of aggregated data, 1992/93* (In Polish). Warsaw: Institute for Social Studies, University of Warsaw.

Clem, R. S. (1977). Recent demographic trends among Soviet nationalities and their implications. In G. W. Simmons (Ed.), *Nationalism in the USSR and Eastern Europe.* Detroit: University of Detroit Press.

Clem, R. S. (1980). Regional patterns of population change in the Soviet Union 1959–1979. *The Geographical Review,* 70, 137–156.

Clements, B. E. (1979). *Bolshevik feminist: The life of Aleksandra Kollontai.* Bloomington: Indiana University Press.

Coale, A. J. (1965). Factors associated with the development of low fertility: An historic summary. In *Proceedings of the World Population Conference,* Belgrade. New York: United Nations, 1967, 205–209.

Coale, A. J. (1969). The decline of fertility in Europe from the French Revolution to World War II. In S. J. Behrman, L. Corsa, and R. Freedman (Eds.), *Fertility and family planning: a world view.* Ann Arbor: University of Michigan Press, 3–24.

Coale, A. J., Anderson, B. A., and Härm, E. (1979). *Human fertility in Russia since the nineteenth century.* Princeton: Princeton University Press.

Coale, A. J., and Watkins, S. C. (Eds). (1986). *The decline of fertility in Europe: The revised Proceedings of the Conference on the Princeton-European fertility project.* Princeton: Princeton University Press.

Cohen, R. (1998). *Hearts grown brutal.* New York: Random House.

Conrad, C., Lechner, M., and Werner, W. (1996). East German fertility: Crisis or adaptation? *Population and Development Review,* 22, 331–358.

Cooper, H. J. (1994). Sex in Albania. *Illyria Newspaper,* 7–9 February.

Coroi, V., and Gorgos, C. (1980). *Social Medicine* (In Romanian). Bucharest: Editura Didactica si Pedagogica.

Corrin, C. (1994). Women's politics in "Europe" in the 1990s. *Women's Studies International Forum,* 2, 289–297.

Council of Europe. (1998). *Recent demographic developments in Europe 1998* and earlier years. Strasbourg: Council of Europe Publications.

Cressin, A. (1977). Past and present sex education in Romania. In M. Kozakiewicz (Ed.), *Family planning and sex education in socialist countries.* Warsaw: Polish Family Planning Association, 96–107.

Cristea, A., Gheorghiu, N. N., Moraru, I., Quai, I., Roman, I., and Fischoff, S. (1961). Early and late sequelae of abortion on demand (In Romanian). *Obstetrica si Ginokolgia,* 8, 489.

Croatia/National Committee for Preparation of International Conference on Population and Development. (1994). National Report by the Republic of Croatia. Unpublished report.

Czech Committee for UNICEF. (1994). The situation of women in the Czech Republic. In B. A. Carlson (Ed.), *Women and gender in countries in transition: A UNICEF perspective.* New York: UNICEF.

Czech Republic. (1994). National Report to the International Conference on Population and Development, Cairo, 5–13 September.

Czech Republic/Czech Statistical Office. (1997). *1995 Statistical yearbook* and earlier years. Prague: Statistical Office.

Czech Republic/Institute for Health Statistics. (1995). *Abortions in 1994* and earlier years (In Czech). Prague: Institute for Health Statistics.

Czech Republic/ Central Statistical Office. (1997). *Demographic developments in the Czech Republic in 1995* and earlier years. (In Czech). Prague: Central Statistical Office.

Czechoslovakia. (1959). Interruption of pregnancy. *International Digest of Health Legislation,* 1959, 10, 283–292; 1962, 13, 491–492.

Czechoslovakia/Federal Statistical Office. (1992). *Survey on family planning, 1991.* Prague: Federal Statistical Office.

Czechoslovakia/Federal Statistical Office. (1993). *Population change in the Czech and Slovak Federal Republic in 1991* and earlier years (In Czech). Prague: Federal Statistical Office.

Dan (Croatia). (1995). 31 December.

Danek, U. (1988). Parenthood and women's health (In Polish). In *Polish women in the 80s.* Warsaw: Niezalezna Oficyna Wydawnicza.

Darsky, L. E., and Dworak, N. B. (1993). Kyrgystan: Fertility indicators and characteristics of the potential market for modern contraceptives. Unpublished report to the U.S. Agency for International Development.

David, H. P. (1970). *Family planning and abortion in the socialist countries of Central and Eastern Europe,* New York: The Population Council.

David, H. P. (1972). Abortion in psychosocial perspective. *American Journal of Orthopsychiatry,* 4, 261–268.

David, H. P. (1974). Abortion and family planning in the Soviet Union: Public policies and private behavior. *Journal of Biosocial Science,* 6, 417–426.

David, H. P. (1981). Abortion policies. In J. E. Hodgson (Ed.), *Abortion and sterilization: Medical and social aspects.* New York: Academic Press, 1–40.

David, H.P. (1982). Eastern Europe: Pronatalist policies and private behavior. *Population Bulletin,* 36(6), 1–47.

David, H. P. (1985). Postabortion and postpartum psychiatric hospitalization. In R. Porter and M. O'Connor (Eds.), *Abortion: Medical progress and social implications.* Ciba Symposium No. 45. London: Pitman, 150–161.

David, H. P. (1990). Romania ends compulsory childbearing. *Population Today,* 18(3), 4–10.

David, H. P. (1991). Poland, abortion and the church. *Population Today,* 9(7/8), 5.

David, H. P. (1992a). Abortion in Europe, 1920–1991: A public health perspective. *Studies in Family Planning,* 23, 1–22.

David, H. P. (1992b). Born unwanted: Long term developmental effects of denied abortion. *Journal of Social Issues,* 48, 163–181.

David, H. P. (1993). Population policy and reproductive behavior: Incentives and disincentives. In L. Severy (Ed.), *Advances in population: Psychosocial perspectives.* London: Kingsley, Vol. 1, 1–33.

David, H. P. (1994a). Abortion legislation in Europe, 1991–1993. *Entre Nous,* (25), 13.

David, H. P. (1994b). Reproductive rights and reproductive behavior: Clash or convergence of private values and public policies. *American Psychologist,* 49, 343–349.

David, H. P. (1997). Postabortion psychological responses. In E. Ketting and J. Smit (Eds.), *Abortion matters.* Utrecht: Stimezo Nederland, 341–347.

David, H. P., and Baban, A. (1996). Women's health and reproductive rights: Romanian experience. *Patient Education and Counseling,* 28, 235–245.

David, H. P., Dytrych, Z., Matejcek, Z., and Schüller, V. (Eds.). (1988). *Born unwanted: Developmental effects of denied abortion.* Prague: Avicenum; New York: Springer.

David, H. P., Fleischhacker, J., and Höhn, C. (1988). Abortion and eugenics in Nazi Germany. *Population and Development Review,* 14, 81–112; *Zeitschrift für Bevölkerungswissenschaft,* 1990, 16, 259–289.

David, H. P., and Friedman, H. L. (1973). Psychosocial research in abortion: A transnational perspective. In H. J. Osofsky and J. D. Osofsky (Eds.), *The abortion experience.* Hagerstown: Harper and Row, 310–337.

David, H. P., and Johnson, R. L. (1977). Fertility regulation in early childbearing years: Psychosocial and psychoeconomic aspects. *Preventive Medicine,* 6, 52–64.

David, H. P., and McIntyre, R. J. (1981). *Reproductive behavior: Central and Eastern European experience.* New York: Springer Publishing Company.

David, H. P., and Pierotti, D. (Eds.). (1993). *Demographic and social effects of population policies in Europe.* Copenhagen: World Health Organization Regional Office for Europe.

David, H. P., and Rademakers, J. (1996). Lessons from the Dutch abortion experience. *Studies in Family Planning,* 27, 341–343.

David, H. P., and Titkow, A. (1994). Abortion and women's rights in Poland, 1994. *Studies in Family Planning,* 25, 239–242.

David, H. P., and Wright, N. H. (1971). Abortion legislation: The Romanian experience. *Studies in Family Planning,* 2, 205–210.

David, H. P., and Wright, N. (1974). Restricting abortion: The Romanian experience. In H. P. David (Ed.), *Abortion research: International experience.* Lexington, MA: Lexington Books, 217–225.

Davis, C., and Feshbach, M. (1980). Rising infant mortality in the USSR in the 1970s. *International Population Reports,* Series P-95, No. 74.

Deak, I. (1994). Post-post-communist Hungary. *New York Review,* 11 August, 33–38.

Demeny, P. (1968). Early fertility decline in Austria-Hungary: A lessen in demographic transition. *Daedalus,* 97, 502–522.

Demoskop. (1991). Research report. Warsaw: Author.

Demoskop. (1992). Research report. Warsaw: Author.

Demosta, 1979, 12(2), 38–40; XVIII.

Dhimgjoka, M. (1996). Albanian party claims win; opposition seeks new vote. *The Washington Post,* 27 May 1996.

Dixon-Mueller, R. (1993). *Population policy and women's rights.* Westport, CT: Praeger.

Djusubalieva, T. (1994). Reproductive health in Kazakhstan: A new approach. *Planned Parenthood in Europe,* 23(3), 17–19.

Djusubalieva, T. (1996). A measurable impact. *Integration,* (49), 22–25.

Dmitrieva, E. (1996). Exercising reproductive rights in Russia. *Choices,* 25(2), 18–19.

Dmitrieva, E. (1999). The power of the media in Russia. *Choices,* 27(1), 14–15.

Dolian, G. (1997). A pilot survey on abortion among young Armenian women. *Entre Nous,* (36/37), 12–13.

Dolian, G., Lüdicke, F., Katchatrian, N., and Morabia, A. (1998). Contraception and abortion in Armenia: A critical need for family planning programs in Eastern Europe. *American Journal of Public Health*, 88, 803–805.

Domanski, H. (1992). *The grateful slave? Inequality between men and women in Poland* (In Polish). Warsaw: Institute of Philosophy and Sociology.

Donkov, K. (1991). Bulgaria. In J. L. Rallu and A. Blum (Eds.), *European population/Demographie europeenne.* Paris: Institut National D'Etudes Demographiques, Vol. 1, 257–275.

Doolotova, A., and Ahmad, J. S. (1996). Increased use of pills. *Integration,* (49), 7–12.

Dorbritz, J. (1992a). The situation of population science in the former GDR and the problem of the prognosis of demographic processes (In German). *Berliner Journal für Soziologie,* (3/4), 431–444.

Dorbritz, J. (1992b). Nuptuality, fertility, and family ways of life in social transformation: Transition to a new demographic regime in East Germany? (In German). *Zeitschrift für Bevölkerungswissenschaft,* 18, 167–196.

Dorbritz, J. (1994a). Trends in births, marriage, and divorces in the former socialist countries in a time of social change (In German). Unpublished paper. Wiesbaden: Federal Institute for Population Research.

Dorbritz, J. (1994b). Changes in the nuptuality and fertility in the former GDR. Paper presented at the American Enterprise Institute Conference on Population Prospects for United Germany, Washington, DC: 15 December.

Dorbritz, J. (1994c). 1994 Report on the demographic situation in Germany (In German). *Zeitschrift für Bevölkerungswissenschaft,* 19, 393–473.

Dorbritz, J. (1995). Demographic shocks in the former German Democratic Republic—the decline in the birth, marriage and divorce rates in the context of demographic changes in East Europe. Unpublished paper. Wiesbaden: Federal Institute for Population Research.

Dorbritz, J. (1997a). Demographic change in the new federal lands and its societal effects (In German). Wiesbaden: Bundesinstitut für Bevölkerungsforschung, 25 February.

Dorbritz, J. (1997b). Demographic change in East Germany: Development and possible explanations (In German). *Zeitschrift für Bevölkerungswissenschaft,* 22, 239–268.

Dorbritz, J. (1998). Personal communication, 2 June.

Dorbritz, J., and Schwarz, K. (1996). Childlessness in Germany—a mass phenomenon? Analyses of manifestations and causes (In German). *Zeitschrift für Bevölkerungswissenschaft,* 21, 231–261.

Dorman, S. (1993). More access to contraception? Russian city surveyed. *Population,* 21(3), 5, 10.

Dornberg, J. (1974). Eastern Europe: Programming the populations. *Ms,* November, 128–130.

Draghici, D. (1997). Building NGO capacity for quality reproductive health programs in Romania. Unpublished paper. Washington, DC: Center for Development and Population Activities.

Dragoti, E. (1996). Ancient crimes return to haunt Albania. *Psychology International,* 7(4), 1–3.

Drakulic, S. (1993). Women and the new democracy in the former Yugoslavia. In N. Funk and M. Mueller (Eds.), *Gender politics and post-communism: Reflections from Eastern Europe and the former Soviet Union.* New York: Routledge, 123–130.

Dreijmanis, J. (1972). The Soviet pro-natalist policy. *Population Review,* 16, 18–31.

Duch, D. (1991). Socioeconomic context of procreational attitudes in Polish society (In Polish). Unpublished paper.

Duch, D. (1994). Marriage, sex, and procreation: Sociological analysis. Warsaw: Institute of Philosophy and Sociology, Polish Academy of Sciences, Unpublished Ph.D. dissertation.

Dumani, B. (1993). Fertility and family planning in Albania. *Planned Parenthood in Europe,* 22(1), 17–19.

Dziewonski, K. (1977). Population problems in Polish regional planning. In H. L. Kostanick (Ed.), *Population and migration trends in Eastern Europe.* Boulder, CO: Westview Press, 81–97.

Earle, J. E. (1994). Unemployment and policies in Romania. *Sfera Politica,* 14, 24–26.

Eberstadt, N. (1981). The health crisis in the USSR. *New York Review of Books,* 20 February.

Eberstadt, N. (1990). Health and mortality in Eastern Europe, 1965–1985. *Communist Economies,* 2, 347–371.

Eberstadt, N. (1993). Mortality and the fate of communist states. *Communist Economies and Economic Transformation,* 5, 499–517.

Eberstadt, N. (1994a). Demographic shocks after Communism: Eastern Germany, 1989–1993. *Population and Development Review,* 20, 137–152.

Eberstadt, N. (1994b). Demographic shocks in Eastern Germany, 1989–1993. *Europe-Asia Studies,* 46, 519–533.

Eberstadt, N. (1994c). Demographic shocks in Eastern Germany, 1989–1994. Paper presented at the AEI Conference on Population Prospects for United Germany, Washington, DC, 15 December.

Ebert, M. (1981). Participation of women in economic development, social policies, and reproductive behaviour of women in the German Democratic Republic. Paper presented at the General Conference of the International Union for the Scientific Study of Population, Manila, Philippines, 1 December.

Economic and Social Commission for Asia and the Pacific Kyrgystan sees fertility declining. *Population Headliners,* March–April, 3.

The Economist. (1994a). 15 January.

The Economist. (1994b). 9 July, 50.

The Economist. (1994c). 12 November.

The Economist. (1995). 12 August.

The Economist. (1996a). 4 May.

The Economist. (1996b). 23 November.

The Economist. (1997a). 4 January.

The Economist. (1997b). 16 August, 40–42.

The Economist. (1997c). 6 December, 52–53.

The Economist. (1998a). A survey of the Balkans. 24 January.

The Economist. (1998b). 31 January.

The Economist. (1998c). 17 October.

The Economist. (1998d), 24 October.

The Economist. (1998e). 12 December.

The Economist. (1999). 16 January.

Einhorn, B. (1993). *Cinderella goes to market; Citizenship, gender and women's movements in East Central Europe.* London: Verso.

Engel, B. A., and Posadskaya-Vanderbeck, A. (1998). *A revolution of their own: Voices of women in Soviet history.* Boulder, CO: Westview Press.

Engels, F. (1972). *The origin of the family, private property and the state.* New York: International Publishers and Pathfinder Press. (First published in German in 1884).

Engels, F. (1983). A letter to K. Kalitsky of 1 February 1881. In A. Drummond (Ed.), *Letters on capital by Karl Marx and Friedrich Engels.* London: New Park Publications.

Engelstein, L. (1992). *The keys to happiness: Sex and the search for modernity in Fin-de-Siecle Russia.* Ithaca: Cornell University Press.

Entwistle, B., and Kozyreva, P. (1997). New estimates of induced abortion in Russia. *Studies in Family Planning,* 28, 14–23.

Erlanger, S. (1992). Medical care in Russia seems near a collapse. *The New York Times,* 9 May.

Farnsworth, B. (1980). *Alexandra Kollontai and the Bolshevik Revolution.* Stanford: Stanford University Press.

Federal Republic of Germany/Federal Statistical Office. (1993). *Special edition with contributions for the territory of the former GDR,* Vol. 3, 1946–1987 (In German). Wiesbaden: Federal Statistical Office.

Federal Republic of Germany/ Federal Statistical Office. (1994). *On the economic and social situation in the new federal states* (In German). Wiesbaden: Federal Statistical Office, March.

Federal Republic of Germany/Federal Statistical Office. (1996). *Population and employment, Special Series I, Vol. 1, district and population, 1994.* Wiesbaden: Federal Statistical Office.

Federation for Women and Family Planning. (1994). *Agenda.* Warsaw: Federation for Women and Family Planning.

Federation for Women and Family Planning. (1996). *Effects of the anti-abortion law: Report No. 2.* Warsaw: Federation for Women and Family Planning.

Feischmidt, M., Magyari-Vincze, E., and Zentai, V. (Eds.) (1997). *Women and men in East European tradition.* Cluj-Napoca: Editura Fundatiei Pentru Studii Europene.

Ferencovic-Nikasinovic, S., Pavlov, M., Radovanovi, Z., and Simic, I. (1975–76). Abortions by women of various nationalities in Vojvodina Province (In Serbian). *Stanovnistvo,* 13/14, 155–157. (Note: actually published in 1978.)

Feshbach, M. (1982a). *Demography and Soviet society.* Armonk, NY: Sharpe.

Feshbach, M. (1982b). The Soviet Union: Population trends and dilemmas. *Population Bulletin,* 37(3), 3–44.

Feshbach, M., and Friendly, A., Jr. (1992). *Ecocide in the USSR: Health and nature under siege.* New York: Basic Books.

Field, D. A. (1996). Communist morality and the meaning of life in post-Stalinist Russia, 1953–1964. Unpublished doctoral dissertation. Ann Arbor: University of Michigan Press.

Field, M. G. (1956). The relegalization of abortion in the Soviet Union. *New England Journal of Medicine,* 255, 421–427.

Field, M. G. (1990). Noble purpose, grand design, flawed execution, mixed results: Soviet socialized medicine after seventy years. *American Journal of Public Health,* 80, 144–145.

Field, M. G. (1994). Postcommunist medicine: Morbidity, mortality, and the deteriorating health situation. In J. R. Millar and S. L. Wolchik (Eds.), *The Social Legacy of Communism.* Washington, DC: W. Wilson Center Press, 178–195.

Filipov, D. (1992). Fertility trends in Bulgaria (1960–1990) (In Bulgarian). *Population* (Sofia) 7, 72–84.

Finkle, J. L., and McIntosh, A. (1980). Policy responses to population stagnation in developed society. In A. A. Campbell (Ed.), *Social, economic, and health aspects of low fertility.* Bethesda: National Institutes of Health, 275–296.

Finn, P. (1999). War clouds Kosovo adversaries' view of the future. *The Washington Post,* 1 January.

Fischer, M. E. (1985). Women in Romanian politics: Elena Ceausescu, pronatalism, and the promotion of women. In S. L. Wolchik and A. G. Meyer (Eds.), *Women state, and party in Eastern Europe.* Durham, NC: Duke University Press, 121–137.

Fischer, M. E., and Harsanyi, D. P. (1994). From tradition and ideology to elections and competition: The changing status of women in Romanian politics. In M. Rueschemeyer (Ed.), *Women in the politics of postcommunist Eastern Europe.* Armonk, NY: Sharpe, 201–223.

Fisher, N. (1992). New on the resume: Sterilization. *The Washington Post,* 27 May.

Fleischhacker, J. (1994). In the West stable—In the East unstable; the East German population in flux (In German). *Forum: Demographie und Politik,* No. 6, 31–52.

Fluss, S. S. (1994). Reproductive health legislation in Central and Eastern Europe: New developments. *Entre Nous,* 25, 9–10.

Fodor, E. (1994). The political woman? Women in politics in Hungary. In M. Rueschemeyer (Ed.), *Women in the politics of post-communist Eastern Europe.* Armonk, NY: Sharpe, 171–199.

Fodor, E. (1997). Gender in transition: Unemployment in Hungary, Poland, and Slovakia. *European Politics and Societies,* 11, 470–500.

Folaron, G. (1994). Suicide after abortion (In Polish). *Zycie Warszawy,* 25 March.

Fong, M. S. (1993). The role of women in rebuilding the Russian economy. *Studies in Economics in Transition,* Paper No. 10. Washington, DC: The World Bank.

Forrai, J. (1996). Adolescent sexuality in Hungary today. *Choices,* 25(3), 33–36.

Frauenreport '90. (1990). Berlin: Verlag der Wirtschaft.

Frejka, T. (1980). Fertility trends and policies: Czechoslovakia in the 1970s. *Population and Development Review,* 6, 65–93.

Frejka, T. (1983). Induced abortion and fertility: A quarter century of experience in Eastern Europe. *Population and Development Review,* 9, 494–520.

Frejka, T. (1985). Induced abortion and fertility. *International Family Planning Perspectives,* 11, 125–129.

Frejka, T. (1998). Personal communication, 18 June.

Frerks, S. (1938). *German population policy, No. 5* (In German). Berlin: Terramare.

Frey, M. (1992). *Women and unemployment* (In Hungarian). Budapest: Kossuth.

Fukalova, D., and Uzel, R. (1975). Attitudes to contraception among women obtaining induced abortions (In Czech). *Ceskoslovenska gynekologie,* 40, 299–310.

Funk, N. (1993). Abortion and German unification. In N. Funk and M. Mueller (Eds.), *Gender politics and post-communism: Reflections from Eastern Europe and the former Soviet Union,* New York: Routledge, 194–200.

Funk, N., and Mueller, M. (Eds.) (1993). *Gender politics and post-communism: Reflections from Eastern Europe and the Former Soviet Union,* New York: Routledge.

Furedi, A. (1993). Review of "Progress postponed: Abortion in Europe in the 1990s. *Planned Parenthood in Europe,* 22(2), 37.

Fuszara, M. (1993). Abortion and the formation of the public sphere in Poland. In N. Funk and M. Mueller (Eds.), *Gender politics and post-communism: Reflections from Eastern Europe and the Former Soviet Union,* New York: Routledge, 241–252.

Fuszara, G., and Tarnowska, M. (1995). Women—the category of "especially sheltered employees in social practice." In A. Titkow and H. Domanski (Eds.), *What does it mean to be a woman in Poland* (In Polish). Warsaw: Institute of Philosophy and Sociology, 119–134.

Gadasina, A. (1997). Struggling to survive in Russia. *Planned Parenthood Challenges,* 1/2, 40–42.

Gal, S. (1994). Gender in the post-socialist transition: The abortion debate in Hungary. *East European Politics and Societies,* 8, 256–286.

Geiger, H. K. (1968). *The family in Soviet Russia.* Cambridge: Harvard University Press.

Gens, A. B. (1927). *Abortion in 1925* (In Russian). Moscow: Statizdat.

Germain, A., and Kyte, R. (1995). *The Cairo consensus: The right agenda for the right time.* New York: International Women's Health Coalition.

German Democratic Republic. (1950). Law on the protection of mother and child and the rights of the woman, 27 September, 1950. *Gesetzblatt der Deutschen Demokratischen Republik,* 1 October, No. 111, 1037–1041. *International Digest of Health Legislation,* 1952, 4, 58–60.

German Democratic Republic. (1965). Instructions for Paragraph 11 of the law for the protection of mother and child and the rights of the woman, 5 June. *Gesetzblatt der Deutschen Demokratischen Republik,* 22 October, 1965. (VU M No. 23/21). Translated in David, H. P. (1970). *Family planning and abortion in the socialist countries of Central and Eastern Europe.* New York: The Population Council, 244–248.

German Democratic Republic. (1972). Law of 9 March 1972, *Berliner Zeitung,* 70, 3. *Gesetzblatt der Deutschen Demokratischen Republik,* 15 March (5), 89–90. *International Digest of Health Legislation,* 23, 767–770.

German Democratic Republic/Central Statistical Office. (1989). *Statistical yearbook* and earlier years. Berlin: Central Statistical Office.

German Democratic Republic/Institute for Social Hygiene. (1979). *Public health of the German Democratic Republic* and earlier volumes (In German). Berlin: Institute for Social Hygiene.

Ghetau, V. (1978). The evolution of fecundity in Romania (In French). *Population,* 33, 426–439.

Ghetau, V. (1979). *Demographic perspectives* (In Romanian). Bucharest: Editura Stiintifica si Enciclopedica.

Ghetau, V. (1998). *Maternal mortality and abortion in Romania. 1990–1997.* Bucharest: United Nations Fund for Population Activities.

Gilberg, T. (1979). Rural transformation in Romania. In I. Volgyes (Ed.), *The peasantry of Eastern Europe.* New York: Pergamon.

Ginic, I. (1971). Fertility and birth control by age of married women (English summary). *Stanovnistvo,* 9, 206–222.

Gjonca, A., Wilson, C., and Falkingham, J. (1997). Paradoxes of health transition in Europe's poorest country: Albania 1950–1990. *Population and Development Review,* 23, 585–609.

Glass, D. V. (1940). *Population: Policies and movements in Europe.* Oxford: Oxford University Press.

Goble, P. (1997). Forget the former Soviet Union. *RFL/RE Newsline,* 98, 19 August.

Goebbels, J. (1929). Cited in Mosse, G. L. (1968: 41).

Gold, R. B. (1996): Poland legalizes abortion again. *Washington Memo,* 20 December.

Goldberg, G. S., and Kremen, E. (Eds.) (1990). *The feminization of poverty: Only in America?* Westport, CT: Greenwood Press.

Goldberg, H. (1997). Personal communication, 29 October.

Goldman, W. Z. (1993). *Women, the state, and revolution: Soviet family policy and social life, 1917–1936.* New York: Cambridge University Press.

Golemanov, N., and Krustev, R. (1973). Social hygiene problems of hormonal contraception. II. The place of hormonal contraception in the prophylaxis of unwanted pregnancy (In Bulgarian). *Khigiena I Zdraveopazvane,* 16, 531–536.

Golod, S. (1993). Sex and young people. In I. Kon and J. Riordan (Eds.), *Sex and Russian Society.* Bloomington: Indiana University Press, 135–151.

Gorbachev, M. (1987a). *Perestroika: New thinking for our country and the world.* New York: Harper and Row, 107.

Gorbachev, M. (1987b). *Perestroika* (In Russian). Moscow: Political Literature, 117.

Grad, J. (1993). Customs and the decay of "real socialism." *Centennial Review,* 37, 73–92.

Grassel, H. (1976). Facilitating cooperation between physician and educator in sex education (In German). *Arztliche Jugendkunde,* 67, 129–145.

Grassel, H., and Bach, K. H. (1979). *Child and adolescent sexuality* (In German). Berlin: Deutscher Verlag der Wissenschaften.

Grassel, H., and Radlbeck, K. G. (1975). Sex education of mentally handicapped children and youth (In German). *Die Sonderschule,* 20, 79–88.

Gray, F. Du Plexis. (1989). *Soviet women: Walking the tight rope,* New York: Doubleday.

Grebesheva, I. (1992). Abortion and the problem of family planning in Russia. *Planned Parenthood in Europe,* 21(2), 8.

Grebesheva, I. (1994). Russian Family Planning Association. *Planned Parenthood in Europe,* 23(2), 35–36.

Grebesheva, I. (1996). Interview with E. Dmitrieva. *Choices,* 25(3), 25–26.

Greenhalgh, S. (1994). Anthropological contributions to fertility theory. New York: The Population Council, Working Paper No. 64.

Griffiths, T. (1996). Bulgaria's opposition candidate wins presidency. *The Washington Post,* 4 November.

Grigoroiu-Serbanescu, M. (1990). How a restrictive abortion policy affected child mental health. *Entre Nous,* (14/15), 6.

Guzevaty, Y. (1966). Population problems in developing countries. *International Affairs,* September.

Guzevaty, Y. (1974). The socio-economic determinants of demographic processes. *Serial Sciences Today,* 1(26), 76–83.

Haberlova, V., Alan, J., and Kovarik, J. (1987). Young spouses after five years of marriage. *Bulletin of the Research Institute of Social Development and Labour,* 4, 27–29.

Hamand, J. (1989). GDR: A model for the East? *People,* 16(3), 7–9.

Harmsen, H. (1936). Critical survey of contraception and contraceptive methods in Germany. In M. Sanger, H. M. Stone, and R. L. Dickenson (Eds.), *The practice of contraception.* Proceedings of the 7th International Birth Control Conference, Zurich, Switzerland, September, 1930. Baltimore: Williams and Wilkins.

Harmsen, H. (1950). Notes on abortion and birth control in Germany. *Population Studies,* 3, 402–405.

Harsanyi, D. P. (1993). Women in Romania. In N. Funk and M. Mueller (Eds.), *Gender politics and post-communism: Reflections from Eastern Europe and the former Soviet Union.* New York: Routledge, 39–52.

Hartmann, H. (1981). The unhappy marriage of Marxism and Feminism: Towards a progressive union. In L. Sargent (Ed.), *Women and revolution.* London: Pluto Press, 1–41.

Hata, K. (1996). Strengthening family planning. *Integration,* (49), 13–15.

Haub, C. (1981). Czechoslovakia. *Intercom,* 9(1), 13.

Haub, C. (1994). Population change in the former Soviet republics. *Population Bulletin,* 49(4), 1–52.

Haub, C. (1995). Demographic change and the status of women in Belarus. Report prepared for the World Bank. Washington, DC: Population Reference Bureau.

Hauser, E., Heyns, B., and Mansbridge, J. (1993). Feminism in the interstices of politics and culture: Poland in transition. In N. Funk and M. Mueller (Eds.), *Gender politics and post-communism: Reflections from Eastern Europe and the former Soviet Union.* New York: Routledge, 257–273.

Hauser, P. M., and Duncan, O. (Eds.) (1959). *The study of population: An inventory and appraisal.* Chicago: University of Chicago Press.

Hausleitner, M. (1993). Women in Romania: Before and after the collapse. In N. Funk and M. Mueller (Eds.), *Gender politics and post-communism: Reflections from Eastern Europe and the Former Soviet Union.* New York: Routledge, 53–61.

Havelka, J. (1974). Population policy. In *Population policy in Czechoslovakia.* Prague: Orbis, 40–50.

Havelka, J. (1978). Population policy. In *Population policy in Czechoslovakia.* Prague: Orbis, 31–35.

Havelkova, H. (1993). A few prefeminist thoughts. In N. Funk and M. Mueller (Eds.), *Gender politics and post-communism: Reflections from Eastern Europe and the former Soviet Union.* New York: Routledge, 62–73.

Havranek, F. (1982). *Interruptio Graviditatis,* Prague: Avicenum.

Heer, D. M. (1965a). Abortion, contraception, and population policy in the Soviet Union. *Demography,* 2, 531–539.

Heer, D. M. (1965b). Abortion, contraception, and population. *Soviet Studies,* 17, 76–83.

Heer, D. M. (1968). The demographic transition in the Russian Empire and the Soviet Union. *Journal of Social History,* Spring, 193–240.

Heer, D. M., and Bryden, J. (1967). Family allowances and fertility in the Soviet Union. *Soviet Studies,* 18, 153–163.

Heilig, G., Büttner, T., and Lutz, W. (1990). Germany's population: Turbulent past, uncertain future, *Population Bulletin,* 45(40), 1–47.

Heitlinger, A. (1976). Pronatalist population policies in Czechoslovakia. *Population Studies,* 30, 123–135.

Heitlinger, A. (1979). *Women and state socialism: Sex inequality in the Soviet Union and Czechoslovakia.* Montreal: McGill-Queen's University Press.

Heitlinger, A. (1980). Marxism, feminism, and sex equality. In T. Yedlin (Ed.), *Women in Eastern Europe and the Soviet Union.* New York: Praeger, 9–20.

Heitlinger, A. (1987). *Reproduction, medicine and the socialist state.* London: Macmillan.

Heitlinger, A. (1993). The impact of the transition from Communism on the status of women in the Czech and Slovak republics. In N. Funk and M. Mueller (Eds.), *Gender politics and post-communism: Reflections from Eastern Europe and the former Soviet Union.* New York: Routledge, 95–108.

Heleniak, T. (1998). Personal communication, 17 July.

Henning, G. (1990). Cited by N. Funk (1993), Abortion and German unification. In N. Funk and N. Mueller (Eds.), *Gender politics and post-communism: Reflections from Eastern Europe and the former Soviet Union*. New York: Routledge, 194.

Henshaw, S. K. (1998). Abortion incidence and services in the United States, 1995–1996. *Family Planning Perspectives*, 30, 263–270, 287.

Henshaw, S. K. (1999). Personal communication, 6 January.

Henshaw, S. K., and Morrow, E. (1990). *Induced abortion: A world review, 1990 Supplement*, Washington, DC: Alan Guttmacher Institute.

Henshaw, S. K., Singh, S., and Haas, T. (1999). Recent trends in abortion worldwide. *Family Planning Perspectives*, 25, 44–48.

Hiatt, F. (1997). Progress, Russian style. *The Washington Post*, 28 September.

Hincu, G., and Gilca, B. (1996). Working with the media in Moldova. *Choices*, 25(3), 29–30.

Hirschler, I. (1958). *In the interest of women* (In Hungarian). Cited in David and McIntyre, 1981: 278.

Hirschler, I. (1961). The abortion situation in the People's Republic of Hungary (In German). In K. H. Mehlan (Ed.), *Internationale Abortsituation, Abortbekämpfung, Antikonzeption*. Leipzig: Thieme, 114–122.

Hitler, A. Cited in Potts et al., 1977: 381–382.

Hoagland, J. (1996). The Czechs' great leap. *The Washington Post*, 12 May.

Hockenos, P. (1996). Controversy over Croatian government's drive for bigger families. *San Francisco Chronicle*, 3 August.

Hockstader, L. (1994). Antiabortion activists backed by Americans launch drive in Russia. *The Washington Post*, 19 May.

Hockstader, L. (1995). For women, new Russia is far from liberating. *The Washington Post*, 1 September.

Hockstader, L. (1997a). Albanian dreams shatter in pyramid schemes' fall. *The Washington Post*, 2 February.

Hockstader, L. (1997b). Bulgaria slides toward economic collapse. *The Washington Post*, 9 February.

Hockstader, L. (1997c). A Balkan winter of discontents. *The Washington Post*, 19 February.

Hockstader, L. (1997d). Young Czechs confront change. *The Washington Post*, 3 December.

Hodann, M. (1934). *Germany*. London: International Group for the Investigation of Contraception, 5th issue, 22–25.

Hodann, M. (1937). *History of modern morals*. London: Heinemann.

Hoderka, J. F. (1988). Czechoslovakia: Abortion and social security reforms. *Journal of Family Law*, 27, 91–100.

Hoffman, G. W., and Hatchett, R. L. (1977). The impact of regional development policy on population distribution in Yugoslavia and Bulgaria. In H. L. Kostanick (Ed.), *Population and migration trends in Eastern Europe*. Boulder, CO: Westview Press, 99–124.

Hollander, D. (1997). In Post-Soviet Russia, fertility is on the decline; marriage and childbearing are occurring earlier. *Family Planning Perspectives*, 29, 92–94.

Hollander, P. (1973). *Soviet and American Society: A comparison*. New York: Oxford University Press.

Holt, A. (Ed.). (1977). *Selected writings of Alexandra Kollontai*. London: Allison and Busby, 148–149; New York: Praeger, 87–114.

Holzer, J. Z. (1984). Demographic highs in Poland: A comparative analysis (In Polish). *Studia Demograficzne*, (4), 55–69.

Holzer, J. Z. (1991). Poland. In J. L. Rallu and A. Blum (Eds.) *European Population, Vol. 1.* Paris: Institut National D'Etudes Demographiques, 177–190.

Holzer, J. Z. (1996). Personal communication, 14 August.

Holzer, J. Z., and Wasilawska-Trenkher, H. (1985). Poland. In V. Bodrova and R. Anker (Eds.), *Working women in socialist countries: The fertility connection.* Geneva: International Labor Office, 129–165.

Homolkova-Krickova (Ed.). (1933). *Mrs. Charlotte G. Masaryk* (In Czech). Brno: Vydavatelsky odbor U.S.J.U., 212.

Höpflinger, F. (1984). Marriages and births among teenagers (In German). *Zeitschrift für Bevölkerungswissenschaft,* 2, 169–192.

Höpflinger, F. (1987). *Change in family formation in Western Europe* (In German). Frankfurt: Campus.

Hopulele-Petri, L., and de Graaf, P. (1992). Current contraceptive use in Romania. *Planned Parenthood in Europe,* 21, 20–22.

Hord, C., David, H. P., Donnay, F., and Wolf, M. (1991). Reproductive health in Romania: Reversing the Ceausescu legacy. *Studies in Family Planning,* 22, 231–240.

Horna, J. L. A. (1978). Current literature on the position and roles of women in Czechoslovakia. *Canadian Slavonic Papers,* 20, 78–90.

Horne, S. (1999). Domestic violence in Russia. *American Psychologist,* 54, 55–61.

Horowitz, R. T. (1979). Jewish immigrants to Israel: Self-reported powerlessness and alienation among immigrants from the Soviet Union and North America. *Journal of Cross-Cultural Psychology,* 10, 366–374.

Huinink, J. (1995). Family development and household establishment in the GDR: From the traditional model to instrumental life planning? (In German). In B. Nauck (Ed.), *Family and life course in societal change* Stuttgart: Enke, 39–55.

Hungary. (1956). Order No. 1047/1956 of the Council of Ministers of 3 June 1956 on the regulation of interruption of pregnancy and Ordinance No. 2/1956 of 24 June 1956. *International Digest of Health Legislation,* 1958, 9, 536–540.

Hungary. (1973). Decision No. 1040 of 18 October, 1973 of the Council of Ministers on the Tasks of Population Policy. *Magyar Kozlongy,* No. 71, 777–778. *International Digest of Health Legislation,* 1974, 25, 332–341.

Hungary. (1977). Instruction No. 6 of 16 February of the Minister of Health on the use and marketing of intrauterine contraceptives. *International Digest of Health Legislation,* 28, 609.

Hungary. (1982). Ordinance No. 3 of 10 February of the Minister of Health on the examination of applications for the termination of pregnancy. *International Digest of Health Legislation,* 33, 494.

Hungary. (1986). Ordinance No. 3 of 23 July of the Minister of Health on the examination of applications or the termination of pregnancies. *International Digest of Health Legislation,* 1987, 38, 56.

Hungary. (1987). Ordinance No. 12 of 19 August of the Minister of Health on artificial sterilization. *International Digest of Health Legislation,* 1988, 39, 80–81.

Hungary. (1988). Ordinance No. 76 of 3 November of the Council of Ministers on the termination of pregnancy. *International Digest of Health Legislation,* 1989, 40, 595–596.

Hungary. (1993). Law No. 79 of 17 December 1992 on the protection of the life of the fetus. *International Digest of Health Legislation,* 44, 614–615.

Hungary. (1994). National report on population. Paper prepared for the International Conference on Population and Development, Cairo Egypt.

Hungary/Central Statistical Office. (1975a). Principal results of the 1974 fertility and family planning (TCS) study (In Hungarian). Budapest: Central Statistical Office.

Hungary/Central Statistical Office. (1975b). *The 1966–1972 years family planning, fertility and birth control attitude of couples married in 1966.* Budapest: Central Statistical Office.

Hungary/Central Statistical Office. (1976). *Family planning in Hungary. Main results of the 1966 fertility and family planning (TCS) study.* Budapest: Central Statistical Office.

Hungary/Central Statistical Office. (1979). *Main results of the 1977 Hungarian fertility, family planning and birth control study. TCS-77, World Fertility Survey.* Budapest: Central Statistical Office.

Hungary/Central Statistical Office. (1984). *Longitudinal marriage surveys in Hungary.* Budapest: Central Statistical Office.

Hungary/Central Statistical Office. (1987). *Summary of the 1986 fertility surveys.* Budapest: Central Statistical Office.

Hungary/Central Statistical Office. (1996). *Main features of the Hungarian demographic situation in the early nineties.* Budapest: Central Statistical Office.

Hungary/Central Statistical Office. (1997). *Statistical yearbook of Hungary, 1994* and earlier years. Budapest: Central Statistical Office.

Hungary/*Documents describing policies concerning women, 1970–1980.* (1980). Budapest: Kossuth.

Husarska, A. (1995). Altared state: Permissive Poland rebels against the Church. *The Washington Post,* 26 November.

Hutter, I. (1996). Induced abortion and contraceptive use in Russia. In IUSSP (Ed.), *Socio-cultural and political aspects of abortion in a changing world.* Liege: International Union for the Scientific Study of Population.

Hyde, G. (1970). Abortion and the birthrate in the USSR. *Journal of Biosocial Science,* 2, 283–292.

Ilea, T., Huszar, R., and Enachescu, D. (1967). Study of the marital fertility of women in Bucharest (In French). *La Sante Publique,* 10, 255–267.

Ilea, T., Muresan, P., Ionescu, I., Pavel, A., Bilegan, M., and Teodorescu, A. (1974). Medico-social research on the fertility of the female population of the Romanian Socialist Republic (In French). *La Sante Publique,* 17, 3–10.

Iliriani, E., and Asllani, P. (1995). Albania's students teach their peers about sexuality and safer sex. *Planned Parenthood Challenges,* 1, 37–38.

Institute of Youth. (1974). *Social and psycho-physical problems of partners' choice* (In Bulgarian). Sofia: Institute of Youth.

Institute for Sociology and Social Policy. (1984). Demographic processes and population policies in the German Democratic Republic, 1970–1983. Berlin: Academy of Sciences.

Ionescu, D. (1987). Abortions in Romania. *Survey,* 29, 113–114.

Isenberg, I. (1963). *Eastern Europe: The Soviet satellites and the other European communist states.* New York: Scholastic Book Service.

Itar-Tass. (1997). 19 August.

Ivanova, E. (1996). Understanding family rejuvenation in Russia: A stable or a fluctuating pattern? Unpublished paper presented at the meetings of the Population Association of America, New Orleans, May.

Jakubowicz, K. (1992). From Party propaganda to corporate speech? Polish journalism in search of a new identity. *Journal of Communication*, 42, 64–73.

Jakubowska, N. (1976). The Polish social movement for sex education and social emancipation. In M. Kozakiewicz (Ed.), *Sex—society—education: Polish experience.* Warsaw: Polish Family Planning Association, 72–85.

Jancar, B. W. (1974). Women and Soviet politics. In H. W. Morton and R. L. Tökes (Eds.), *Soviet politics and society in the 1970s.* New York: Free Press.

Jancar, B. W. (1978). *Women under communism.* Baltimore: Johns Hopkins University Press.

Janikowska, J. (1994). Nonsense about sex (In Polish). *Gazeta Wyborcza,* No. 87.

Janowska, Z., Martini-Fiwelz, J., and Goral, Z. (1992). *Unemployment among women in Poland* (In Polish). Warsaw: Ebert Stiftung.

Johnson, B. R., Horga, M., and Andronache, L. (1993). Contraception and abortion in Romania. *The Lancet,* 341, 875–878.

Johnson, B. R., Horga, M., and Andronache, L. (1996). Women's perspective on abortion in Romania, *Social Science and Medicine,* 42, 521–530.

Jones, M. (1992). Abortions up as Russia runs out of condoms. *New Scientist,* 1 August, 10.

Jozan, P. (1977). Planned parenthood in Hungary. *IPPF Europe Regional Information Bulletin,* 6(4), 8.

Jozan, P. (1978). Hungary. *IPPF Europe Regional Information Bulletin Supplement,* 7(1), 4–5.

Juchasz, M. (1929). A question of births—sexual counseling (In German). Arbeiterwohlfart, 730–734.

Jurukovski, J., Antonovski, L., and Ljatkova, K. (1972). The anatomy of the serial abortion (English summary). *Godisen Zbornik n'Medicinskiot Fakultet vo Skopje,* 18, 61–69.

Jurukovski, J., and Sukarov, L. (1971). A critical review of legal abortion. *International Journal of Gynaecology and Obstetrics,* 9, 111–117.

Juveler, P. H. (1978). Women and sex in Soviet law. In D. Atkinson, A. Dallin, and G. W. Lapidus (Eds.), *Women in Russia.* Hassocks: Harvester Press, 243–265.

Kallajxhiu, F., and Gliozheni, O. (1994). Albania: Breaking chains of the past. *Entre Nous,* 26–27, 17.

Kalnicka, V. (1996). Attitudes of the Czech public towards induced abortion (In Czech). *Demografie,* 38 (2).

Kamaras, F. (1995). European fertility and family planning survey in Hungary. *Demografia,* 38, 309–339.

Kamaras, F., and Kowalska, I. (1995). Matrimonial and procreational attitudes and behaviors in Hungary and Poland: Comparative study. Paper prepared for the European PopulationConference, Milano, Italy, 4–8 September.

Kamsiouk, L. (1998). Summary of report presented to N.I.S. and Eastern Europe Consultation, Hannover, Germany, 12–16 October. Washington, DC: Population Institute.

Kapor-Stanulovic, N. (1980). Motivation for parenthood among adolescents in Yugoslavia. Paper presented at the International Congress of Psychology, Leipzig, GDR, July.

Kapor-Stanulovic, N. (1985). *The psychology of parenthood* (In Serbian). Belgrade: Nolit.

Kapor-Stanulovic, N. (1988a). Psychological aspects of population growth. In N. Milosavljevic (Ed.), *Natural population growth in Vojvodina and the factors that influence it* (In Serbian). Novi Sad: Medical Faculty, 67–78.

Kapor-Stanulovic, N. (1988b). Current investigations of the psychological variables relevant to population policy in Vojvodina. In N. Milosavljevic (Ed.), *Natural population growth in Vojvodina and the factors that influence it* (In Serbian). Novi Sad: Medical Faculty, 157–166.

Kapor-Stanulovic, N. (1988c). Pronatalist population policy-stimulating measures. In N. Milosavljevic (Ed.), *Natural population growth in Vojvodina and the factors that influence it* (In Serbian). Novi Sad: Medical Faculty, 195–202.

Kapor-Stanulovic, N. (1989a). Liberal approach for a patchwork population. *People,* 16(3), 16–18.

Kapor-Stanulovic, N. (1989b). Operationalization of population policy. In M. Macura (Ed.), *Population renewal in Serbia* (In Serbian). Belgrade: Serbian Academy for Science, 371–374.

Kapor-Stanulovic, N. (1991). Why sex education is not effective. *Entre Nous,* 17, April, 11.

Kapor-Stanulovic, N. (1996). Thirty years follow-up study on knowledge and practices of family planning in Yugoslavia; Stagnation or development? Unpublished paper presented at the symposium on Problems of Demographic Development of AP Vojvodina (In Serbian). Novi Sad, 10–11 November.

Kapor-Stanulovic, N., and Beric, B. (1973). Family planning in Yugoslavia with special reference to the Province of Vojvodina. *Journal of Biosocial Science,* 5, 179–185.

Kapor-Stanulovic, N., and Beric, B. (1979). Repeated abortions. In G. I. Zatuchni, J. J. Sciarra, and J. J. Speidel (Eds.), *Pregnancy termination.* Hagerstown, MD: Harper and Row, 333–342.

Kapor-Stanulovic, N., and Friedman, H. L. (1978). Studies of choice behavior in Yugoslavia. In H. P. David, H. L. Friedman, J. Van der Tak, and M. J. Sevilla (Eds.), *Abortion in psychosocial perspective: trends in transnational research.* New York: Springer, 119–144.

Kapor-Stanulovic, N., and Shabanah, N. (1990/1991). Inequalities in health status of women in Europe (In Serbian). *Stanovnistvo,* 3–4, 1–2, 209–216.

Karcher, H. L. (1992). Sterilizations up in eastern Germany. *British Medical Journal,* 305, 75.

Karro, H. (1997). Abortion in the framework of family planning in Estonia. *Acta Obstetrica et Gynecologica Scandinavica,* Supplement No. 164, 45–50.

Karro, M., Klimas, V., and Lazdane, G. (1997). Reproductive health in the Baltic countries. *Choices,* 26(1), 13–17.

Kaser, M. (1976). *Health care in the Soviet Union and Eastern Europe.* London: Croom Helm.

Katkova, I. P. (1968). Some sociohygienic aspects of the childbearing functions among young women (In Russian). Moscow State University: Dissertation.

Katus, K., and Zakharov, S. (1997). Demographic adaptation to socioeconomic changes in the USSR successor states. Paper presented at the meetings of the International Union for the Scientific Study of Population, Beijing, October 11–17.

Kaupova, N. A. (1996). Ensuring safe motherhood. *Integration,* (49), 17–21.

Kazakstan/National Institute of Nutrition, Academy of Preventive Medicine and Macro International. (1996). *Kazakstan demographic and health survey, 1995.* Summary in *International Family Planning Perspectives,* 1997, 23, 92–94.

Ketting, E., Visser, A. P., Uzel, R., and Lehert, P. (1993). Oral contraception in the former Czech and Slovak Federal Republic: Attitudes and use. *Advances in Contraception,* 9, 141–152.

Khachikian, M. (1998). Summary of report presented to the N.I.S. and Eastern Europe Consultation, Hanover, Germany, 12–16 October. Washington, DC: Population Institute.

Khalatbari, P. (1992). Population science in the former GDR—past, present, and future (In German). *Gesellschaft für Demographie, Beiträge zur Demographie,* 7–23.

Kharchev, A. G. (1964). *Marriage and family in the USSR* (In Russian). Moscow: Mysl.

Khomassuridze, A. G. (1985). The main aspects of hormonal contraception (In Russian). *Medical and social aspects of birth rate.* Tbilisi: Zhordania Institute.

Khomassuridze, A. G. (1993). Abortion and contraception in the USSR. In K. Newman (Ed.), *Progress postponed: Abortion in Europe in the 1990s.* London: International Planned Parenthood Federation/Europe Region, 78–91.

Kinzer, S. (1994). $650 a baby: Germany to pay to stem decline in births. *The New York Times,* 25 November.

Kissling, F. (1991). The Church's heavy hand in Poland. *Conscience,* September/October, 20–21.

Klein-Schonnefeld, S. (1994). Germany. In B. Rolston and A. Eggert (Eds.), *Abortion in the new Europe: A comparative handbook.* Westport, CT: Greenwood, 113–138.

Kligman, G. (1985). The rites of women. In S. L. Wolchik and A. G. Meyer (Eds.), *Women, state and party in Eastern Europe.* Durham, NC: Duke University Press, 323–343.

Kligman, G. (1988). *The wedding of the dead.* Berkeley: University of California Press.

Kligman, G. (1991). Women and reproductive legislation in Romania: Implications for the transition. In G. Breslauer (Ed.), *Dilemmas of transition in the Soviet Union and Eastern Europe.* Berkeley: University of California Press, 141–166.

Kligman, G. (1992). The politics of reproduction in Ceausescu's Romania: A case study in political culture. East European Politics and Societies, 6, 364–418.

Kligman, G. (1994). The social legacy of communism: Women, children, and the feminization of poverty. In J. Millar and S. Wolchik (Eds.), *The social legacy of communism.* Washington, DC: Woodrow Wilson Center and Cambridge Presses, 252–270.

Kligman, G. (1995). Political demography: The banning of abortion in Ceausescu's Romania. In F. D. Ginsburg and R. Rapp (Eds.), *Conceiving the new world order: The global politics of reproduction.* Berkeley: University of California Press, 234–255.

Kligman, G. (1998). *The politics of duplicity: Controlling reproduction in Ceausescu's Romania.* Berkeley: University of California Press.

Kliment, M. (1994). Reproductive health needs in the Slovak Republic. *Planned Parenthood in Europe,* 23(2), 10–11.

Kliment, M. (1998). Slovak Republic. Summary of report presented to the N.I.S. and Eastern Europe Consultation, Hanover, Germany, 12–16 October. Washington, DC: Population Institute.

Kliment, M., and Cupanik, M. (1997). Ethics in reproduction: Health versus religion. *Choices,* 26(2), 20–22.

Klinger, A. (1966). Abortion programs. In B. Berelson (Ed.), *Family planning and population programs.* Chicago: University of Chicago Press, 405–476.

Klinger, A. (1971). Demographic aspects of abortion. In *International Population Conference London 1969.* Liege: International Union for the Scientific Study of Population, 1153–1165.

Klinger, A. (1974). Hungary. In B. Berelson (Ed.), *Population policy in developed countries.* New York: McGraw-Hill, 225–269.

Klinger, A. (1977). Fertility and family planning in Hungary. *Studies in Family Planning,* 8, 166–176.

Klinger, A. (1979). Birth control and family planning in Hungary in the last decades. *World Health Statistics,* 32, 257–268.

Klinger, A. (1984). *The impact of policy measures, other than family planning programmes, on fertility.* Budapest: Demographic Research Institute.

Klinger, A. (1988). Hungary. In P. Sachdev (Ed.), *International handbook on abortion.* Westport, CT: Greenwood, 218–234.

Klinger, A. (1991). Hungary. In J. L. Rallu and A. Blum (Eds.), *European population. Vol. 1 Country analysis.* Paris: Institut National D'Etudes Demographiques, 209–223.

Klinger, A. (1995). 1945–1995: Half a century of population research; past achievements and experiences in Hungary (In Hungarian). *Demografia,* 38, 253–274.

Klinger, A., and Szabady, E. (1978). Patterns of abortion and contraceptive practice in Hungary. In H. P. David, H.L. Friedman, J. van der Tak and M. J. Sevilla (Eds.), *Abortion in psychosocial perspective: Trends in transnational research.* New York: Springer, 168–198.

Knopp, S. (1980). Sexism in the pictures of children's readers: East and West Germany compared. *Sex Roles,* 189–205.

Knothe, M. (1994). Warsaw as a labor market for women (In Polish). In *Kobiety: dawne I nowe role.* European Centrum, Warsaw University, Bulletin No. 1.

Kobieta I Zycie. (1994). No. 30. Editorial.

Kohak, E. (1992). Ashes, ashes . . . Central Europe after forty years. *Daedalus,* 121, 197–215.

Kolinsky, E. (1993). *Women in contemporary Germany.* Providence: Berg Publications.

Kollontai, A. (1971a). *Autobiography of a sexually emancipated Communist woman.* New York: Herder and Herder.

Kollontai, A. (1971b). *Communism and the family.* Sydney: Young.

Kollontai, A. (1972). *Sexual relations and the class struggle: Love and new morality.* Bristol: Falling Wall.

Kollontai, A. (1977). *Selected writings.* New York: Lawrence Hill.

Komyssova, N. (1992). Family planning in the Russian Federation. *Planned Parenthood in Europe,* 21(2), 7.

Kon, I. S. (1981). *Culture and sexology.* Budapest: Kossuth (In Hungarian). Revised in 1985 as *Introduction to sexology.* Berlin: Deutscher Verlag der Wissenschaften (In German); Cologne: Pahl-Rugenstein Verlag (1985). Moscow: Meditsina (1988) (In Russian).

Kon, I. S. (1992). Culture and sexuality in the former USSR. *Planned Parenthood in Europe,* 21(2), 2–4.

Kon, I. S. (1993). Sexuality and culture. In I. S. Kon and J. Riordan (Eds.), *Sex and Russian Society.* London: Pluto Press, 15–44.

Kon, I. S. (1995). *The sexual revolution in Russia.* New York: Free Press.

Kon, I. S. (1997). Sexuality and politics in Russia (1700–2000). Moscow: Unpublished paper.

Kon, I. S., and Riordan, J. (Eds.) (1993). *Sex and Russian society.* Bloomington: Indiana University Press.

Konarska, I. (1993a). Playing games in bed (In Polish). *Kobieta I Zycie,* No. 18.

Konarska, I. (1993b). Women you make a mistake (In Polish). *Kobieta I Zycie,* No. 50.

Konarska, I. (1994). Questionable pleasure (In Polish). *Kobieta I Zycie,* No. 25.

Köö, B. (1992). Societatea de Educate Contraceptiva si Sexuala (SECS). *Planned Parenthood in Europe,* 21, 34–35.

Köö, B. (1998). Summary of report presented to the N.I.S. and Eastern Europe Consultation, Hanover, Germany, 12–16 October. Washington, DC: Population Institute.

Koonz, C. (1987). *Mothers in the fatherland.* New York: St. Martin's Press.

Koop, C. E. (1989). Cited in *The federal role in determining the medical and psychological impact of abortion on women.* Washington, DC: 10th report by the Committee on Government Operations, 11 December. House Report 101-392.

Kornai, J. (1992). *The socialist system: The political economy of communism.* Princeton: Princeton University Press.

Kornet, E. (1972). Motives for termination of pregnancy. In *Tartu seminar on family problems I.* Tartu: 181.

Kosinski, L. A. (1977a). Population trends in Poland after World War II. In L. A. Kosinski (Ed.), *Demographic developments in Eastern Europe.* New York: Praeger, 309–338.

Kosinski, L. A. (1977b). Sources of demographic statistics in East-Central Europe. In L. A. Kosinski (Ed.), *Demographic developments in Eastern Europe.* New York: Praeger, 64–75.

Kostanick, H.L. (1977). Characteristics and trends in Southeastern Europe: Romania, Yugoslavia, Bulgaria, Albania, Greece, and Turkey. In H.L. Kostanick (Ed.), *Population and migration trends in Eastern Europe.* Boulder, CO: Westview Press, 11–22.

Kostova, D. (1994). Similar or different? Women in postcommunist Bulgaria. In M. Rueschemeyer (Ed.), *Women in the politics of postcommunist Eastern Europe.* Armonk, NY: Sharpe, 117–132.

Kot, N. (1994). Polish frustration (In Polish). *WPROST,* No. 6.

Kowalska, H. (1991). *Family transformation and fertility patterns.* Warsaw: Central Statistical Office.

Kowalska, L. (1993a). Sex education in Polish language and sports classes (In Polish). *Zycie Warszawy,* 21–22 August.

Kowalska, L. (1993b). Alone without help from school (In Polish). *Zycie Warszawy,* 21–22 August.

Kozakiewicz, M. (1972). Sex education in Poland. *Polish Perspectives,* 1972, 15(9), 21–28.

Kozakiewicz, M. (Ed.). (1976). *Sex—society—education: Polish experience.* Warsaw: Polish Family Planning Association.

Kozakiewicz, M. (Ed.). (1977). *Family Planning and sex-education in socialist countries.* Warsaw: Polish Family Planning Association.

Kozakiewicz, M. (1978). Sex education in Poland. *IPPF Europe,* 7, 4–5.

Kozakiewicz, M. (1980). Planned Parenthood in non-member countries. *IPPF/Europe Regional Information Bulletin,* 9(2), 9–12.

Kozakiewicz, M. (1997). Poland—the struggle for "free choice" continues. *Choices,* 26(1), 18–19.

Kozlowska, E., and Wojtyla, J. (1976). Law and fertility in Poland. In M. Kirk, M. Livi Bacci, and E. Szabady (Eds.), *Law and fertility in Europe.* Dolmain: Ordina, 495–519.

Kubicka, L., Matejcek, Z., David, H. P., Miller, W. B., and Roth, Z. (1995). Prague children from unwanted pregnancies revisited at age thirty. *Acta Psychiatrica Scandinavica,* 91, 361–369.

Kucera, M. (1974). Population and its development. In *Population policy in Czechoslovakia.* Prague: Orbis, 5–16.

Kucera, M. (1994). *Population of the Czech Republic, 1918–1991* (In Czech). Prague: Czech Demographic Association.

Kucera, M., and Fialova, L. (1996). *Demographic behavior of the Czech population in the course of social transformation after 1989* (In Czech). Working Paper 96:1, Prague: Institute of Sociology.

Kuciarska-Ciesielska, M., Nowak, L., and Smolinski, Z. (1988). *Family life and procreational attitudes of newly married couples.* (In Polish). Warsaw: Central Statistical Office.

Kulakov, V. I. (1995). Abortion and infertility in Russia. *Planned Parenthood in Europe,* 24(1), 10–12.

Kulcsar, K. (1984). Possibilities for influencing demographic trends. In D. Biro, P. Jozan, and K. Miltenyi (Eds.), *Population and population policy in Hungary.* Budapest: Akademia: Kiado, 13–26.

Kulczycki, A. (1995). Abortion policy in postcommunist Poland. *Population and Development Review,* 21, 471–505.

Kulczycki, A. (1996). Demographic research and abortion policy: The limits to statistics. In IUSSP (Ed.), *Socio-cultural and political aspects of abortion in a changing world.* Liege: International Union for the Scientific Study of Population.

Kulczycki, A., Potts, M., and Rosenfield, A. (1996). Abortion and fertility regulation. *The Lancet,* 347, 1663–1668.

Kuliesyte, E. (1998). Summary of report presented to the N.I.S. and Eastern Europe Consultation, Hanover, Germany, 12–16 October. Washington, DC: Population Institute.

Kuratowska, Z. (1992). Present situation of women in Poland. In *The impact of economic and political reform on the status of women in Eastern Europe: Proceedings of a United Nations Regional Seminar.* New York: United Nations.

Kuzibaeva, G. (1998). Personal communication, 7 July.

Kyrgyz Republic/Ministry of Health. (1997). *Kyrgyz Republic Demographic and Health Survey 1997; Preliminary Report.* Bishkek: Ministry of Health and Calverton: Macro International.

Kyrgyz Republic/National Statistical Committee. (1997). Communication to Population Reference Bureau, 11 August.

Labuda, B. (1994). *Kobieta I Zycie,* 10 February.

Lagadinova, E. (1974). Women and the population: The experience of the People's Republic of Bulgaria. Report presented to the International Forum of the United Nations' Organization on the Role of Women in Population and Development, March.

Lalovic, M. (1971). The fecundity of married women age 15 to 49 years and birth control (French summary). *Stanovnistvo,* 9, 237–253.

Lammich, S., and Nagy, F. (1988). Country Report Hungary (In German). In A. Esser and G. Koch (Eds.), *Abortion in international comparison.* Baden-Baden: Nomos, 1665–1744.

The Lancet. (1994). 10 September, Vol. 344, 745.

Langmeier, J., and Matejcek, Z. (1975). *Psychological deprivation in childhood.* New York: Halstead.

Lapidus, G. W. (1978). *Women in Soviet society: Equality, development, and social change.* Berkeley: University of California Press.

Lapidus, G. W. (Ed.) (1982). *Women, work, and family in the Soviet Union.* Armonk. NY: M. E. Sharpe.

Latuch, N. (1973). Elements of population policy in Poland. In *Population policy-contemporary problem* (In Polish). Warsaw: Institute of Social Economy, 106–139.

Lazdane, G. (1997). *Report prepared for Council of Europe.* Strasbourg: COE, EG/TAL (97)8.

Lazdane, G. (1998).Summary of report presented to the N.I.S. and Eastern Europe Consultation, Hanover, Germany, 12–16 October. Washington, DC: Population Institute.

League of Nations. (1940). *Bulgaria.* Geneva: European Conference on Rural Life, Monograph No. 28.

Lech, M. (1974). Estimation of the proportion of pregnant women living in towns (In Polish). *Zdrowie Publiczne,* No. 11.

Leetz, I. (1977). On the development of the number of live born children in the years 1970–1977 (In German). Unpublished paper, presented in Rostock, October.

Legge, J. S. (1985). *Abortion policy: An evaluation of the consequences for maternal and infant health.* Albany: State University of New York Press.

Lehert, P., Pavlenko, I., Remennick, L., and Visser, A. (1992). Contraception in the former USSR: Recent survey results on women's behavior and attitudes. *Planned Parenthood in Europe,* 21(2), 9–11.

Lehfeldt, H. (1986). Felix A. Theilhaber—pioneer sexologist. *Archives of Sexual Behavior,* 15, 1–12.

Lenin, V. I. (1913). The working class and neo-Malthusianism. *Pravda,* 16 June; also published in Lenin, V. I. (1963). *Collected works,* Vol. 19. Moscow: Foreign Languages Publishing House, 235–237. (Cited in Heitlinger, 1979; and Besemeres, 1980.)

Lenin, V. I. (1951). The woman question. In *Selections from the writings of K. Marx, F. Engels, V. I. Lenin, and J. Stalin.* New York: International Publishers.

Lenin, V. I. (1963). The female workers (1920) (In Russian). In *Collected Works.* Vol. 40. Moscow: Gosizdat, 197–205.

Lenin, V. I. (1966a). *The emancipation of women.* New York: International Publishers.

Lenin, V. I. (1966b). Correspondence with I. Armand. In *On the emancipation of women, from the writings of V. I. Lenin.* New York: International Publishers, 97.

Lenin, V.I. (1976). A great beginning: Heroism of the workers in the rear; Communist Subbotniks. In V.I. Lenin, *Selected works in three volumes.* Moscow: Progress Publishers, Vol. 3.

Lenz, K. (1983). Population science in the Third Reich (In German). *Materialien zur Bevölkerungswissenschaft* No. 35. Wiesbaden: Federal Institute for Population Research.

Leon, D. A., and Shkolnikov, V. M. (1998). Social stress and the Russian mortality crisis. *Journal of the American Medical Association,* 279, 790–793.

Leshi, A. (1994). *Popolli Po,* 20 February.

Leskaj, V. (1996). Reaching people in Albania. *Choices,* 25(3), 31.

Leskaj, V. (1998). Summary of report presented at the N.I.S. and Eastern Europe Consultation, Hanover, Germany, 12–16 October. Washington, DC: Population Institute.

Lew-Starowicz, Z. (1994). Cited in David and Titkow (1994).

Liegle, L. (1975). *The family's role in Soviet education.* New York: Springer.

Liiceanu, A. (1994). Adolescence in Romania; developing future orientation in a changing world. Unpublished report. Bucharest: Society for Education in Contraception and Sexuality.

Lipovskaya, O. (1994). Sisters of stepsisters. *Women's Studies International Forum,* 2, 273–276.

Lithuanian frontline. (1994). *Planned Parenthood in Europe,* 23(3), 35–36.

Litinaviciene, S. (1996). Family and fertility in Lithuania. Paper presented at the Conference of European statisticians. Geneva, 29–31 January.

Littmann, E., and Szewczyk, H. (1977). The influence of "the pill" on the sexual behavior of students (In German). In A. Burghardt and H. Szewczyk (Eds.), *Sexuality—facts norms and societal responsibility.* Berlin: Volk und Gesundheit.

Lorimer, F. (1946). *The population of the Soviet Union: History and prospects.* Geneva: League of Nations.

Lorimer, F. (1958). Population policy and politics in the Communist world. In P. M. Hauser (Ed.), *Population and world politics.* Glencoe, IL: Free Press, 214–236.

Los Angeles Times. (1991). 9 December.

Lubell, I. (1979). The IUD and other methods of family planning in the Soviet Union. Paper presented at the International Symposium on IUDs and Biodegradable Delivery Systems, Amsterdam, June 1979. Summarized in *Ob. Gyn. News,* 14, No. 8.

Luker, K. (1975). *Taking chances: Abortion and the decision not to contracept.* Berkeley: University of California Press.

Lungwitz, K. (1981). Principles, purposes and results of population policies in the German Democratic Republic. Paper presented at the General Conference of the International Union for the Scientific Study of Population, Manila, Philippines, December.

Lunin, I., Hall, T. L., Mandel, J. S., Kay, J., and Hearst, N. (1995). Adolescent sexuality in Saint Petersburg, Russia. *AIDS,* 9 (Suppl. 1), 553–560.

Lunin, I., Krizanskaya, J., Meliktiova, L., Light, L., and Brandt-Sorheim. (1997). Use of the Internet for sex education in Russia. *The Journal of Sex Education and Therapy,* 22(4), 74–78.

Lydolph, P. E., Johnson, R., Mintz, J., and Mills, M. E. (1978). Recent population trends in the USSR. *Soviet Geography: Review and Translation,* 19, 505–539.

Mace, D., and Mace, V. (1963). *The Soviet family.* London: Hutchinson.

Macedonia/Statistical Office. (1995). *Statistical yearbook of the Republic of Macedonia 1995.* Skopje: Statistical Office of Macedonia.

Macedonia/Statistical Office. (1997). Information provided to Population Reference Bureau.

Macura, M. (1974a). Socioeconomic conditions and policy related to development and population. In D. Breznik (Ed.), *The population of Yugoslavia.* Belgrade: Demographic Research Center, 75–89.

Macura, M. (1974b). Population policies in socialist countries of Europe. *Population Studies,* 28, 369–379.

Macura, M. (1981). Evolving population policies. In H. P. David and R. J. McIntyre (Eds.), *Reproductive behavior: Central and Eastern European experience.* New York: Springer, 30–52.

Macura, M. (1989). *Population renewal in Serbia* (In Serbian). Belgrade: Serbian Academy for Science.

Macura, M. (1995). Fertility and nuptiality changes in Central and Eastern Europe: 1982–1993. *Studia demograficzne,* 122, 9–34.

Maddock, J. W., Hogan, J. M., Antonov, A. I., and Matskovsky, M. S. (Eds.). (1994). *Families before and after Perestroika: Russian and U.S. perspectives.* New York: Guilford Press.

Madison, B. Q. (1968). *Social welfare in the Soviet Union.* Stanford: Stanford University Press.

Magas, B. (1993). *The destruction of Yugoslavia: Tracking the breakup 1980–1992.* New York: Verso.

Maggs, P. B. (1977). Law and population in Eastern Europe. *Law and Population Monograph series,* No. 3 (revised).

Mahler, K. (1997). In Kazakstan, contraceptive knowledge is nearly universal but abortion rates remain high. *International Family Planning Perspectives,* 23, 92–94.

Makusheva, V. P. (1983). Problems of contraception: Literature review (In Russian). *Problems of Motherhood and Childhood Protection,* 28(12), 37–39.

Mala, E. (1974). Women and culture. *Society and Leisure,* 4, 172.

Malacic, J. (1990). Political changes in Yugoslavia and family planning. *Planned Parenthood in Europe,* 19(2), 18–20.

Malanowski, J. (1960). *Societal attitudes toward an increase of reproductivity.* Warsaw: Public Opinion Polling Center.

Malewska, H. (1966). Sex norms and their correlates (In Polish). *Studia Socjologiczne,* No. 3(22).

Malewska, H. (1968). Cultural and psychological determinants of sexual behavior. *Social Science and Medicine,* 2, 319–335.

Malysheva, M. (1992). Feminism and Bolshevism: Two worlds, two ideologies. In S. Rai, M. Pilkington, and A. Phizacklea (Eds.), *Women in the face of change.* New York: Routledge, 186–199.

Malysheva, M., Ballaeva, M., Klimenkova, T., Konstantinova, V., and Tyuryukanova, E. (1998). *Russia—1997: Women's rights in the context of social and economic reforms.* Moscow: Moscow Center for Gender Studies.

Malysheva, M., and Bertaux, D. (1996). The social experiences of a country woman in Soviet Russia. In S. Leydesdorff, L. Passerini, and P. Thompson (Eds.), *International yearbook of oral history and life stories,* Vol. IV, Gender and memory. Oxford: Oxford University Press, 31–43.

Mamonova, T. (Ed.) (1984). *Women and Russia: Feminist writings from the Soviet Union.* Boston: Beacon Press.

Mandel, W. (1975). *Soviet women.* Garden City, NY: Doubleday.

Manuilova, I. (1991a). Family planning in the USSR: The role of the Soviet Family and Health Association. *Planned Parenthood in Europe,* 19(2), 9–13.

Manuilova, I. (1991b). A major challenge. *Integration* (JOICFP), (29), 4–5.

Mares, P., Musil, L., and Rabusic, L. (1991). Social change in perception of the Czech public (In Czech). *Sociologicky casopis,* 27, 702–714.

Marinova, J. (1997). Bulgarian medical students' attitudes to abortion. *Choices,* 26(2), 15–17.

Markowska, A. (1998). Summary of report presented to the N.I.S. and Eastern Europe Consultation, Hanover, Germany, 12–16 October. Washington, DC: Population Institute.

Marmullaku, R. (1975). *Albania and the Albanians.* London: Hurst.

Marvin, G. M. (1995). Two steps back and one step forward: East German women since the fall of the wall. *Humanity and Society,* 19, 37–52.

Marx, K. (1936). *Capital.* New York: Modern Library.

Marx, K., and Engels, F. (1934). *Correspondence, 1846–1895.* London: Lawrence and Wishart.

Marx, K., Engels, F., and Lenin, V. (1973). *On the woman question* (In Czech). Prague: Nakladatezstvi Svoboda.

Matejko, A. J. (1991). Civic culture and consumption in Eastern Europe. *Sociologia Internationalis,* 29, 75–102.

Matioc, A., and Costa, E. (1963). The problem of pregnancy interruption on demand (In Romanian). *Obstetrica si Ginokolgia,* 10, 193–202.

Matskovksi, M., and Bodrova, V. (1990). *Value of the family as perceived by various strata of the population* (In Russian). Moscow: Institute of Sociology, USSR Academy of Sciences.

Matuchniak-Krasuska, A. (1995). The abortion discussion—what it really was (In Polish). In A. Titkow and H. Domanski (Eds.), *What does it mean to be a woman in Poland.* Warsaw: Institute of Philosophy and Sociology, 189–212.

Mazur, P. (1981). Contraception and abortion in Poland. *Family Planning Perspectives,* 13, 195–198.

McIntosh, C. A., and Finkle, J. L. (1995). The Cairo conference on population and development: A new paradigm? *Population and Development Review,* 21(2), 223–260.

McIntyre, R. J. (1975). Pronatalist programmes in Eastern Europe. *Soviet Studies,* 27, 336–380.

McIntyre, R. J. (1980). The Bulgarian anomaly: Demographic transition and current fertility. *Southeastern Europe,* 7, 147–170.

McNicoll, G. (1994). Institutional analysis of fertility. New York: Population Council Working Paper, No. 62.

Mecklinger, L. (1968). Co-responsibility and tasks of the physician in family planning (In German). In K. H. Mehlan (Ed.), *Physician and family planning.* Berlin: Verlag Volk und Gesundheit, 11–30.

Meekers, D., Holscher, M., and Munteanu, A. (1997). Sexual and reproductive health behavior among Romanian adolescents: An exploratory narrative research analysis. Unpublished paper. Washington, DC: Population Services International Research Division, Working Paper No. 11.

Mehlan, K. H. (1955). Abortion statistics and fertility in the German Democratic Republic (In German). *Das Deutsche Gesundheitswesen,* 10, 1648–1659.

Mehlan, K. H. (1961). The abortion situation in the German Democratic Republic (In German). In K. H. Mehlan (Ed.), *International Abortsituation, Abortbekämpfung, Antikonzeption.* Leipzig: Thieme, 52–63.

Mehlan, K. H. (1965). Legal abortions in Romania, *Journal of Sex Research,* 1, 31–38.

Mehlan, K. H. (1966). The socialist countries of Europe. In B. Berelson (Ed.), *Family planning and population programs.* Chicago: University of Chicago Press, 207–226.

Mehlan, K. H. (1967). Reducing the abortion rate and increasing fertility by social policy in the German Democratic Republic. In Proceedings, *World Population Conference,* 1965. New York: United Nations, Vol. II, 223–227.

Mehlan, K. H. (1969). Teaching family planning in Eastern Europe: The significance of the high abortion rate. In J. Leban (Ed.), *Teaching family planning.* Report of an International Macy Conference. New York: Josiah Macy, Jr. Foundation, 61–76.

Mehlan, K. H. (1970). *Wanted children? Family planning, contraception and the fight against abortion in our time* (In German). Rudolstadt: Greifenverlag.

Mehlan, K. H. (1974). Effects of the legalization of abortion in the Rostock District (In German). *Das Deutsche Gesundheitswesen,* 29, 2216–2219.

Mehlan, K. H. (1978). The problem of adolescent pregnancy (In German). Cited in *Pro Familia Informationen, (3)* November 17–18.

Mehlan, K. H. (1988). German Democratic Republic. In P. Sachdev (Ed.), *International handbook on abortion.* Westport, CT: Greenwood.

Mehlan, K. H., and Akkermann, S. (1978). Adolescent fertility in the German Democratic Republic. Paper presented at the WHO Meeting on Service Oriented Research in Adolescent Fertility, Warnemünde, April.

Mehlan, K. H., and Geissler, A. (1973). Family planning in the GDR (In German). *Das Deutsche Gesundheitswesen,* 28, 1136–1140.

Meier, C. S. (1997). *Dissolution: The crisis of Communism and the end of East Germany.* Princeton: Princeton University Press.

Mesaros, E. (1972). The socioeconomic basis of demographic behavior and family planning (In Romanian). *Viitorul Social,* 1, 1245–1255.

Mesaros, E. (1975). Social demographic considerations on the family (In Romanian). *Viitorul Social,* 4, 683–688.

Meszaros, A. (1990). Birth control in Hungary: Changing legal conditions. *Planned Parenthood in Europe,* 19(2), 14–16.

Meyer, A. G. (1985). Feminism, socialism, and nationalism in Eastern Europe. In S. L. Wolchik and A. G. Meyer (Eds.), *Women, state, and party in Eastern Europe.* Durham, NC: Duke University Press, 13–30.

Meyerfeldt, M. (1996). Demographic transformation processes in Bulgaria (In German). *Europa Regional,* 4(1), 24–31.

Meznaric, S., and Ule, M. (1994). Women in Croatia and Slovenia. In M. Rueschemeyer (Ed.), *Women in the politics of postcommunist Eastern Europe.* Armonk, NY: Sharpe, 153–170.

Michnik, A. (1994). *The church and the left.* Chicago: University of Chicago Press.

Mikhnyuk, D. (1994). Belarus. Country Report prepared for the WHO/Europe Conference on Women's Health Counts, Vienna, 16–18 February.

Milewicz, E., and Nowakowska, A. (1993). Abortion forbidden (In Polish). *Gazeta Wyborcza,* No. 6.

Milic, A. (1993). Women and nationalism in the former Yugoslavia. In N. Funk and M. Mueller (Eds.), *Gender politics and post-communism: Reflections from Eastern Europe and the former Soviet Union.* New York: Routledge, 109–122.

Miller, W. B. (1983). Chance, choice, and the nature of reproduction. *American Psychologist,* 38, 1198–1205.

Milosavljevic, N. (Ed.) (1988). *Natural population growth in Vojvodina and the factors that influence it* (In Serbian). Novi Sad: Medical Faculty.

Mitchell, B. R. (1975). *European historical statistics, 1750–1970.* London: Macmillan.

Mitlyng, V. (1997). Russia prepares for sex education. *Moscow Times,* 13 May.

Mlynarczyk, J. (1994). Strike of storks (In Polish). *Kobieta I Zycie,* No. 26/2243.

Moffet, J. (1987). Sex and birth control in the USSR. In V. Mastney (Ed.), *Soviet/East European survey, 1985–1986.* Durham, NC: Duke University Press, 323–330.

Mogilevkina, I., Markote, S., Avakyan, Y., Mrochek, L., Liljestrand, J., and Hellberg, D. (1996). Induced abortions and childbirths: Trends in Estonia, Latvia, Lithuania, Russia, Belarussia, and the Ukraine during 1970–1994. *Acta Obstetrica et Gynecologia Scandinavica,* 75, 908–911.

Mojic, A. (1969). Fluctuation in abortion and births in the health establishments of the Yugoslavian Socialist Federal Republic, 1963–1967. Belgrade: Unpublished Report to the Federal Institute for the Protection of Health.

Molnar, E. S. (1976). Some problems of interpretation of the number of children considered as ideal per family (In Hungarian). *Demografia,* 19, 212–227.

Moore, E. C. (1971). Induced abortion and contraception: Sociological aspects. In *Abortion obtained and denied: Research approaches.* New York: The Population Council, 131–155.

Moore, E. C. (1972). To S. F.: My god, we've been telling you! *Family Planning Perspectives,* 4, 2–4.

Morris, K. (1998). Polish abortion law reversed. *The Lancet,* 351, 3 January.

Morris, S. J. (1997). The new face of Bulgaria. *The Wall Street Journal,* 18 June.

Moscow Center for Gender Studies. (1996). *Institutionalization project, 1994–1996.* Moscow: MCGS.

Moscow Times. (1997). 29 May.

Moshin, V., Bodrug, V., and Blaja, M. (1996). Adolescent sexual health in Moldova. *Choices,* 25(2), 22–23.

Moskoff, W. (1978). Sex discrimination, commuting, and the role of women in Romanian development. *Slavic Review,* 37, 440–456.

Moskoff, W. (1980). Pronatalist policies in Romania. *Economic Development and Cultural Change,* 23, 597–614.

Muckle, J. (1987). The new Soviet child: Moral education in Soviet schools. In G. Avis (Ed.), *The making of the Soviet citizen: Character formation and civic training in Soviet education.* London: Croom Helm, 1–22.

Mullaj, E. (1994). Abortion Unit, Tirana Women's Hospital. Personal Communication, November 1994.

Mundigo, A. (1998). The role of men in improving reproductive health: The direction research should take. In J. Khenna and P.F.A. Van Look (Eds.), *Reproductive health research: The new directions.* Geneva: World Health Organization, 124–131.

Münz, R., and Ulrich, R. (1994). Depopulation after unification? The demographic future of East Germany. Occasional Paper, Center for German and German Studies, University of California at Berkeley.

Muresan, C. (1996). The demographic evolution in Romania. Past trends (1948–1994) and future perspectives (1995–2030) (In French). *Population,* 51, 813–814.

Muresan, P. (1991). Cited in R. Rochat (1991).

Muresan, P., and Copil, I. M. (1974). Romania. In B. Berelson (Ed.), *Population policy in developed countries.* New York: McGraw-Hill, 355–384.

Murphy, D. E. (1997). Polish doctors reject abortion. *Los Angeles Times,* 2 February.

Murphy, E. (1998). A quick history of health communication. In Michau, L. (Ed.), *Report from the meeting on changing communication strategies for reproductive health and rights.* Takoma Park, MD: Health and Development Policy Project, 3–5.

Murstein, B. I. (1974). *Love, sex, and marriage through the ages.* New York: Springer.

Musil, J. (1971). Some aspects of social organization of the contemporary Czechoslovak family. *Journal of Marriage and the Family,* 33, 196–206.

Mutler, A. (1994). Confusion about sex, birth control in Romania. *Chicago Tribune,* 17 May.

Nachtwey, Y. (1990). Romania's lost children. *The New York Times Magazine*, 24 June: 29–33.

Nagorski, A. (1993). Trying to say "sex" in Polish. *Newsweek,* 21 June, 44.

Narodny Novine, 1973, No. 10.

National Abortion Federation (1997). *The role of physician assistants, nurse practitioners, and nurse-midwives in providing abortion: Recommendations from a national symposium.* Washington, DC: National Abortion Federation.

Neier, A. (1997). Impasse in Kosovo. *The New York Review,* 25 September.

Neubert, R. (1967). *The new marriage manual* (In German). Rudolstadt: Greifenverlag.

Neubert, R. (1969). *What do I tell my child?* (In German). Rudolstadt: Greifenverlag.

Neumann, H. G. (1987). On the position of irreversible contraception (sterilization) within the context of family planning in the GDR (In German). *Pro Familia Magazin,* 15(6), 20–21.

Neumann, H. G. (1990). Thoughts on the present day family planning situation in the German Democratic Republic. *Planned Parenthood in Europe,* 19(2), 2.

Neumann, H. G. (1991). Farewell to Ehe und Familie. *Planned Parenthood in Europe,* 20(2), 24–25.

Newman, B. (1994). Sex education leaves much to be desired in liberated Poland. *The Wall Street Journal,* 22 November.

Newman, L. F. (1985). *Women's medicine: A cross-cultural study of indigenous fertility regulation.* New Brunswick: Rutgers University Press.

The New York Times. (1990). 9 June.

Nickel, H. M. (1993). Women in the German Democratic Republic and in the new federal states: Looking backward and forward. In N. Funk and M. Mueller (Eds.), *Gender politics and post-communism: Reflections from Eastern Europe and the former Soviet Union.* New York: Routledge, 138–150.

Nolan, D. (1998). Abortion: Should men have a say? In E. Lee (Ed.), *Abortion, law and politics today.* New York: St. Martin's Press, 216–230.

Notzon, F. C., Komarov, Y. M., Emakov, S. P., Sempos, C. T., Marks, J. S., and Sempos, E. V. (1998). Causes of declining life expectancy in Russia. *Journal of the American Medical Association,* 279, 793–800.

Novak, F. (1968). *Instructions for usage of vacuum aspirator VE-2.* Wilmington: Lalor Foundation.

Novak-Reiss, A. (1978). The implementation of sex education in Yugoslavia. *IPPF/Europe Regional Information Bulletin,* 7(2), 6–7.

Nowak, S. (1981). Values and attitudes of the Polish people. *Scientific American,* 245, 23–31.

Nowakowska, A. (1993). Faulty law (In Polish). *Gazeta Wyborcza,* No. 70.

Nowakowska, A., Domostaski, A., and Nielowiejska, D. (1994). Abortion still forbidden (In Polish). *Gazeta Wyborcza,* No. 205.

Nowicka, W. (1993). The new abortion law in Poland. *Planned Parenthood in Europe,* 22(3), 21–24.

Nowicka, W. (1995). Consequences of the anti-abortion law. Warsaw: Federation for Women and Family Planning, Report No. 1, May.

Nowicka, W. (1996a). The effects of the 1993 anti-abortion law in Poland. *Entre Nous,* 34–35, December, 13–15.

Nowicka, W. (1996b). Personal communication. Amsterdam, 28 March.

Nukusheva, S. (1996). An encouraging effect. *Integration,* (49), 16.

Nurt, L. (1989). On the legal protection of the conceived child (In Polish). Discussion. Warsaw: NUVT.

Obersnell-Kveder, D. (1992). The study of use of abortion as a fertility regulating method or an emergency procedure among women in Ljubljana. Unpublished research report to the World Health Organization, Geneva. Ljubljana, Slovenia: University Institute for Public Health and Social Welfare.

Obregon, R. (1995). Bulgaria. *Population Today*, 23(2), 7.

Oddens, B. J., Vemer, H. M., Visser, A. P., and Ketting, E. (1993). Contraception in Germany: A review. *Advances in Contraception,* 9, 105–116.

Oddens, B. J., Visser, A. P., Vemer, H. M., and Everaerd, W. T. A. M. (1994). Contraceptive use and attitudes in reunified Germany. *European Journal of Obstetrics and Gynecology,* 57, 201–208.

Okolski, M. (1983). Abortion and contraception in Poland. *Studies in Family Planning,* 14, 263–274.

Okolski, M. (1988). *Reproduction and modernization: Polish syndrome* (In Polish). Warsaw: Ksiazka I Wiedza.

Olczyk, E. (1991). To go out from the kitchen (In Polish). *Zycie Warszawy,* 4 April.

Omeragic, F. (1996). Reproductive health and family planning in Bosnia and Herzegovina. *Entre Nous,* (34/35), 17.

Omeragic, F. (1998). Bosnia-Herzegovina: A case study in service collapse. *Entre Nous,* (38), 10.

Open File/IPPF. (1998). Social marketing project launched in Lithuania. March/April, 10.

Orlova, N. (1973). Women in Soviet society. *Soviet Life,* March.

Ostner, I. (1994). Back to the fifties: Gender and welfare in unified Germany. *Social Politics,* 1, 32–59.

Pain, E. (1998). Interview with O. Moroz. *Literaturnaya Gazeta*, 9 December.

Panova, R., Gavrilova, R., and Merdzanska, C. (1993). Thinking gender: Bulgarian women's impossibilities. In N. Funk and M. Mueller (Eds.), *Gender politics and post-communism: Reflections from Eastern Europe and the former Soviet Union.* New York: Routledge, 15–21.

Papic, Z. (1993). Nationalism, patriarchy, and war. Paper presented at Network of East-West women public policy forum, Washington, DC, October 26–27.

Paton, C. R. (1989). Perestroika in the Soviet Union's health system. *British Medical Journal,* 299, 45–46.

Paulickova, M. (1976). Development of kindergartens and day care centers (In Czech). *Socialna Politika,* No. 8.

Pavlik, Z. (1964). *An outline of the world population development* (In Czech). Prague: Czechoslovak Academy of Sciences.

Pavlik, Z. (1978). Baby boom in Czechoslovakia. *Intercom*, 6(10), 8–9.

Pavlik, Z. (Ed.). (1994). *Population change in the Czech Republic* (In Czech). Prague: Charles University Department of Demography and Geodemography.

Pavlik, Z. (Ed.). (1996). *Human development report: Czech Republic, 1996.* Prague: Faculty of Science, Charles University.

Pavlik, Z., and Kucera, M. (Eds). (1997). *Population development in the Czech Republic.* Prague: Department of Demography and Geodemography, Charles University.

Pavlik, Z., and Wynnyczuk, V. (1974). Czechoslovakia. In B. Berelson (Ed.), *Population policy in developed countries.* New York: McGraw-Hill, 319–354.

Peers, J. (1985). Workers by hand and womb—Soviet women and the demographic crisis. In B. Holland (Ed.), *Soviet sisterhood.* Bloomington: Indiana University Press, 116–144.

Peil, I. (1996). Acceptance of family-policy measures in the Federal Republic of Germany—An East-West Comparison (In German). *Materialien zur Bevölkerungswissenschaft* No. 65. Wiesbaden: Federal Institute for Population Research.

Penev, G. (1995a). Cohort fertility of Serb and ethnic Albanian women in former Yugoslavia (English summary). *Stanovnistvo,* 33, 1–4, 5–19.

Penev, G. (1995b). The natural population increase in Serbia by settlements (English summary). *Demografske Sveske,* 24, 38–41.

Perevedentsev, V. I. (1972). Why not two? A demographic commentary. *Soviet Sociology,* 10, 386–389.

Perlez, J. (1994a). No love and little care: Romania's sad orphans. *The New York Times,* 27 October.

Perlez, J. (1994b). Uprising or coup? Romanians ask 5 years later. *The New York Times,* 25 December.

Perlez, J. (1995a). A painful case tests Poland's abortion ban. *The New York Times International,* 2 April.

Perlez, J. (1995b). Sex barely gets a foot in the classroom door. *The New York Times International,* 19 July.

Perlez, J. (1996). Non-communist elected Romania's leader. *The New York Times,* 17 November.

Perlez, J. (1997). Pope renews call for Poles to fight abortion. *The New York Times,* 5 June.

Pescaru, A. (1966). Birth rate, nuptuality, and work (In Romanian). *Obstetrica si Ginokolgia,* 14, 173–180.

Peterson, W. (1964). *The politics of population.* London: Gollancz.

Petric, N. (1969). Family planning in Yugoslavia (English summary). *Stanovnistvo,* 7, 209–218.

Petrova, D. (1993). The winding road to emancipation in Bulgaria. In N. Funk and M. Mueller (Eds.), *Gender politics and post-communism: Reflections from Eastern Europe and the former Soviet Union.* New York: Routledge, 22–29.

Petrovic, M. (1995). Research in parenthood: Value and empiric aspects. *Demografske Sveske,* (24), 11–14.

Pierotti, D. (1991). From abortion to contraception in Romania. *World Health,* November–December.

Pierotti, D. (1993). Padlocked societies. *Entre Nous,* (22/23), 10–11.

Pierotti, D., and Blayo, C. (1994). The long march from abortion to contraception. *World Health,* 47(3), 18–19.

Pine, P. (1991). Personal interviews with women in Tirana Hospital. Unpublished.

Pine, R.N., and Mertus, J. A. (1993). *Meeting the health needs of women survivors of the Balkan conflict.* New York: Center for Reproductive Law and Policy.

Plakwicz, J., and Zielinska, E. (1994). Poland. In B. Rolston and A. Eggert (Eds.), *Abortion in the new Europe; A comparative handbook.* Westport, CT: Greenwood Press, 199–213.

Podyashchikh, P. (1968). Impact of demographic policy on the growth of population. In E. Szabady, G. Acsadi, J. Tamasy, and G. Vukovich (Eds.), *World views of population problems.* Budapest: Akademiai Kiado, 230–251.

Poland. (1958). Law No. 61 of 27 April 1956. *Dziennik Ustaw,* 8 May, No. 12, p. 71. *International Digest of Health Legislation,* 9, 319–323.

Poland. (1960). Law No. 61 of 1956. Instruction No. 5/59 and 52/59 of 11 December 1959 relating to the interruption of pregnancy, *Dziennik Urzedowy,* 15 January.

Poland/Central Statistical Office. (1997). *Demographic yearbook 1997* and earlier years. Warsaw: Central Statistical Office.

Poland/Government Population Commission. (1984).

Poland/Government Population Commission. (1988).

Poland/Government Population Commission. (1992).

Poland/Government Population Commission. (1993).

Poland/Ministry of Health. (1990). Health behavior of teenage schoolgoers in Poland. Unpublished paper. Warsaw: Ministry of Health.

Poland/Ministry of Health. (1993). Report on the implementation of the act on family planning. Warsaw: Ministry of Health.

Polchanova, S. A. (1986). Popularization of contraception (In Russian). *Nurse and Midwife,(*5), 38–42.

Pollo, S., and Pluto, A. (1981). *The history of Albania.* London: Routledge and Kegan Paul.

Pongracz, M. (1976). Report of Working Group on Social Demography, Zakopane, 1976. New York: United Nations, 49–50. (SOA/ESDP/1976/2).

Pongracz, M. (1991). Induced abortion in Hungary today: Results of a public opinion poll. *Planned Parenthood in Europe,* 20(2) 8.

Pongracz, T., and Molnar, E. S. (1975). Some preliminary results of the public opinion poll concerning demographic questions (In Hungarian). *Demografia,* 18, 435–461.

Pongracz, T., and Molnar, E. S. (1976). *Public opinion survey concerning population questions.* Budapest: Central Statistical Office/Demographic Research Institute, No. 43.

Popov, A. A. (1980). Medical-demographic and socio-hygienic reasons and factors for induced abortions; a review of the literature (In Russian). *Zdravookhraneniye Rossiyskoy Federatsii,* (9), 27–31.

Popov, A. A. (1990). Sky-high abortion rates reflect dire lack of choice. *Entre Nous,* (16), 5–7.

Popov, A. A. (1991). Family planning and induced abortion in the USSR: Basic health and demographic characteristics. *Studies in Family Planning,* 22, 368–377.

Popov, A. A. (1993). A short history of abortion and population policy in Russia. *Planned Parenthood in Europe,* 22(2), 23–25.

Popov, A. A. (1994a). Family planning and induced abortion in the Post Soviet Russia of the early 1990s: The unmet needs in information supply. *Working Papers,* 16 August. Moscow: Center of Demography and Human Ecology.

Popov, A. A. (1994b). The USSR. In B. Rolston and A. Eggert (Eds.), *Abortion in the new Europe.* Westport, CT: Greenwood Press, 267–297.

Popov, A. A. (1994c). Sterilization in Post-Soviet Russia. *Planned Parenthood in Europe,* 23(3), 25–26.

Popov, A. A. (1995a). Family planning and induced abortion in Post-Soviet Russia of the early 1990s: Unmet needs in information supply. In J. Da Vanzo (Ed.), *Russia's demographic "crisis."* Santa Monica, CA: Rand, 84–108.

Popov, A. A. (1995b). Family planning in Russia: The role of NGOs in demonopolizing population policy, *Planned Parenthood in Europe,* 24(2), 26–30.

Popov, A. A. (1996). Family planning and induced abortion in Post-Soviet Russia of the early 1990s: Unmet needs in information supply. In J. DaVanzo (Ed.), *Russia's demographic "crisis".* Santa Monica, CA: Rand, 84–108.

Popov, A. A., Visser, A. P., and Ketting, E. (1993). Contraceptive knowledge, attitudes, and practice in Russia during the 1980s. *Studies in Family Planning,* 24, 227–235.

Popova, V. J. (1992). Youth work initiatives in Moscow. *Planned Parenthood in Europe,* 21(2), 1–2.

Popova, V. J. (1996). Sexuality education moves forward in Russia. *SIECUS Report,* 24(3), 14–15.

Posadskaya, A., and Azhgikhina, N. (1995). Women's issues in Russia, 1986–1995. Moscow: Unpublished paper.

Posadskaya, A. et al. (Eds.). (1995). *Women in Russia; A new era in Russian feminism.* London: Verso.

Potts, M. (1967). Legal abortion in Eastern Europe. *Eugenics Review,* 59, 232–250.

Potts, M., Diggory, P., and Peel, J. (1977). *Abortion.* Cambridge: Cambridge University Press.

Prager, K. M. (1987). The critical condition of Soviet health care. *Wall Street Journal*, 29 January, reprinted in *Survey,* 29: 145–147.

Pressat, R. (1979). The demographic position of the USSR on the eve of its fifth census (In Russian). *Population*, 34, 845–868.

Presser, H. B. (1997). Demography, feminism, and the science-policy nexus. *Population and Development Review*, 23, 295–331.

Prifti, P. (1978). *Socialist Albania since 1944: Domestic and foreign developments.* Cambridge: MIT Press.

Prisacaru, C., Buese, D., and Goilea, F. (1992). Evolution of birth rate and fertility (In Romanian). *Romanian Journal of Statistics,* 12, 7–25.

Pro Familia Magazine. (1987). (6), 11–13.

Pro Familia Magazine (1995). (5), 30–31

Prokopec, J. (1963). A married woman in the family and in employment (In Czech). *Demografie,* 5, 17–29; 109–120.

Prokopec, J. (1965). Investigation of young people before marriage in rural areas (In Czech). *Bulletin of the Population Commission,* 5, 30–37.

Prokopec, J. (1966). Young people in towns and cities before marriage (In Czech). *Bulletin of the Population Commission*, 6, 3–32.

Prokopec, J. (1968). The sexual life of young urban married couples. *Czechoslovak Population Problems,* 16–39.

Prozhanova, V., and Tantchev, S. (1995). Bulgaria: Adolescent pregnancy on the increase. *Entre Nous*, (30/31), 16–17.

Public Opinion Research Center. (1990). *Newsletter.* (In Polish).

Public Opinion Research Center. (1993). *Newsletter.* (In Polish).

Puia, S., and Hirtopeanu, C. (1990). Coming out of the dark: Family planning in Romania. *Planned Parenthood in Europe,* 19(2): 5–6.

Quinn, F. (1998). *Democracy at dawn: Notes from Poland and points East.* College Station, TX: Texas A and M University Press.

Racioppi, L., and O'Sullivan, K. (1995). Organizing women before and after the fall. *Signs,* 4, 818–850.

Raduly, J. (1997). Women and birth in Romania. In M. Feischmidt, E. Magyari-Vincze, and V. Zentai (Eds.), *Women and men in East European transition.* Cluj-Napoca: Editura Fundatiei Studii Europene, 171–180.

Rahman, A., Katzive, L., and Henshaw, S. K. (1998). A global review of laws on induced abortion, 1985–1997. *International Family Planning Perspectives,* 24, 56–64.

Randolph, E. (1996). *Waking the tempests: Ordinary life in the new Russia.* New York: Simon and Schuster.

Ransel, D. L. (Ed.), (1979). *Family in Imperial Russia.* Urbana: University of Illinois Press.

Rasevic, M. (1971). Influence of education on population fertility and birth control (In Serbian). *Stanovnistvo,* 9, 255–265.

Rasevic, M. (1990a). Abortion as a method of birth control. *Yugoslav Survey,* 31(3), 99–112.

Rasevic, M. (1990b). Yugoslavia: Too many abortions. *Entre Nous,* (16) 11.

Rasevic, M. (1990c). Thirty years of induced abortion in Yugoslavia. Paper presented at the U.N. Conference "From abortion to contraception," Tbilisi, Georgia, July.

Rasevic, M. (1993). *Towards an understanding of abortion in Serbia* (English summary). Belgrade: University of Belgrade Demographic Research Center.

Rasevic, M. (1993/94). Women's knowledge and induced abortion: A survey in Metropolitan Belgrade, 1990. *Stanovnistvo,* 32(3/4), 1–12.

Rasevic, M. (1994). Yugoslavia: Abortion as a method of birth control. *Reproductive Health Matters,* 3, 68–74.

Rasevic, M. (1995a). Population policy acceptance among women who decided to terminate pregnancy (English summary). *Stanovnistvo,* 33(1–4), 41–54.

Rasevic, M. (1995b). Research results relevant for the family planning program (English summary). *Demografske Sveske,* 24, 23–26.

Rasevic, M. (1998). Personal communication, 27 July.

Rasevic, M., and Petrovic, N. (1995). *Case studies in population policy: An overview.* Belgrade: Demographic Research Center.

Regulska, J. (1994). Transition to local democracy: Do Polish women have a chance?. In M. Rueschemeyer (Ed.), *Women in the politics of postcommunist Eastern Europe,* Armonk, NY: Sharpe, 35–62.

Remennick, L. I. (1991). Epidemiology and determinants of induced abortion in the U.S.S.R. *Social Science and Medicine,* 33, 841–848.

Remennick, L. I. (1993). Patterns of birth control. In I. Kon and J. Riordan (Eds.), *Sex and Russian Society.* Bloomington: Indiana University Press, 45–63.

Remennick, L. I. (1996). Terra Incognita of Russian sex: Seven decades of socialism and the morning after. *Journal of Sex Research,* 33, 383–385.

Reporteri. (1993). 20 October.

Reproductive Freedom News. (1996). 22 November.

RFE/RL Newsline. (1997). 30 December, No. 187, Part II.

Rich, V. (1992). Albania: Family planning available. *The Lancet,* 17 October, 964.

Rich, V. (1994). Polish abortions. *The Lancet,* 343, 1090.

Riley, N. E. (1997). Gender, power, and population change. *Population Bulletin,* 52(1), 2–48.

Ristic, J. (1994). Medical staff and family planning. *Stanovnistvo* (3/4), 27–52.

Rivkin-Fish, M. (1996). Improving women's health in urban Russia: A cultural analysis of international and local efforts for change. Unpublished report presented to the MacArthur Foundation.

Rochat, R. (1991). *Women's health, family planning, and institutionalized children in Romania.* Atlanta, GA: USAID/Trust Through Health.

Rogow, D. (1993). From abortion to contraception: Getting to the Emerald City. In K. Newman (Ed.), *Progress postponed: Abortion in Europe in the 1990s.* London: International Planned Parenthood Federation.

Roloff, J. (1997). Pregnancy termination in West and East Germany: Analyses of backgrounds, facts, and acceptance with special consideration of the results of the German Fertility and Family Survey (In German). *Materialien zur Bevölkerungswissenschaft,* No. 27. Wiesbaden: Federal Institute for Population Research.

Romania. (1966). Decrees No. 770 and No. 771 of 29 September 1966. *Buletinul Oficial al Republicii Socialiste Romania,* Part I, 1 October, No. 60. *International Digest of Health Legislation,* 1967, 18, 822–837 (includes instructions No. 819).

Romania. (1967). Instruction No. 819 of 19 October 1966. *International Digest of Health Legislation,* 18, 822–837 (includes Decrees No. 770 and 771).

Romania. (1976). Revision of Penal Code of 23 April 1973. *International Digest of Health Legislation,* 27, 803.

Romania. (1986). Decree No. 185 of 2 June. Translated in *Annual Review of Population Law,* 1987, 14, 39.

Romania/Central Board of Statistics. (1969). Census of the population of Romania. March 15, 1966. Bucharest: Central Board of Statistics.

Romania/Central Board of Statistics. (1993). *Statistical Yearbook, 1993* and prior years. Bucharest: Central Board of Statistics.

Romania/Ministry of Health. (1957). Decree No. 463, 25 November.

Romania/Ministry of Health. (1968). General Report of the Ministry of Health. *Munca,* 27 October, 2–5.

Romania/Ministry of Health. (1970). Social investigation of fertility 1967–68 (In Romanian). Bucharest: Ministry of Health.

Romania/Ministry of Health. (1979). *Health Statistics Bulletin* and earlier years. Bucharest: Ministry of Health.

Romania/Ministry of Health. (1989). Decree No. 605 (In Romanian), 26 December; *International Digest of Health Legislation,* 1992, 43, 530.

Romania/Ministry of Health. (1990). *Reports on abortion;* various years. Bucharest: Ministry of Health.

Romania/Ministry of Health. (1993). *Health statistics yearbook,* 1992 and earlier editions. Bucharest: Ministry of Health.

Romania/Ministry of Health. (1994). *Official Bulletin,* Bucharest: Ministry of Health.

Romania/Ministry of Health. (1995). *Official Bulletin,* Bucharest: Ministry of Health.

Romania/Ministry of Health/Computing and Health Statistics Centre. (1995). Reports; various years. Bucharest: Ministry of Health.

Romania/Ministry of Labor. (1993), *Official Monitor;* various years. Bucharest: Ministry of Labor.

Romania/National Center for Statistics (NCS). (1997). *Romanian Statistical Yearbook* and earlier editions. Bucharest: National Center for Statistics.

Romania/National Center for Statistics. (1998). Report of 9 January.

Romania/National Commission of Demography. (1971). Law concerning the founding organization and functioning of the National Commission of Demography (In Romanian). *Buletinul Oficial al Republicii Socialiste Romania,* 18 March.

Romania/National Commission of Demography. (1976). Report to the Working Group on Social Demography. Geneva: United Nations, European Social Development Programme, 59–61.

Romania/National Committee for the United Nations International Conference on Population and Development, Cairo, Egypt. (1994). *National Report.* Bucharest: National Committee.

Ross, J., and Frejka, T. (1998). Paths to subreplacement fertility: Overview and ten country studies. Paper prepared for the conference on the Global Fertility Transition, Bellagio, Italy, May 18–22.

Rosset, E. (1973). *Polish demography in the service of social progress, 1946–1971* (In Polish). Warsaw: Instytut Wydawniczy CRZZ.

Rothe, J. (1968). Health protection and irreversible contraception (In German). *Das Deutsche Gesundheitswesen,* 23, 1721.

Rothe, J. (1973). Irreversible contraception in the German Democratic Republic. Paper presented at the International Conference on Voluntary Sterilization, Geneva.

Royston, E., and Armstrong, S. (1989). *Preventing maternal deaths.* Geneva: World Health Organization.

Rozenfeld, B. A. (1996). The crisis of Russian health care and attempts at reform. In J. Da Vanzo (Ed.), *Russia's demographic "crisis."* Santa Monica, CA: Rand, 163–174.

Rueschemeyer, M. (Ed.) (1994a). *Women in the politics of postcommunist Eastern Europe.* Armonk, NY: Sharpe.

Rueschemeyer, M. (1994b). Women in the politics of Eastern Germany: The dilemmas of unification. In M. Rueschemeyer (Ed.), *Women in the politics of postcommunist Eastern Europe.* Armonk, NY: Sharpe, 87–116.

Ruevekamp, D. (1994). The urgent need for quality improvement in Russia. *Planned Parenthood Challenges,* (2), 22–23.

Rusinow, D. I. (1972). Yugoslavia. In H. Brown and A. Sweezy (Eds.), *Population: Perspective, 1971.* San Francisco: Freeman Cooper, 269–288.

Rusinow, D. I. (1973). Slovenia: Modernization without urbanization? In H. Brown, J. P. Holdren, A. Sweezy, and B. West (Eds.), *Population: Perspective 1973.* San Francisco: Freeman Cooper, 138–157.

Russia/Goskomstat. (1997) and earlier years. *The demographic yearbook of the Russian Federation* (In Russian). Moscow: Goskomstat.

Russia/Ministry of Health. (1988). On organizing of commercial clinics (departments, rooms) and free-of-charge clinics (departments, rooms) with state financing for patients who require permanent care and commercial gynecologic clinics for provision of operations for artificial interruption of pregnancy (abortion). Order No. 250, 29 March. Moscow: M2RF.

Russia/Ministry of Health. (1992a). On the abortion situation and organizing family planning services (In Russian). Directive No. 6-15/7-15, 30 August. Moscow: M2RF.

Russia/Ministry of Health. (1992b). On the collection and presentation of monthly medical records (In Russian). Directive No. 171-4, 3 March. Moscow: M2RF.

Russia/Ministry of Health. (1992c). On the collection and presentation of monthly medical records (In Russian). Directive No. 171-4, 3 March. Moscow: M2RF.

Russia/Ministry of Health. (1993a). On the confirmation of the list of medical indicators for artificial interruption of pregnancy (In Russian). Order No. 302, 26 December. Moscow: M2RF.

Russia/Ministry of Heralth. (1993b). On providing medical sterilization of citizens. Order No. 303, 28 December. Moscow: Ministry of Health.

Russia/Ministry of Health. (1994). National report of the Russian Federation on population problems. Moscow: Ministry of Health.

Russia/Ministry of Health. (1998). Statement released 29 November. *RFE/RL Newsline.* 30 November.

Russia/State Committee on Statistics. (1996). *Russian statistical yearbook.* Moscow: State Committee on Statistics, 93.

Russian Centre for Public Opinion and Market Research, Centers for Disease Control and Prevention, and U.S. Agency for International Development. (1997). *1996 Russia women's reproductive health survey: A study of three cities. Preliminary report.* Atlanta: Centers for Disease Control and Prevention.

Russian Centre for Public Opinion and Market Research, Centers for Disease Control and Prevention, and U.S. Agency for International Development. (1998). *1996 Russia women's reproductive health survey: A study of three sites. Final report.* Atlanta: Centers for Disease Control and Prevention.

Russian Federation/Goskomstat. (1997). *Demographic yearbook, 1997 and earlier years.* Moscow: Goskomstat.

Russo, N. F. (1992). Psychological aspects of unwanted pregnancy and its resolution. In J. D. Butler and D. F. Walbert (Eds.), *Abortion, medicine, and the law (4th ed).* New York: Facts on File, 593–626.

Russo, N. F., and Zierk, K. L. (1992). Abortion, childbearing, and women's well-being. *Professional Psychology; Research and Practice,* 23, 269–280.

Ryabushkin, T. (1978). Social policy and demography in the Soviet Union. *Population and Development Review,* 4, 715–720.

Ryan, M. (1987). Illegal abortions and the Soviet Health Service. *British Medical Journal,* 294, 525–527.

Ryan, M. (1990). *Doctors and the state in the Soviet Union.* New York: St. Martin's Press.

Rychtarikova, J. (1994). Czech and Slovak families in the European context. *Journal of Family History,* 19, 131–147.

Rychtarikova, J. (1995). Recent changes in fertility and families in the Czech Republic. *Studia Demograficzne,* 122(4), 35–46.

Rychtarikova, J. (1996). Current changes of the reproduction character in the Czech Republic and the international situation (In Czech). *Demografie,* 38(2), 72–89.

Sabatello, E. F. (1992). Estimates of demand for abortion among Soviet immigrants in Israel. *Studies in Family Planning,* 23, 268–273.

Sacks, K. (1975). Engels revisited: Women, the organization of production, and private property. In R. R. Reiter (Ed.), *Toward an anthropology of women.* London: Monthly Review Press.

Sacks, M. P. (1976). *Women's work in Soviet Russia: Continuity in the midst of change.* New York: Praeger.

Sadik, N. (1998). Human rights: Women have special needs. *Populi,* 25(1), 16–17.

Sadvokasova, E. A. (1969). *Social hygiene aspects of the regulation of family size* (In Russian). Moscow: Meditsina.

Sahatci, E. (1993). Albania discovers the human right to family planning. *Planned Parenthood in Europe,* 22(3), 25–26.

Savage, M. (1988). The law of abortion in the Union of Soviet Socialist Republics and the People's Republic of China: Women's rights in two socialist countries. *Stanford Law Review,* 40, 1027–1117.

Schehl, M. (1996). Contraceptive use increases in Russia. *AVSC News,* 34(4), 1, 8.

Scheuflerowa, R. (1977). Some problems of sex education in CSSR. In M. Kozakiewicz (Ed.), *Family planning and sex education in socialist countries.* Warsaw: Polish Family Planning Association, 61–69.

Schlesinger, R. (Ed.) (1949). *The family in the USSR: Changing attitudes in Soviet Russia.* London: Routledge and Kegan, Paul.

Schmemann, S. (1998). How can you have a bust if you never had a boom? *The New York Times Magazine,* 27 December.

Schnabl, S. (1970). *Man and woman intimate* (In German). Rudolstadt: Greifenverlag.

Schneck, P. (1985). The fate of social gynecology in the time of fascism in Germany (In German). In A. Thon and H. Spaar (Eds.), *Medizin im Faschismus.* Berlin: Volk und Gesundheit.

Schneider, N. (1994). *Family and private life in West and East Germany: A comparative analysis of family life 1970–1992* (In German). Stuttgart: Enke.

Schulz, G. E. (1971). German Democratic Republic. In L. T. Lee and A. Larson (Eds.), *Population and law: A study of the relations between population problems and law.* Leiden: Sijhoff, 221–231.

Schwarz, K. (1992). Development of births and family policy in the former GDR: An example of a pronatalist policy (In German). *Zeitschrift für Familienfoschung,* 4, 248–262.

Scott, H. (1974). *Does socialism liberate women?* Boston: Beacon.

Scott, H. (1976). *Women and socialism: Experiences from Eastern Europe.* London: Alison and Busby.

Scott, H. (1977). Women's place in socialist society: The case of Eastern Europe. *Social Biology,* 7, 32–35.

SECS. (1993). *1993 activities* (In Romanian). Bucharest: Society for Education on Contraception and Sexuality, 1993.

Segodnya. (1997). 24 December.

Senturia, K. D. (1996). Maternal and child health in Albania. *Social Science and Medicine,* 43, 1097–1107.

Senturia, K. D. (1997). A woman's work is never done: Women's work and pregnancy outcome in Albania. *Medical Anthropology Quarterly,* 11, 375–395.

Serbanescu, F. (1994). Personal communication, 3 May.

Serbanescu, F. (1998a). Reproductive health services in Eastern Europe and the Newly Independent States: Country cases. Unpublished paper presented at the ECE Regional Population Meeting, Budapest, 7–9 December. Atlanta, GA: Centers for Disease Control and Prevention.

Serbanescu, F. (1998b). Personal communication, 18 December.

Serbanescu, F., and Morris, L. (1995). *Reproductive health survey Romania 1993. Report.* Bucharest and Atlanta, Institute for Mother and Child Care, Ministry of Health, Bucharest and Division of Reproductive Health, Centers for Disease Control and Prevention.

Serbanescu, F., and Morris, L. (1997). *Young adult reproductive health survey Romania 1996. Preliminary report.* Bucharest and Atlanta: International Foundation for Children and Families, National Institute for Mother and Child Health Care, National Commission for Statistics, Centers for Disease Control and Prevention, United States Agency for International Development, and Centre for Development and Population Activities.

Serbanescu, F., Morris, L., Stratila, M., and Bivol, O. (1998). *Reproductive health survey Moldova 1997.* Chisinau, Moldova: Moldovan Ministry of Health and Atlanta, GA: Centers for Disease Control and Prevention.

Serbanescu, F., Morris, L., Stupp, P., and Stanescu, A. (1995). The impact of recent policy changes on fertility, abortion, and contraceptive use in Romania. *Studies in Family Planning,* 26, 76–87.

Serbia. (1991). *Official Gazette,* No. 45.

Serbia. (1992). *Official Gazette,* No. 27.

Serbia. (1994). *Official Gazette,* No. 28, 47.

Serbia. (1995). *Official Gazette*, No, 16.
Serbia/Bureau of Statistics. (1997). *Vital events in Serbia in 1995 and 1996.* Belgrade: Republic Bureau of Statistics.
Sexological Institute. (1994). *Sexual behavior of the inhabitants of the Czech Republic* (In Czech). Prague: Sexological Institute.
Shalin, D. N. (1978). The development of Soviet sociology, 1956–1976. *Annual Review of Sociology,* 4, 171–191.
Shkolnikov, V. M., and Mesle, F. (1996). The Russian epidemiological crisis as mirrored by mortality trends. In J. Da Vanzo (Ed.), *Russia's demographic "crisis."* Santa Monica, CA: Rand, 113–162.
Shkolnikov, V. M., Leon, D. A., Adamets, S., Andreev, E., and Deev, A. (1998). Educational level and adult mortality in Russia: An analysis of routine data 1979 to 1994. *Social Science and Medicine,* 47, 357–369.
Shlindman, S. I., and Zvidrinsh, P. P. (1973). *The study of fertility* (In Russian). Moscow: Statistika.
Sigrist, H. E. (1947). *Medicine and health in the Soviet Union.* New York: Gollancz.
Siklova, J. (1997). Different region, different women: Why feminism isn't successful in the Czech Republic. In M. Feischmidt, E. Magyari-Vincze, and V. Zentai (Eds.), *Women and men in Eastern European transition.* Cluj-Napoca: Editura Fundatiei Pentru Studu Europene, 34–41.
Silber, L., and Little, A. (1997). *The death of Yugoslavia.* New York: Penguin.
Silver, B. D. (1987). Political beliefs of the Soviet citizen: Sources of support for regime norms. In J. R. Millar (Ed.), *Politics, work, and daily life in the USSR: A survey of former Soviet citizens.* Cambridge: Cambridge University Press, 100–141.
Simoneti, S. K., Breznik, D., Radovan, A., and Stevanov, M. (1976). Law and fertility in Yugoslavia. In M. Kirk, M. Livi Bacci, and E. Szabady (Eds.), *Law and fertility in Europe,* Vol. 2. Dolhain: Ordina, 665–693.
Simpson, P. (1995). Poland's morning after. *On the Issues,* 4(Spring), 36–39.
Singh, S., and Henshaw, S. (1996). The incidence of abortion: A worldwide overview. In IUSSP (Ed.), *Socio-cultural and political aspects of abortion in a changing world.* Liege: International Union for the Scientific Study of Population.
Sjoberg, O. (1991). *Rural change and development in Albania.* Boulder, CO: Westview Press.
Skipietrow, N. (1993a). Conference on sexual education (In Polish). *Gazeta Wyborcza,* 12 May, No. 109.
Skipietrow, N. (1993b). Secret report about the refusal of abortion (In Polish). *Gazeta Wyborcza,* 21/22 August, No. 176.
Skipietrow, N. (1994). Troublesome pregnancy (In Polish). *Gazeta Wyborcza,* No. 23.
Sklar, J. L. (1974). The role of marriage behavior in the demographic transition: The case of Eastern Europe 1900. *Population Studies,* 28, 231–247.
Skwiecinska, O. (1993). Let's go to live longer (In Polish). *Gazeta Wyborcza,* No. 303.
Skwiecinska, O. (1994a). Abortion officially (In Polish). *Gazeta Wyborcza,* No. 93.
Skwiecinska, O. (1994b). Neglected pregnant (In Polish). *Gazeta Wyborcza,* No. 122.
Skwiecinska, O. (1994c). Place for reports—wastepaper basket (In Polish). *Gazeta Wyborcza,* No. 126.
Slovak Rep/Statistical Office. (1997). *Statistical yearbook of the Slovak Republic,* 1997. Bratislava: Statistical Office.

Slovenia/Ministry of Labour, Family and Social Affairs. (1994). Slovenian National Report to International Conference on Population and Development. Ljubljana: Ministry of Labor.

Slovenia/SO. (1995). *Statistical yearbook of the Republic of Slovenia 1994* and earlier years. Ljubljana: Statistical Office of the Republic of Slovenia.

Smith, H. (1977). *The Russians.* New York: Ballantine.

Smolinski, Z. (1980). *Fertility of Polish women* (In Polish). Warsaw: Central Statistical Office.

Smolinski, Z., and Kuciarska-Ciesielska, M. (1989). *Procreation in the family.* Warsaw: Central Statistical Office, Vol. 18.

Snitow, A. (1993). The church wins, women lose. *The Nation,* 26 April, 556–559.

Sokolov, L. (1974). Labor force (economically active population). In D. Breznik (Ed.), *The population of Yugoslavia.* Belgrade: Demographic Research Center, 56–67.

Sokolova, N. S. (1970). Statistical analysis of the outcomes of pregnancies (In Russian). *Zdravookhraneniye Rossiyskoy Federatsy,* 14, 38–40.

Sokolowska, M. (1977). Poland: women's experience under socialism. In J. Z. Giele and A. C. Smock (Eds.), *Women; roles and status in eight countries,* New York: Wiley, 349–381.

Sokolowska, M. (1980). Family policy in Poland (In Polish). *Przeglad Humanistyczny,* No. 11/12.

Solnickova, L. (1979). Position of women in Czechoslovakia. *Demosta,* 12, 68–71.

Solomon, S. G. (1989). David and Goliath in Soviet public health; the rivalry of Soviet hygienists and psychiatrists for authority other than Bytovi alcoholic. *Soviet Studies,* 41, 254–275.

Sora, V., Hristache, L., and Despa, M. P. (1983). *Demography* (In Romanian). Bucharest: Editura Didactica si Pedagogica.

Sougareva, M. (1992). Cohabitation: A new form of the family or a crisis of the family (In Bulgarian). *Population* (Sofia), 7(3), 79–90.

Specter, M. (1998). Traffickers' new cargo: Naive Slavic women. *The New York Times,* 11 January.

Speigner, W. (1992). Consequences of a reassessment of Marxist ideology and practice for demography: On the theory of the transition (In German). *Gesellschaft für Demographie, Beiträge zur Demographie,* 24–41.

Spolar, C. (1996a). Romania speeds reforms. *The Washington Post,* 19 May.

Spolar, C. (1996b). Albania reverts to one-party government. *The Washington Post,* 22 June.

Spolar, C. (1996c). Bulgaria pays price for lagging on market reforms. *The Washington Post,* 7 July.

Spolar, C. (1996d). Romania chooses change. *The Washington Post,* 18 November.

Spolar, C. (1997). Hungary turns austerity into growth. *The Washington Post,* 28 December.

Spolar, C. (1998). Conservative parties win power in Hungary. *The Washington Post.* 25 May.

Sracek, J. (1988). Czechoslovakia. In P. Sachdev (Ed.), *International handbook on abortion.* Westport, CT: Greenwood Press, 127–137.

Srb, V. (1979). World Fertility Survey in Czechoslovakia. *Demosta,* 12, 25, II–XIII.

Srb, V. (1993). Population movement in the territory of Czechoslovakia, 1938–1944 (In Czech). *Demografie,* 35, 73–94.

Srb, V. (1995). *Demographic profile of Czechoslovakia.* Prague: Czech Family Planning Association.

Srb, V., and Kucera, M. (1959). Survey on parenthood (In Czech). Prague: State Statistical Office.

Stampar, D. (1971a). Symposium on Family Planning, Zagreb, June (English summary). *Stanovnistvo,* 9, 101–105.

Stampar, D. (1971b). The effect of the liberalization of abortion legislation in the Republic of Croatia estimated on the basis of existing statistical data (English summary). Unpublished Dissertation.

Stampar, D. (1971c). Some aspects of natality trend in Croatia (English summary). *Lijecnicki Vjesnik,* 93, 729–737.

Stampar, D. (1971d). Demographic and other effects of the liberalization of abortion legislation in the Republic of Croatia. Paper presented at the International Seminar on Family Planning in the Public Health Systems of Socialist Countries, Rostock, GDR.

Stampar, D. (1972a). Abortion trends in Croatia in the period from 1960 to 1970 (English summary). *Arhiv za Zastitu Majke I Dejeteta,* 16, 3–13.

Stampar, D. (1972b). Outcome of pregnancy in women whose application for approval of termination of pregnancy was refused by the commissions of the first instance and by commissions of appeal in Zagreb. Unpublished paper.

Stampar, D. (1973a). Abortion mortality in the Socialist Republic Croatia (English summary). *Jugoslovenska Ginekologija: Opstetricija,* 13, 113–123.

Stampar, D. (1973b). Croatia: Outcome of pregnancy in women whose requests for legal abortion have been denied. *Studies in Family Planning,* 4, 267–269.

Stampar, D. (1973–74). Study of reliability of data concerning the incidence of induced abortion (English summary). *Stanovnistvo,* (11/12), 273–277.

Stampar, D. (1975–76). Data on repeat abortions in Croatia (English summary). *Stanovnistvo,* (13/14), 146–154. (Note: Actually published in 1978).

Stampar, D. (1976). Fertility regulated behavior in adolescents and adults in Croatia. Unpublished paper.

Stampar, D. (1995). Personal communication of data from Ministry of Public Health.

Stampar, D., and Beluhan, A. (1972). Survey of knowledge, attitude and practice of family planning in the population of women visiting family planning clinic of the Department for Family Planning for the first time (English summary). *Arhiv za Zastitu Majke I Djeteta,* 16, 67–78.

Stampar, D., and Kapor-Stanulovic, N. (1990). New population law gives region a choice. *Entre Nous,* (16), 12.

Standish, D. (1998). From abortion on demand to its criminalization: The case of Poland in the 1990s. In E. Lee (Ed.), *Abortion law and politics today.* New York: St. Martin's Press, 116–129.

Stanley, A. (1994). Sexual harassment thrives in the new Russian climate. *The New York Times,* 17 April.

Starkalev, I., Papasov, B., and Stoimenov, G. (1961). The abortion situation in the People's Republic of Bulgaria (In German). In K. H. Mehlan (Ed.), *Internationale Abortsituation, Abortbekämpfung, Antikonzeption.* Leipzig: Thieme, 26–31.

Starke, K. (1987). Love and sexuality of youth (In German). *Pro Familia Magazin,* 15(6), 2–5.

Starke, K., and Friedrich, W. (1984). *Love and sexuality to age 30* (In German). Berlin: Verlag der Wissenschaften.

Starke, K., and Visser, A. P. (1994). Sexuality, sexual behaviour, and contraception in East Germany. *Patient Education and Counseling,* 23, 217–226.

Stefanov, I., and Naumov, N. (1974). Bulgaria. In B. Berelson (Ed.), *Population policy in developed countries,* New York: McGraw-Hill, 149–170.

Steinem, G. (1980). Feminist notes; the Nazi connection. *Ms,* October.

Stephenson, J. (1975). *Women in Nazi society.* New York: Barnes and Noble.

Stephenson, P., and Porter, R. (1994). Preventing maternal mortality in St. Petersburg. Unpublished report to the World Health Organization.

Stephenson, P., Wagner, M., Badea, M., and Serbanescu, F. (1992). Commentary: The public health consequences of restricted induced abortion—lessons from Romania. *American Journal of Public Health,* 82, 1328–1331.

Stern, M., and Stern, A. (1980). *Sex in the USSR.* New York: Times Books.

Sternadel, Z. (1982). *Research on effectiveness of basic fertility regulating methods* (In Polish). Poznan.

Stites, R. (1978). *The women's liberation movement in Russia: Feminism, nihilism, and Bolshevism, 1860–1930.* Princeton: Princeton University Press.

Stloukal, L. (1993). The politics of population policy: Abortion in the Soviet Union. *Working Papers in Demography,* No. 43. Canberra: Australian National University, Research School of Social Sciences.

Stloukal, L. (1995). Demographic aspects of abortion in Eastern Europe: A study with special reference to the Czech Republic and Slovakia. A thesis submitted for the degree of Doctor of Philosophy at the Australian National University, Canberra.

Stloukal, L. (1996a). Eastern Europe's abortion culture: Puzzles of interpretation. Paper presented at the IUSSP Seminar on Socio-Cultural and Political Aspects of Abortion from an Anthropological Perspective, Trivandrum, India, 25–28 March.

Stloukal, L. (1996b). Personal communication, 12 August.

Stloukal, L. (1996c). Abortion. In Z. Pavlik and M. Kucera (Eds.), *Population development in the Czech Republic 1996.* Prague: Charles University, Department of Demography and Geodemography, 35–41.

Stloukal, L. (1997). Changing patterns of extramarital conceptions in the Czech Republic, 1960–1993. *Journal of Biosocial Science,* 29, 471–489.

Stloukal, L. (1998a). Personal communication, 15 June.

Stloukal, L. (1998b). An APC analysis of demographic responses to population policy measures: The case of the Czech and Slovak republics. 1960–1990. *Genus,* 51, 87–121.

Stloukal, L. (1998c). Induced abortion in the Czech Republic from a cohort perspective (In Czech). *Demografie,* 40, 81–92.

Stoimenov, G. (1976). Planned parenthood in Bulgaria. *IPPF Europe Regional Information Bulletin,* 5(2), 5.

Stoimenov, G., Sepetliev, D., and Dimitrov, R. (1965). *Birth rate in Bulgaria* (In Bulgarian). Sofia: Medicine and Physculture.

Stoimenov, G., Vassilev, D., and Tzankov, L. (1977). Social psychological factors influencing the family for the determination of wanted number of children (In Bulgarian). In A. Latov (Ed.), *The woman: Work and way of life.* Sofia: Bulgarian Academy of Sciences, 265–281.

Storey, J. D., Ilkhamov, A., and Saksvig, B. (1997). Perceptions of family planning and reproductive health issues: Focus group discussions in Kazakhstan, Turkmenistan,

Kyrgyzstan, and Uzbekistan. Baltimore: Johns Hopkins School of Public Health, Reproductive Health Services Expansion Program, Field Report, November 10.

Stotland, N. L. (1992). The myth of the abortion trauma syndrome. *Journal of the American Medical Association,* 268(15), 2078–2079.

Stoyanov, P. (1998). Bulgaria's courage. *The Washington Post,* 10 February.

Studio Gazeta. (1992). Poles about abortion (In Polish). *Gazeta Wyborcza,* No. 270.

Sucholinska, A. (1993). Women's unemployment (In Polish). *Statistical Newsletter No. 17.* Warsaw: Central Statistical Office.

Swiatkowska, B. (1994). Public prosecutors criticize anti-abortion law, *Zycie Warszawy,* April 20.

Szabady, E., and Klinger, A. (1970). Report from Budapest: Studies in abortion-seeking behavior. In H. P. David and J. Bernheim (Eds.), *Proceedings of the Conference on Psychosocial Factors in Transnational Family Planning Research,* Geneva, Switzerland, April 1970. Washington, DC: American Institutes for Research, 31–43.

Szabady, E., and Klinger, A. (1972). Pilot surveys of repeat abortion seeking. *International Mental Health Research Newsletter,* 14, 6–9.

Szalai, J. (1988). Abortion in Hungary. *Feminist Review,* 29, 98–100.

Taaffe, R. N. (1977). The impact of rural-urban migration on the development of Communist Bulgaria. In H. L. Kostanick (Ed.), *Population and migration trends in Eastern Europe.* Boulder, CO: Westview Press, 157–180.

Tagliabue, J. (1997). December 23, 1989. *The New York Times Magazine,* 15 June.

Taniguchi, H. (1991). The agenda for family planning in the USSR. *Integration* (29), 2–3.

Tare, V. (1992). Family planning and health. Unpublished report to a national seminar, Tirana, June.

Tare, V. (1993). Unpublished paper on abortion in Albania. Tirana Hospital Family Planning Unit.

Tarifa, F. (1994). Disappearing from politics: Social change and women in Albania. In M. Rueschemeyer (Ed.), *Women in the politics of postcommunist Eastern Europe.* Armonk, NY: Sharpe, 133–151.

Teitelbaum, M. S. (1972). Fertility effects of the abolition of legal abortion in Romania, *Population,* 26, 405–417.

Teitelbaum, M. S. (1974). The delegalization of abortion in Romania. *Family Planning* (U.K.), 23, 38–41.

Tesarova, D. (1979). Assertion of Czechoslovak women in employment. *Demosta,* 12, 72–75.

Tesarova, D. (1981). Labour force in Czechoslovakia after the World War II. *Demosta,* 14, 14–18.

Tess, E. (1994). The grateful slave (In Polish). *Femina,* 4 June.

This Week in Germany. (1995). 3 March.

This Week in Germany. (1997). 10 October.

This Week in Germany. (1998). 30 January.

Thomas, L. (1994). Establishing FPAs in the Central Asian Republics. *Planned Parenthood in Europe,* 23(3), 16.

Tichonova, L., Borisenko, K., Ward, H., Mehaus, A., Gormyko, A., and Renton, A. (1997). Epidemics of syphilis in the Russian Federation: Trends, origins, and priorities for control. *Lancet,* 350, (9072), 210–213.

Tietze, C. (1983). *Induced abortion: A world review. (5th ed.).* New York: The Population Council.

The Times. (London). (1992). 8 April.

The Times. (London). (1996). 21 June.

Tirpak, M., and Adamica, F. (1995). Population change in the Slovak Republic in 1994 (In Slovak). *Demografie,* 37, 254–266.

Titkow, A. (1984). *Child and values.* Warsaw: Institute of Philosophy and Sociology.

Titkow, A. (1992). Status evolution of Polish women. Unpublished paper prepared for the Ford Foundation.

Titkow, A. (1993a). Political change in Europe: Cause, modifier, or barrier to gender equality. In N. Funk and M. Mueller (Eds.), *Gender politics and post-communism: Reflections from Eastern Europe and the former Soviet Union.* New York: Routledge, 252–256.

Titkow, A. (1993b). Stress and social life: Polish experience. Warsaw: State Publishing Institute.

Titkow, A. (1994a). Polish women in politics: An introduction to the status of women in Poland. In M. Rueschmeyer (Ed.), *Women in the politics of postcommunist Eastern Europe,* Armonk, NY: Sharpe, 29–34.

Titkow, A. (1994b). Status evolution of Polish women—the paradox and chances. In M. Alesteld, E. Allardt, A. Rychard, and W. Wesotowski (Eds.), *The transformation of Europe, social conditions and consequences,* Warsaw: Institute of Philosophy and Sociology.

Titkow, A. (1995). Women under pressure (In Polish). In A. Titkow and H. Domanski (Eds.), *What does it mean to be a woman in Poland?* Warsaw: Institute of Philosophy and Sociology, 9–39.

Tobias, S. (1997). *The faces of feminism: An activist's reflections on women's movements.* Boulder, CO: Westview Press.

Todhe, T. (1994). *Policies supporting women's health: Social and legal problems related to abortion in Albania.* Tirana: Ministry of Health and Environment.

Todorova, M. (1993). The Bulgarian case: Women's issues or feminist issues? In N. Funk and M. Mueller (Eds.), *Gender politics and post-communism: Reflections from Eastern Europe and the former Soviet Union.* New York: Routledge, 30–38.

Tomala, G. (1994). Budget and money market will feed the poor (In Polish). *Zycie Warszawy,* 17 June.

Tomilin, S. A. (1927). The problem of population. In S. A. Tomilin (Ed.), *Demography and public health* (In Russian). Moscow: Statistika, 180.

Tomsic, V. (1973). The status of women and family planning in Yugoslavia. *Faits et Tendances,* (10), 57–80.

Toth, O. (1993). No envy, no pity. In N. Funk and M. Mueller (Eds.), *Gender politics and post-communism: Reflections from Eastern Europe and the former Soviet Union.* New York: Routledge, 213–223.

Trebici, V. (1968). Organization of scientific research in demography (In Romanian). *Revista de Statistica,* 17, 29–39.

Trebici, V. (1971). *The population of Romania and economic development* (In Romanian). Bucharest: Editura Politica.

Trebici, V. (1976). Law and fertility in Romania. In M. Kirk, M. LiviBacci, and E. Szabady (Eds.), *Law and fertility in Europe*, Vol. 2. Dolhain: Ordina, 520–543.

Trebici, V. (1978). The value of children in Romania (In German). In L. Franke and H. W. Jürgens (Eds.), *Keine Kinder-Keine Zukunft?* Boppard: Boldt, 59–66.

Trebici, V. (1979). *Demography* (In Romanian). Bucharest: Editura Stiintifica si Enciclopedica.

Trebici, V. (1991). *Genocide and demography* (In Romanian). Bucharest: Editura Humanitas.

Trenc, P., and Beluhan, A. (1971). Experience and attitudes of secondary school pupils concerning sexual life (English summary). *Arhiv za Zastitu Majke I Djeteta,* 15, 275–339; 1973, 17, 269–320.

Tucholska-Zaluska, H. (1975). Abortions of married women (In Polish). *Studia Demograficzne,* No. 39.

Tulchinsky, T. H., and Varavikova, E. A. (1996). Addressing the epidemiologic transition in the former Soviet Union: Strategies for health system and public health reform in Russia. *American Journal of Public Health,* 86, 313–320.

Turner, R. (1992). Russian fertility is low, despite early age at first birth and lack of effective contraceptive methods. *Family Planning Perspectives,* 24, 236–237.

Turnock, D. (1989). *Eastern Europe: An economic and political geography.* London: Routledge.

Tzankov, C., and Vassilev, D. (1976). Some peculiarities in the sexual life of women during their reproductive years (In Bulgarian). *Problems of Obstetrics/Gynecology,* 4, 125–131.

Ukraine. (1994). Demographic situation and population policy in Ukraine in the 80s and early 90s of the 20th century. Summary of National Paper of Ukraine for the International Conference on Population and Development, Cairo.

United Nations. (1987). *World population policies,* Vol. 1. New York: United Nations.

United Nations. (1991). *World population prospects 1990.* New York: United Nations.

United Nations. (1992). *Abortion policies: A global review.* Vol. I, 19–20. New York: United Nations.

United Nations. (1995a). *Demographic yearbook 1993 and earlier years.* New York: United Nations.

United Nations. (1995b). *Report of the International Conference on Population and Development.* New York: United Nations A/Conf. 171/13/Rev. 1.

United Nations. (1996). *The Beijing declaration and the platform for action: Fourth World Conference on Women: Beijing, China: 4–15 September 1995.* New York: UN/DPI/1766/Wom.

United Nations Fund for Population Activities (1990). Report of UNFPA mission to Romania 5–15 March 1990. Unpublished.

United Nations Fund for Population Activities. (1991). *Albania country report.* New York: UNFPA.

United Nations Fund for Population Activities. (1995). *World population report.* New York: UNFPA.

United Nations Fund for Population Activities/ World Health Organization European Regional Office. (1995). *Family planning and reproductive health in CCEE/NIS.* Copenhagen: World Health Organization European Regional Office.

UNICEF. (1994). *Women and gender in countries in transition: A UNICEF perspective.* New York: Regional Office for Central and Eastern Europe, Commonwealth of Independent States, and Baltic States.

UNICEF. (1996a). *News in Brief,* June/July

UNICEF. (1996b). Situation analysis of women and children in the Federa Republic of Yugoslavia. Belgrade: UNICEF.

UNICEF. (1996c). Situation analysis of women and children in Former Republic of Macedonia. Skopje: UNICEF.

UNICEF. (1996d). Situation analysis of women and children in Croatia. Zagreb: UNICEF.
UNICEF. (1998). Transmonee Data Base. New York: UNICEF.
United Nations/Population Division. (1995). *Abortion: A global review, Vol. III.* New York: United Nations.
United Nations/Population Division. (1996). Reproductive rights and reproductive health: A concise report. New York: United Nations.
United States/Department of State. (1995). *The Dayton peace agreement on Bosnia Herzegovina.* Washington, DC: Department of State.
United States/Bureau of the Census. (1996). *Health in Eastern Europe.* Washington, DC: U.S. Bureau of the Census.
United States/Bureau of the Census. (1998). International Data Base, International Programs Center. Washington, DC: Bureau of the Census.
Urlanis, B. T. (1968). The problem of demography and planning. In E. Szabady (Ed.), *World views of population problems.* Budapest: Akademiai Kiado, 343–349.
Urlanis, B. T. (1974). Current problems of the Soviet demographic policy. *Social Sciences Today,* 26, 30–35.
Ushborne, C. (1992). *The politics of the body in Weimar Germany.* London: Macmillan.
USSR. (1920). Supreme Soviet Decree of 18 November (In Russian). Translated in. Field, M. G. (1956). The relegalization of abortion in the Soviet Union. *New England Journal of Medicine,* 255, 421–427.
USSR. (1936). Central Executive Committee and Council of People's Commissars' Decision of 27 June (In Russian). Translated in Sigerist, H. E. (1947). *Medicine and health in the Soviet Union.* New York: Citadel.
USSR. (1955a). Supreme Soviet Decree of 23 November. *Bulletin of the Institute for the Study of the USSR* (Munich), 2(12), 44; Also in World Health Organization, *Abortion laws: a survey of current world legislation.* Geneva: WHO, 1971, 57.
USSR. (1955b). Ministry of Health. Instructions of 28 December. In H. Harmsen (Ed.), *Monograph of the Academy for State Medicine in Hamburg,* 1961, 16, 64–66.
USSR/Central Executive Committee and Council of People's Commissars of the Soviet Union. (1936). About increasing production of rubber products and other production for personal prophylactics. Decree No. 1348, 31 July. *The Medical Herald,* 1994, No. 1, 8.
USSR/Goskomstat. (1988 and earlier years). *Population of the USSR* (In Russian). Moscow: Goskomstat of the USSR.
USSR/Goskomstat. (1990). *Demographic yearbook of the USSR, 1990* (In Russian). Moscow: Goskomstat, 640.
USSR/Ministry of Health. (1975–1991). Official statistical data published by the Central Data Processing Center. Moscow: Ministry of Health.
USSR/Ministry of Health. (1990a). On permission for surgical sterilization of women (In Russian). Order No. 484. Moscow: Ministry of Health, 14 December.
USSR/Peoples' Commissariat of Health. (1936). Instructions of women's consultations for the fight against abortion (In Russian). Moscow: Herald of the Narcomzdav of the USSR, No. 10, 15–16 and later years.
USSR/Peoples' Commissariat of Health. (1939). On the prohibition of the operation of cutting or ablation of healthy fallopian tubes of the uterus (In Russian). Order No. 303. Moscow: Peoples' Commissariat of Health, 7 August.
Uzbekistan/Goskomstat. (1996). *Demographic yearbook* (In Russian). Tashkent: Goskomstat.

Uzbekistan Institute of Obstetrics and Gynecology and Macro International. (1997). *Uzbekistan demographic and health survey.* Calverton, MD: Macro International. (Summarized in *International Family Planning Perspectives,* 1998, 24, 43–44.)

Uzel, R. (1994). Czech and Slovak Republics. In B. Rolston and A. Eggert (Eds.), *Abortion in the new Europe: A comparative handbook.* Westport, CT: Greenwood Press, 55–65.

Uzel, R. (1998). Czech Republic. Summary of report presented to the N.I.S. and Eastern Europe Consultation, Hanover, Germany, 12–16 October. Washington, DC: Populaton Institute.

Uzel, R., Ketting, E., Visser, A., and Lehert, P. (1992). *Contraceptive practices and attitudes of Czech and Slovak women.* Brussels: International Health Foundation.

Uzel, R., and Wynnyczuk, V. (1992). Family planning and sex education for national minorities in the Czech and Slovak Federal Republic. *Planned Parenthood in Europe,* 21(3), 20–21.

Uzel, R., and Wynnyczuk, V. (1994). Private fee no barrier to family planning in the Czech Republic. *Planned Parenthood in Europe,* 23(1), 10–11.

Valentei, D., and Kwasha, A. (1974). Demographic and population policy. In *Population policy: Modern problems.* Warsaw: 138–141.

van Dam, A. (1993). Preventive methods badly needed. *Planned Parenthood Challenges,* 1, 22–24.

van der Walle, E., and Knodel, J. (1980). Europe's fertility transition: New evidence and lessons for today's developing world. *Population Bulletin,* 34(6),1–43.

vanden Heuvel, K. (1993). Sex in the time of revolution. *The Nation,* 21 June, 870–873.

Vasilevsky, L. A., and Vassilevsky, L. M. (1924). Abortion as a social phenomenon (In Russian). Moscow: Izdatelstvo Frenkelia.

Vaskovics, L. A., Garhammer, M., Schneider, N. F., and Kabat vel Job, O. (1994). Family and household structures in the former GDR and FRG, 1980–1989: A comparison (In German). *Materialien zur Bevölkerungswissenschaft, Supplement No. 24.* Wiesbaden: Federal Institute for Population Research.

Vassilev, D. (1972). *Medico-social study on the causes of abortion on request and problems of medical prophylaxis* (In Bulgarian). Sofia: Postgraduate Medical Institute.

Vassilev, D. (1978). Service oriented research in adolescent fertility in the People's Republic of Bulgaria. Paper presented at WHO Seminar, Warnemünde, GDR.

Vassilev, D. (1979). *Contraception* (In Bulgarian). Sofia: Medizina c Phykultura.

Vassilev, D. (1982). Obstetrical aspects of the reproductive function of the family (In Bulgarian). Sofia: Unpublished Doctor of Science dissertation.

Vassilev, D. (1986). Orientation toward a two children family model in Bulgaria (In Bulgarian). *Population* (Sofia), 1, 67–82.

Vassilev, D. (1988). *In the intimate world* (In Bulgarian). Sofia: Ministry of Education.

Vassilev, D. (1989a). Birth control and declining birth rates: The Bulgarian experience. *Planned Parenthood in Europe,* 18(2), 9–13.

Vassilev, D. (1989b). *Sexuality of adolescents and youths* (In Bulgarian). Sofia: Narodna Prosveta.

Vassilev, D. (1989c). *Sexual consultation for students.* Sofia: Narodna Prosveta.

Vassilev, D. (1994). Bulgaria. In B. Rolston and A. Eggert (Eds.), *Abortion in the new Europe: A comparative handbook.* Westport, CT: Greenwood Press, 43–54.

Vassilev, D., Golemanov, N., and Mishkov, K. (1974). *The problems of social hygiene in hormonal contraception* (In Bulgarian). Sofia: National Information Center of Medical Science.

Vassilev, D., and Rushkarska, S. (1990). Defloration: A medico-sociological study (In Bulgarian). *Akusherstvo: Gynecologia,* 29(4), 56–63.

Velkoff, V.A. (1993). Hungary. In V.A. Velkoff and K. Kinsella (Eds.), *Aging in Eastern Europe and the Former Soviet Union.* Washington, DC: Bureau of the Census.

Velkoff, V. A. (1995). *Women in Poland.* Washington, DC: Bureau of the Census, WID/95

Verbenko, A. (1968). Only desired children (In Russian). *Literaturnaya Gazeta,* 31 July.

Verdery, K. (1994). From parent-state to family patriarchs. *Eastern European Politics and Societies,* 2, 225–255.

Vickers, M., and Pettifer, J. (1997). *Albania: From anarchy to Balkan identity.* New York: New York University Press.

Vidlakova, O. (1976). Law and fertility in Czechoslovakia. In M. Kirk, M. Livi Bacci, and E. Szabady (Eds.), *Law and Fertility in Europe,* Vol. 1. Dolhain: Ordina, 174–198.

Vikhlayeva, E. M., and Nikolaeva, E. (1996). Epidemiology of abortions in Russia. *Entre Nous,* (34/35), 18.

Vinereanu, D. (1991). Romanian Family Planning Association (SECS) starts sexual and contraceptive consultations on the Black Sea Coast. *Planned Parenthood in Europe,* 20(1), 24–25.

Vishnevsky, A. G. (1996). Family, fertility, and demographic dynamics in Russia: Analysis and forecast. In J. Da Vanzo (Ed.), *Russia's demographic "crisis".* Santa Monica: Rand, 1–35.

Visser, A. P., Bruyniks, N., and Remennick, L. (1993). Family planning in Russia: Experiences and attitudes of gynecologists. *Advances in Contraception,* 9(2), 93–104.

Visser, A. P., Pavlenko, I., Remennick, L., and Lehert, P. (1992). Contraception in the former USSR: Recent survey results on women's behavior and attitudes. *Planned Parenthood in Europe,* 21(2), 9–11.

Visser, A. P., Pavlenko, I., Remennick, L., Bruyniks, N., and Lehert, P. (1993). Contraceptive practices and attitudes in former Soviet women. *Advances in Contraception,* 9, 13–23.

Visser, A. P., Remennick, L., and Bruyniks, N. (1993). Contraception in Russia: Attitude, knowledge, and practice of doctors. *Planned Parenthood in Europe,* 22(2), 26–29.

Visser, A. P., Uzel, R., Ketting, E., Bruyniks, N., and Oddens, B. J. (1993). Attitudes of Czech and Slovak gynecologists toward family planning. *Advances in Contraception,* 9, 351–362.

Visser, A. P., Uzel, R., Ketting, E., Bruyniks, N., and Oddens, B. (1994). Practice, attitude and knowledge of Czech and Slovak gynaecologists concerning contraception. *Planned Parenthood in Europe,* 23, 19–23.

Vjatere, D. (1995). Family planning in Latvia: A blossoming flower. *Choices,* 24(2), 31–33.

Volgyes, I., and Volgyes, N. (1977). *The liberated female; Life, work, and sex in socialist Hungary.* Boulder, CO: Westview Press.

Volk, I. (1995). Cited in *International Dateline,* January 4.

von Baross, J. (1993). German Constitutional Court rejects abortion compromise. *Planned Parenthood in Europe,* 22(3), 14–16.

von Soden, K. (1988). *The sexual advisory centers in the Weimar Republic,* 1919–1933. (In German). Berlin: Edition Heinrich Berlin.

Vornik, B., and Govorlin, T. (1996). Challenges in the Ukraine. *Choices,* 25(2), 13–15.

Voronina, O. A. (1988). Women in a man's society. *Soviet Sociology*, (2), 104–110.

Vostrikova, A. M. (1964). Examination of fertility, marriages, and the family in the USSR. In E. Szabady (Ed.), *Studies on fertility and social mobility*. Budapest: Akademiai Kiado, 214–228.

Vukovich, E. (1989). *Case studies in population policy: Hungary*. New York: United Nations Population Policy Paper No. 19.

Weinert, A. (1976). Psychological-pedagogical problems of psychosexual development and education of blind persons (In German). *Wissenschaftliche Zeitschrift der W. Pieck Universität Rostock*, 25, 659–663.

Weiss, P., and Zverina, J. (1996). Risk factors in the sexual behavior of Czechs: Results from the national survey. Paper presented at the International Conference on HIV/AIDS, Vancouver, Canada, July.

Wendt, H. (1993). Family formation and family policy in the forner GDR (In German). *Materialien zur Bevölkerungswissenschaft*, Supplement No. 22. Wiesbaden: Federal Institute for Population Research.

Wertheimer-Baletic, A. (1979). Women's employment in Yugoslavia. In M. Macura (Ed.), *The effect of current demographic change in Europe on social structure*. Belgrade: Ekonomski Institut, 43–50.

West, R. (1996). *Tito and the rise and fall of Yugoslavia*. New York: Carroll and Graf.

Westoff, C.F., Calot, G., and Foster, A.D. (1983). Teenage fertility in developed nations, 1971–1980. *Family Planning Perspectives*, 15, 105–110.

Westoff, C. F., Sharmanov, A. T., and Sullivan, J. M. (1998). The replacement of abortion by contraception in three Central Asian Republics: Digest. Calverton, MD: Macro International

Whitaker, I. (1981). "A sack for carrying things." The traditional role of women in Northern Albanian society. *Anthropological Quarterly*, 54, 146–156.

Willekins, F. J., and Scherbov, S. (1995). Demographic trends in Russia. In H. van den Brekel and F. Deven (Eds.), *Population and family in the low countries, 1994: Selected current issues*. Norwell, MA: Kluwer Academic, 177–230.

Williams, C. (1994). Sex education and the AIDS epidemic in the Former Soviet Union. *Sociology of Health and Illness*, 16, 81–102.

Williams, D. (1997). Just say nyet: Talk show breaks taboo. *The Washington Post*, 6 December.

Williams, D. (1999). In Russia, a host of new year "presents." *The Washington Post*, 3 January.

Winkler, G. (1980). *Theory and practice of social policy* (In German). Berlin: Akademie Verlag.

Wissenschaftlicher Rat für Sozialpolitik und Demographie. (1979). Problems of demographic development in the developing socialist society in the GDR (In German). *Wirtschaftswissenschaft*, (7), 775–806.

Witte, J. C., and Wagner, G. E. (1995). Declining fertility in East Germany after unification: A demographic response to socioeconomic change. *Population and Development Review*, 21, 387–397.

Wolchik, S. L. (1981a). Elite strategy toward women in Czechoslovakia: Liberation or mobilization? *Studies in Comparative Communism*, 15(2/3), 140–141.

Wolchik, S. L. (1981b). Ideology and equality; The status of women in Eastern and Western Europe. *Comparative Political Studies*, 13, 445–476.

Wolchik, S. L. (1989). Women and the state in Eastern Europe and the Soviet Union. In E. M. Charlton, J. Everett, and K. Staudt (Eds.), *Women, the state, and development.* New York: State University of New York Press, 44–65.

Wolchik, S. L. (1991). Women and work in Communist and Post-Communist Central and Eastern Europe. In H. Kahne and J. Giele (Eds.), *Women's work in modernizing and industrial societies.* Boulder, CO: Westview Press.

Wolchik, S. L. (1994a). Women and the politics of transition in the Czech and Slovak republics. In M. Rueschemeyer (Ed.), *Women in politics of postcommunist Eastern Europe.* Armonk, NY: Sharpe, 3–27.

Wolchik, S. L. (1994b). Women's issues in Czechoslovakia in the communist and post-communist periods. In B. Nelson and N. Chowdhury (Eds.), *Women and politics worldwide.* New Haven: Yale University Press.

Wolf, E., and Hammer, K. (1978). *Cyankali: A documentation* (In German). Berlin: Aufbau.

Women's Global Network for Reproductive Rights Newsletter. (1992). No. 3, April–June.

World Bank. (1992). *Romania: Human resources and the transition to a market economy.* Washington, DC: World Bank.

World Fertility Survey. (1980). *The Hungarian Fertility Survey, 1977: A summary of findings.* London: World Fertility Survey, June.

World Health Organization. (1971). *Abortion laws: a survey of current world legislation.* Geneva: World Health Organization.

World Health Organization. (1990). *Abortion: A tabulation of available data on the frequency and mortality of unsafe abortion.* Geneva: World Health Organization Maternal and Child Health Unit and Family Planning Division of Family Health.

World Health Organization. (1992). *Reproductive health: A key to a brighter future. Biennial Report 1990–1991.* Geneva: World Health Organization.

World Health Organization. (1993). *The narrative research method: Studying behavior patterns of young people—by young people. A guide to its use.* Geneva: World Health Organization.

World Health Organization. (1995). *Complications of abortions: Technical and managerial guidelines.* Geneva: World Health Organization.

World Health Organization. (1998). *Unsafe abortion, 3rd edition.* Geneva: World Health Organization Division of Reproductive Health.

World Health Organization/EURO. (1991). Unpublished report. Copenhagen: World Health Organization European Regional Office.

World Health Organization/EURO. (1992a). Country Highlights: Azerbaijan. Copenhagen: World Health Organization European Regional Office.

World Health Organization/EURO. (1992b). Country Highlights: Kazakhstan. Copenhagen: World Health Organization European Regional Office.

World Health Organization/EURO. (1992c). Country Highlights: Kyrgystan. Copenhagen: World Health Organization European Regional Office.

World Health Organization/EURO. (1992d). Country Highlights: Tajikistan. Copenhagen: World Health Organization European Regional Office.

World Health Organization/EURO. (1992e). Country Highlights: Turkmenistan. Copenhagen: World Health Organization European Regional Office.

World Health Organization/EURO. (1992f). Country Highlights: Uzbekistan. Copenhagen: World Health Organization European Regional Office.

World Health Organization/EURO (1993). Highlights on health in Bulgaria. Copenhagen: World Health Organization European Regional Office.

World Health Organization/EURO. (1994). Data provided for the 1994 Conference on Women's Health Counts. Copenhagen: World Health Organization European Regional Office.

World Health Organization/EURO. (1997). *Family planning and reproductive health in CCEE and CIS.* Copenhagen: World Health Organization European Regional Office.

Wright, N. H. (1973). Romania: Some consequences of restrictive abortion legislation. In S. Lewit (Ed.), *Advances in Planned Parenthood,* Vol. 8. Amsterdam: Excerpta Medica, International Congress Series, No. 271, 200–208.

Wright, N. H. (1975). Restricting legal abortion: Some maternal and child health effects in Romania. *American Journal of Obstetrics and Gynecology,* 121, 246–256.

Wulf, D. (1980). The Hungarian fertility survey, 1977. *Family Planning Perspectives,* 12, 44–46.

Wynnyczuk, V. (1969). *Social-hygienic aspects of the employment of women* (In Czech). Prague: Ministry of Labour and Social Affairs.

Wynnyczuk, V. (1974). Parenthood education. In *Population policy in Czechoslovakia.* Prague: Orbis, 69–76.

Wynnyczuk, V. (1978). Parenthood education. In *Population policy in Czechoslovakia.* Prague: Orbis, 63–70.

Wynnyczuk, V. (1980). *Introduction to the study and systems of parenthood education* (In Czech). Prague: Czech Research Institute of Labour and Social Affairs.

Wynnyczuk, V., and Sracek, J. (1976). Young family in the industrial region of Ostrava—a survey. *Demosta,* 9, 68–73.

Wynnyczuk, V., and Sracek, J. (1977). Young family in a rural area/1970–1975: A comparative study. *Demosta,* 10, 55–61.

Yedlin, T. (1980). *Women in Eastern Europe and the Soviet Union.* New York: Praeger.

Yugoslavia. (1952). Decree No. 27 of 11 January 1952 relating to justifiable interruption of pregnancy. *International Digest of Health Legislation,* 4, 450–453.

Yugoslavia. (1969). General law on the interruption of pregnancy of 26 April 1969. *International Digest of Health Legislation,* 20, 573–575.

Yugoslavia. (1974). Article 191 of the constitution of the Socialist Federal Republic of Yugoslavia. *Official Bulletin,* (9), 21 February.

Yugoslavia/Croatia. (1979). Law on abortion of 21 April 1978. *Narodne Novine,* 4 May 1978. *International Digest of Health Legislation,* 30, 329–333.

Yugoslavia/Federal Assembly. (1969). Resolution on family planning of 25 April 1969. *Yugoslav Survey,* 10, 103–106.

Yugoslavia/Federal Statistical Office. (1991). *Statistical yearbook of the Socialist Federal Republic of Yugoslavia 1991* and earlier years. Belgrade: Federal Statistical Institute.

Yugoslavia/Federal Statistical Office. (1997). *1997 Statistical yearbook of Yugoslavia* and earlier years. Belgrade: Federal Statistical Institute.

Yugoslavia/Slovenia. (1977). Law of 20 April 1977 on medical measure to implement the right to a free decision regarding the birth of children. *International Digest of Health Legislation,* 28, 1112–1115.

Yvert-Jalu, H. (1981). The history of divorce in the Soviet Union: Relationship to family policy and social realities (In French). *Population,* 36, 41–62.

Zakharov, S. V. (1994). Changes in spatial variation of demographic indicators in Russia. In W. Lutz, S. Scherbov, and A. Volkov (Eds.), *Demographic trends and patterns in the Soviet Union before 1991.* London: Routledge, 113–130.

Zakharov, S. V. (1997). Personal communication, 1 November.

Zakharov, S. V. (1998a). Personal communication, 5 March.

Zakharov, S. V. (1998b). Personal communication, 8 July.

Zakharov S. V. (1999) Personal communication, 8 January.

Zakharov, S. V., and Ivanova, E. L. (1996a). Fertility decline and recent changes in Russia: On the threshold of the second demographic transition. In J. Da Vanzo (Ed.), *Russia's demographic "crisis."* Santa Monica, CA: Rand, 36–83.

Zakharov, S. V., and Ivanova, E. I. (1996b). Regional fertility differentiation in Russia: 1959–1994. *Studies on Russian Economic Development,* 354–368.

Zakharova, N., Posadskaya, A., and Rimashevskaya, N. (1989). How we are resolving the Women Question? (In Russian). *Kommunist,* (4), 56–65.

Zaplatic, J. (1972). Survey on knowledge, attitudes and practice of birth control of women applying for legal abortion in the General Hospital. "Dr. J. Kajfes" in Zagreb (English summary). *Arhiv za Zastitu Majke I Djeteta,* 16, 79–82.

Zarkovic, G. (1972). Natural increase and population planning in Yugoslavia (Serbo-Croatian). *Narodno Zdravlje,* 28(2), 41.

Zarkovic, G. (1973). *Factors affecting the displosion of the natural population growth in the SFR Yugoslavia* (English summary). Sarajevo: Institute for Hygiene and Social Medicine, Medical Faculty at Sarajevo, Publication No. 72.

Zelinskaya, D. (1994). In A. Bernard, Russia: A risky place to be pregnant. *The Moscow Times,* 3 November.

Zetkin, K. (1929). *Reminiscences of Lenin.* London: Modern Books. Also New York: International Publishers, 1934.

Zetkin, K. (1958). *On the history of the women's proletarian movement in Germany* (In German). Berlin: Dietz.

Zickel, R., and Iwaskiw, W. (Eds.) (1994). *Albania: A country study.* Washington, DC: Library of Congress.

Zielenska, E. (1987). European socialist countries. In S. J. Frankowski and G. F. Cole (Eds.), *Abortion and protection of the human fetus: Legal problems in a cross-cultural perspective.* Dordrecht: Martinus Nijhoff Publishers, 241–334.

Zielenska, E. (1993). Recent trends in abortion legislation in Eastern Europe, with particular reference to Poland. *Criminal Law Forum,* 4, 47–93.

Zielinski, T. (1994). By the rule of the jungle (In Polish). *Wprost,* No. 36.

Ziolkowski, J. A. (1974). Poland. In B. Berelson (Ed.), *Population policy in developed countries.* New York: McGraw-Hill, 445–488.

Ziolkowski, M. (1993). Group interests and group consciousness in the process of system transformation: The case of the Polish intelligentsia. *Centennial Review,* 37, 115–134.

Zupancic, K. (1978). Legal abortion in Europe: Yugoslavia. *IPPF Europe Regional Information,* No. 1 Supplement, 7–8.

Zwerenz, I. (1980). *Women: The history of Paragraph 218* (In German). Frankfurt: Fischer.

Index

Note: Additional guidance is provided in the Contents and in the Tables and Figures chapters.

About the Contributors

ADRIANA BABAN is Associate Professor of Health Psychology at Babes-Bolyai University in Cluj-Napoca, Romania.

HENRY P. DAVID is a psychologist and Founder/Director of the Transnational Family Research Institute in Bethesda, MD, United States.

JÜRGEN DORBRITZ, a demographer formerly residing in the GDR, is a senior researcher at the Federal Institute for Population Research in Wiesbaden, Germany.

AMILDA DYMI is an Albanian journalist and Fulbright Scholar living in the United States.

JOCHEN FLEISCHHACKER is a senior researcher at the Max Planck Institute for Demographic Research in Rostock, Germany.

NILA KAPOR-STANULOVIC is a University Professor, the first chair of the Department of Psychology at the University of Novi Sad (Vojvodina), Yugoslavia, and a consultant to the World Health Organization, UNICEF, and UNESCO.

PAMELA PINE is an American MPH in International Health and a Ph.D. candidate in health education at the University of Maryland who has worked extensively in Albania.

ANDREJ A. POPOV was a gifted medical demographer and respected public health researcher. He served as Research Associate of the Center of Demography and Human Ecology, Russian Academy of Sciences, and Director of the Transnational Family Research Institute, Moscow, Russian Federation.

ANASTASIA POSADSKAYA-VANDERBECK is an economist who in 1990 co-founded and served as first director of the Moscow Centre for Gender Studies, established within the Institute for Socio-Economic Population Studies of the Russian Academy of Sciences. Since 1997 she has directed the Open Society Institute Network Women's Program (Soros Foundation), in New York, NY, United States.

JOANNA SKILOGIANIS is an American Ph.D. candidate in anthropology at Case Western Reserve University who has conducted abortion research in Southern Europe.

LIBOR STLOUKAL is a Czech demographer presently on the staff of the Demography Department, Charles University, Prague (Czech Republic) and the Population Programme Service, Women and Population Division, Food and Agriculture Organization, Rome, Italy.

ANNA TITKOW is Associate Professor and Chair of the Research Unit on Women and Family, Institute of Philosophy and Sociology, Polish Academy of Sciences, Warsaw, Poland.

RADIM UZEL, a gynecologist and sexologist, is Executive Director of the Czech Family Planning Association, Prague, Czech Republic.

DIMITER VASSILEV, an obstetrician-gynecologist who founded the Bulgarian Family Planning Association in 1973 and was its president for many years, currently resides in the United States.

VLADIMÍR WYNNYCZUK, a demographer and former member of the Secretariat of the State Population Commission, is Research Adviser to the Czech Family Planning Association, Prague, Czech Republic.

www.ingramcontent.com/pod-product-compliance
Ingram Content Group UK Ltd.
Pitfield, Milton Keynes, MK11 3LW, UK
UKHW041641031026
15587UKWH00001B/432

9 780313 305870